Well-Oiled Diplomacy

SUNY series in Global Politics
James N. Rosenau, editor

*A complete listing of books in this series
can be found at the end of this volume.*

Well-Oiled Diplomacy

Strategic Manipulation and
Russia's Energy Statecraft in Eurasia

Adam N. Stulberg

STATE UNIVERSITY OF NEW YORK PRESS

Published by
State University of New York Press, Albany

© 2007 State University of New York

For information, contact State University of New York Press, Albany, NY
www.sunypress.edu

Production by Diane Ganeles
Marketing by Anne M. Valentine

Portions of chapters 5 and 7 are from *Geopolitics Journal*, 2005, 10 (1). Reproduced
by permission of Taylor & Francis, Inc., http://www.taylorandfrancis.com

The "Outline Maps" of Central and Southwest Asia are from the Houghton Mifflin
Education Place web site (http://www.eduplace.com/ss/maps/pdf/cent_swasia_nl.pdf).
Copyright © Houghton Mifflin Company. Reprinted by permission of
Houghton Mifflin Company.

Library of Congress Cataloging in Publication Data

Stulberg, Adam N., 1963-
  Well-oiled diplomacy : strategic manipulation and Russia's energy
statecraft in Eurasia / Adam N. Stulberg.
    p. cm. — (SUNY series in global politics)
  Includes bibliographical references and index.
  ISBN-13: 978-0-7914-7063-3 (hardcover : alk. paper)
  ISBN-13: 978-0-7914-7064-0 (pbk. : alk. paper)
  1. Energy policy—Russia (Federation) 2. Petroleum industry and
trade—Political aspects—Russia (Federation) 3. Gas
industry—Political aspects—Russia (Federation) 4. Nuclear
industry—Political aspects—Russia (Federation) 5. Russia
(Federation)—Foreign relations—Asia, Central. 6. Asia,
Central—Foreign relations—Russia (Federation) 7. Geopolitics—Russia
(Federation) I. Title. II. Series.

HD9502.R82S75 2007
333.790947—dc22

2006014432

10 9 8 7 6 5 4 3 2 1

To Cara,
whose love, understanding, keen wit,
and aesthetic touch simply fill my tank

Contents

Illustrations

Figures

Maps

Preface

I started this project when directions in the study of international politics and Russian foreign policy were diverging. The end of the Cold War and bipolarity provoked a welter of scholarly inquiry into the significance of globalization. The common good of energy security was generally treated as both a byproduct of new intimate forms of cooperative engagement, as well as a potential panacea for the numerous regional conflicts precipitated by global transition. This coincided with focused exploration of the evolving dimensions and importance of soft power as practiced by the envied few in the West, instead of by imperial anachronisms, like Russia, that seemed doomed to clutch to crude and waning vestiges of hard military and economic power. At the same time, scholars of post-Soviet affairs remained steeped in statist approaches, emphasizing the importance of realpolitik, domestic institutional capacity, and/or national identity to explain the many idiosyncrasies of Russian and newly independent states (NIS) international behavior. The regional "energy boom" was seen largely as a magnet for a renewed "Great Game" in the Caspian Basin, an instrument for reasserting Moscow's imperial hubris, and a "disease" that encouraged insatiable rent-seeking and indefinite enfeeblement of Eurasian institutions.

By completion of the manuscript, mainstream scholarly and policy landscapes had fundamentally converged. The aftermath of the September 11th terrorist attacks and the 2003 war in Iraq, coupled with the Chinese government's bids for foreign-based oil and gas companies, punctuated a shift from infatuation with globalization to analysis of new forms of statecraft. No longer was Russia relegated to the margins of international politics, as it too seemed determined to wield soft power, with energy as a pillar of its resurgence. At home, Putin's political ascendance, the economy's overreliance on swelling energy revenues, and the imprisonment on tax evasion of one of the country's leading oil barons, Mikhail Khordokovsky, situated energy at the crux of Russia's turn toward "managed democracy," with attendant implications for reconsidering state direction of the "commanding heights," private sector behavior, civil liberties, and the trajectory of national development. With recovery of the oil and gas sectors, Russia was poised to leverage its energy superpower status to influence political transition throughout the NIS, to reintegrate the region as part of a "liberal empire," and to punch above its weight in asserting national interests and setting the terms for cooperation with energy-dependent great powers of Asia, Europe, and North America. Global anxiety heightened with Moscow's 2006 gas war with Ukraine and Europe's awakening to growing dependence on Russian gas. Accordingly, debate shifted to how and to what ends, not whether, energy would provide the base for Russia's rising global stature.

This volume contributes to "squaring the circle" in the study of international relations and Russian foreign policy by stretching our thinking about state power and interdependence. It begins by noting that soft power, like hard power, is not a sure thing, as evidenced by Russia's often overlooked mixed success at exerting its energy dominance of Eurasia throughout the first decade of independence. The book then develops arguments to explain the persistent sources and limitations to Russia's soft energy diplomacy that speak broadly to the literature on international security, political economy, and energy security by increasing our understanding of interfaces between strategic and domestic conditions and between statecraft and globalization. It supplements traditional analyses of statecraft, highlighting indirect market and regulatory mechanisms for altering the behavior of foreign and subnational actors, as well as demonstrates the usability of soft power and global networks, both by expanding our appreciation for how national leaderships can manipulate the risk as well as utility of compliance for targets. The conclusion suggests new directions for the study of international coercion and inducements, as well as offers practical guidelines for engaging a new Russia in the coming decades.

Acknowledgments

I am indebted to many patrons and colleagues for the scholarship and interest that fueled this project. I owe deep gratitude to my mentors, Professors Arthur Stein, Arnold Horelick, David Lake, Richard Rosecrance, and Richard Anderson at UCLA and RAND who taught me to think big and deductively about Eurasian international behavior without succumbing to artificial disciplinary divides or compromising appreciation for the region. Senator Sam Nunn and Dr. Richard Combs, with support from the Carnegie Corporation of New York, piqued my interest in Eurasian energy security and urged me to "push the limits" of conventional wisdom to explore prospective strategic and policy implications. I thank Drs. William Potter, Amy Sands, Clay Moltz, Nikolai Sokov, and Scott Parrish, as well as their colleagues Kenley Butler, Christina Chuen, and Elena Sokova at the Center for Nonproliferation Studies for sharing their vast expertise and resources, and for affording such a stimulating and fraternal postdoctoral setting to study nuclear developments in the former Soviet Union. Professors Doug Blum, Mark Katz, and Blair Ruble, as well as fellow participants at an IREX symposium dedicated to the Caspian Basin offered invaluable critiques of my initial research. I also thank Professors Scott Sagan and Chip Blacker at Stanford University, and Professor Steven Lamy at USC for their hospitality and for providing access to the extensive research facilities at their institutions. Scholars, students, and policy analysts at UCLA, RAND, the University of Michigan, International Studies Association, Association for the Study of Nationalities, and CISAC offered very constructive comments on successive drafts of the main arguments. Many officials, policy "insiders," and business executives in Russia, Azerbaijan, Georgia, Kazakhstan, and Kyrgyzstan were extremely generous with their time and candid insights, and rendered crucial empirical correctives to my theoretical musings. My colleagues at Georgia Tech, especially Bill Long, graciously provided encouragement and indispensable relief that enabled me to cross the finish line. Christa Gray, Nadia Marinova,

Gene Germanovich, and Omer Khan were relentless in hunting down materials, and together with Marilu Suarez rescued me from my computer travails. I also thank the editors and publishers at *Geopolitics* and *The Carl Beck Papers* (University of Pittsburgh) for allowing me to reprint excerpts from previous work, as well as to Houghton Mifflin for permission to adapt their map outlines. My deep gratitude extends to the editor, reviewers, and production staff at SUNY Press for their assistance in bringing this manuscript to fruition. Of course, I alone accept responsibility for any shortcomings.

This book would not have been possible without the indefatigable support and nourishment by friends and family. Stephen McMillan lent an especially deft editorial hand at several critical junctures. As always, my parents deserve special thanks for inspiring me and for their unwavering support for all of my "academic" pursuits. Finally, my greatest thanks goes to Cara for sustaining me in more ways than she can imagine, and for her willingness to put so many things on hold for this project.

Introduction

How and under which conditions can states leverage economic interdependence for purposes of security? These questions are hotly contested, as globalization and the political liberalization unleashed at the end of the Cold War created new vulnerabilities and opportunities for great power politics. While attention to date has been on whether and how trade affects conflict, there has been a surge of interest among scholars to understand how political leaders can exploit economic interconnectivity and the attendant curbs on sovereign autonomy to advance security policy goals short of war.[1] These concerns are especially relevant for preponderant states situated at either the global or regional level that face a new paradox of power: as their relative military and economic advantages are growing, their capacity to formulate and implement foreign economic policies are increasingly circumscribed by global markets and the activities of companies and subnational agencies that are the primary producers, distributors, regulators, and consumers of strategic resources.

This book addresses the political economy of these international security concerns. It offers a new way of thinking about how national leaderships can and have used global energy markets and regulatory institutions as instruments for influence and nonmilitary intervention on vital energy security issues. I argue in these pages that much can be gained from broadening consideration of statecraft from traditional elements of coercive diplomacy to include issues of "strategic manipulation" when considering the "soft" dimensions to international security. Whereas the former entail using threats, inducements, or limited force to alter the expected utility of specific outcomes, strategic manipulation involves restructuring a target's decision situation, alignment choices, and risks to maximize the appeal of a favorable outcome or minimize the appeal of an unfavorable one.

A systematic inquiry into the "manipulative" dimensions to strategic energy policy is important for scholarly and practical reasons.

1

Students of international political economy typically regard energy as an instrument of economic power used to address matters of "low politics." It is treated as a form of either positive or negative trade, aid, or financial inducements that supplier states brandish for specific political and economic purposes against vulnerable customers. Often overlooked, however, is how energy levers are used strategically in relations between rival supplier states. By contrast, the subfield of energy economics regards the supply of strategic goods as a national security concern, and applies market analysis to explain cooperation and efficient responses among producer states. Energy security, defined as "protection against the loss of welfare that may occur as a result of a change in price or availability of a strategic resource," derives from the interplay of exogenous forces. National leaderships are presumed to have both little choice and marginal effect on complex energy markets, with state policies confined mostly to influencing the efficiency of domestic adjustment. Scant attention is devoted to exploring links between the nature and consequences of supply disruption, or to the relative effectiveness of foreign strategies for deterring, containing, and protecting against these shocks.[2] Finally, students of the dominant paradigm in security studies, realism, are generally preoccupied with how threats of using military force solve issues of "high politics." They tend not to subject assumptions about the strategic fungibility of asymmetrical energy interdependence to critical analysis, despite acknowledging the security consequences of economic statecraft. They also typically limit the study of coercion to the military realm, treating positive inducements as a form of appeasement. Thus, this book sits at the nexus of security studies, international political economy, and energy economics, and is part of a growing body of research on the causal connections between security considerations and foreign economic policy.[3]

Similarly, the study lies at the intersection of debates over globalization and coercion. The former refers to dense networks of transnational interdependence and their implications for the continued relevance of nation-state centered concepts of international relations. Most current research does not presume that globalization has rendered the state obsolete; instead, debates center around whether, how, and to what extent international institutions, markets, firms, and other subnational actors bend the rules of state sovereignty and governance.[4] At the same time, the literature on coercion explores the effectiveness of national governments at exploiting policy instruments, including threats of force, sanctions or inducements, to get adversaries to act in a certain way. While there has been long-standing empirical debate over the relationship between power and interdependence, scholars are only at the precipice of thinking critically about how and under which conditions central government actors can exploit "soft power" and extraterritorial networks—for example, manipulating alignment options, interdependent markets, outsourcing arrangements, regulatory regimes, and commercial risks—to further national security objectives.[5] The energy sector is a prime arena for exploring these analytical links, as there is a long history of foreign oil interests and investments serving as precipitants and instruments of war, imperial expansion, and coercive diplomacy.[6] The intertwining of national and firm policy decisions concerning production, investment, pricing, transport, subsidies, tariffs, consumption, conservation, and substitution within and across energy sectors creates situations where one state's policies can have strategic repercussions for another. The book contributes to this research program by systematically bringing statecraft into the study of globalization, while broadening analysis of international coercion to world energy markets and distribution networks.

The significance of this study extends beyond the realm of theory. Henry Kissinger once quipped, "aside from military defense, there is no project of more central importance to national security and indeed independence as a sovereign nation than energy security."[7] Notwithstanding obvious market concerns, energy security is fundamentally "politicized," as states allow foreign ambitions to alter their behavior in energy markets; employ political instruments to advance their position in energy markets; and exploit this standing to influence the strategic behavior of target states.[8] In the post–Cold War era, these dimensions to diplomacy are assuming an even higher profile in econocentric security policies. The pivotal importance of energy to national economic growth, military power, and private consumption, combined with the intense volatility of supply and demand, contested ownership, and critical geographic chokepoints and constraints on access to energy, simultaneously raise the stakes of resource competition and vulnerability across the globe. One has only to survey foreign activism in the Persian Gulf, Europe's growing energy business with Russia, strategic interaction between Israel and its Arab neighbors, regional stability in the Caucasus and Central Asia, and the strategic trajectory of southwest and northeast Asia to appreciate the growing significance of natural resource and pipeline politics for geopolitics and national security. Whether one state's manipulation of another's resource scarcity, foreign markets, equity stakes, and transit options fuels belligerence or opens windows for international cooperation, however, remains an open question.[9]

States face similar complexities concerning the practice of coercive diplomacy across multiple issue areas. With the end of the Cold War, the United States threatened (then used) force repeatedly in the Persian Gulf, Afghanistan, Haiti, Somalia, and the Balkans. No longer consumed with countering Soviet influence, Washington also relied on economic statecraft to deter weapons proliferation, stem sensitive technology transfers and the hosting of terrorist cells, promote nation-building, and punish human rights violations, drug trafficking, and egregious forms of environmental degradation.[10] Yet, the United States achieved only mixed success, notwithstanding favorable gross asymmetries of power. The relevance of specific policy instruments, as well as the credibility and effectiveness America's coercive strategy, were altered by the reaction of small states and transnational entities. Allies and adversaries alike exercised veto authority in multilateral settings, exploiting the high value that the United States placed on other dimensions of interconnectedness to soften or confound Washington's coercive diplomacy.[11] As illustrated by the countermeasures to American sanctions on Iran, Libya, and Cuba in 1996, failure to comprehend the dynamics and potential blowback of practicing statecraft within a dense international networks can be extremely costly.

At the same time, globalization has created opportunities for, if not rewarded, coercive diplomacy. Statistical studies show that states are effective at exploiting complex interdependence to signal and pressure foreign rivals short of military violence.[12] Notwithstanding the difficulty, Washington was able to compel and entice extraterritorial compliance by asserting control over the distribution of strategic goods in other states and altering the opportunity costs for political actors and firms of neglecting U.S. interests.[13] Similarly, global markets often acquiesce to rather than constrain coercive diplomacy. The United States' March 2003 48-hour ultimatum for Iraq's Saddam Hussein to resign or face "severe consequences," prompted oil prices to drop and stocks to soar in the anticipation of a world economic recovery in the aftermath of Washington's successful military operations.[14] Thus, a careful examination of how, when, and to what effect states can leverage power and resources in densely connected networks is clearly warranted.

Finally, a new take on statecraft is warranted to explicate the central paradox of power in post-Soviet international relations. On the one hand, Russia emerged from the Soviet breakup as the undisputed hegemon in Eurasia. The residual political and economic dependencies of the newly independent states provided Moscow with relative and "relevant" power advantages that fortified geopolitical impulses to cast a long shadow over its immediate periphery.[15] On the other hand, enfeebled state institutions and fragmented decision making, coupled with increasing levels of privatization and globalization, steadily undermined Moscow's capacity to mobilize resource advantages in support of effective statecraft. Footloose private firms, subnational agencies, and transnational forces stymied the Russian government's ability to restore imperial domination in its own backyard, and on occasion generated pressure for closer integration into global markets and institutions for NIS targets.[16]

Nowhere was tension between Russia's regional dominance, diffusion of interests, and domestic weakness more transparent than with its energy diplomacy. The natural gas, oil, and electricity sectors offered potentially cheap avenues of influence due to the combination of Russia's superior resource endowments, the Soviet legacy of integrated supply and distribution infrastructures, and the intense dependence on energy subsidies among transitioning NIS. Situated at the hub of regional energy networks, Moscow exited the Soviet Union poised either to threaten appropriation outright or to control prices, supply, pipelines access, and equity stakes for residual energy assets to coerce its neighbors and advance great power ambitions. Yet it was precisely in these energy sectors where the Kremlin confronted the greatest challenges to state autonomy and influence. The emergence of world-class quasi-state and private energy companies, and the de facto control over energy policy exerted by loosely supervised federal and regional agencies weakened the state's grip over Russia's energy strategies and statecraft. Simultaneously, global energy demand and market pressures fueled the independent ambitions of new NIS suppliers and lured multinational corporations and outside powers into the post-Soviet space as potential counterweights to Russia. These factors considerably reduced the vulnerability of the NIS and mitigated Moscow's coercive power, while creating opportunities for Russia to expand participation in international supply networks.

Russia's post-Soviet energy security dilemma raises a crucial question: What is the scope of its strategic leverage in Eurasia? Scholars of realist and liberal theories of power politics advance contending propositions about the success of statecraft, generating expectations that Russia's energy leverage in Eurasia would be either awesome or ineffectual, respectively. In repeated confrontations with Eurasian suppliers over the terms of reintegration and reclamation of residual assets, however, Russia's record of coercion was both more effective than expected by integrationists and less potent than predicted by realists. Counter to integrationist expectations, Russia adroitly seized on the trade of natural gas to wrangle concessions from potential competitors, Turkmenistan and Kazakhstan, concerning regional energy ownership, control, and reintegration. Yet Moscow flaunted realist predictions by failing miserably to capitalize on its power advantages and petroleum pipeline politics to coerce similar concessions from both Azerbaijan and Kazakhstan, as well as from its own oil companies operating in the region.

The Argument

Why and under which conditions have Russian policy-makers succeeded at exploiting relative advantages to secure postimperial energy security initiatives in Eurasia on some issues and not others? Addressing these questions as a window into understanding the analytical links between globalization and statecraft, I present an argument for "strategic manipulation" that challenges the conventional wisdom on coercion and inducements. Traditional accounts best understand statecraft in relational terms, described as the ability of one state to get another to do something that it would not do otherwise. Success rests on a state's ability ex ante to threaten credibly to punish, reverse, or reward a target state's behavior ex post. Coercion obtains when the anticipated costs of defying a threat or limited application of direct pressure outweigh the expected gains from defiance. Positive inducements work by increasing the payoffs of compliance via concessions. In these pages, I take a different tact, arguing that a state can influence a target's policy choices by altering its decision-making situation. Whereas success of direct forms of statecraft turns on issuing credible and explicit threats/promises during dramatic showdowns, states can manipulate a target indirectly by altering the opportunity costs and risks of compliance without precipitating a crisis.

The logic of strategic manipulation derives directly from the empirical observation that decision makers must contend with risk as well as with uncertainty. Policy choices are based on the prospects for achieving gains and avoiding losses, as opposed to simply maximizing expected utility. Drawing on insights from prospect theory that policy makers are generally risk-acceptant when facing sure losses and risk averse in choosing among gains, I argue that the key to strategic manipulation rests with a state's power to set the decision-making agenda for targets. This is accomplished by affecting the baselines used by target states to assess their domain and to make trade-offs between compliance and noncompliance. A state that can determine the value of an exchange, including the difference between potential positive and negative outcomes, is in position to shape a target's decision frame and risk-taking propensity. To the extent that a state like Russia can make compliance attractive either as a safe bet relative to other options, or as a potentially high-value outcome among other losing prospects, it can be expected to realize preferred energy security initiatives without having to threaten, punish, or reverse a target's behavior.

A state's capacity to manipulate energy security decisions rests on two basic conditions. The first relates to the state's power in a global energy market. Absent third party alternatives or domestic options for adjustment, the more vulnerable a target is to unilateral decisions by the state concerning the scope, sequence, and opportunity costs of available policy choices. Vulnerability alone does not guarantee compliance, however, as targets can lash out in different directions—gambling on compliance or defiance—in response to waning prospects. A manipulator also must ensure that domestic actors with direct responsibility for controlling energy resources and extraterritorial activities line up behind its statecraft. This requires that a national leadership possesses discrete regulatory authority to mobilize national resources so that domestic energy firms pursue policies that align the substantive appeal of compliance with a target's risk-taking propensity. This does not necessitate the political capacity to impose or enforce compliance at home, as much as the authority to shape the commercial and political incentives for domestic agents and firms in ways that make upholding national interests more rewarding.

Together market and domestic institutional conditions are crucial for determining the Russian government's capacity to set decision frames and recast the value and riskiness of compliance for both NIS target states and domestic energy lobbies. In those sectors where Moscow can use its market power and domestic regulatory authority to manipulate the substantive appeal of alternative policy options for risk-averse and risk-acceptant NIS targets, it should be poised to discourage defection and guide targets toward favored energy security outcomes. Conversely, Russia's attempts at securing foreign energy security policies should fall short precisely on those issues where the smaller NIS targets are less commercially vulnerable and the Kremlin lacks authority to offset the opportunity costs of domestic compliance. Finally, in those sectors where Russia enjoys *either* market power *or* clearly specified regulatory authority (not both), the best that Moscow can hope for is to wrangle minimally acceptable regional energy security policies from foreign targets that are overwhelmingly commercial not political in nature. Given Russia's status during the first post–Cold War decade as a suppliant in the global political economy and protracted internal weakness, these findings should be especially poignant and readily applicable to other states that enjoy greater market and institutional stature both at home and abroad.

Plan of the Book

Part I situates Russia's diplomacy in the context of the debate over energy supply. Chapter 1 discusses the puzzle of Russia's energy statecraft in light of gaps in the extant theoretical literature. In addition to reviewing the empirical challenges to rival explanations of energy diplomacy, I critique the "hard power" myopia of mainstream scholarship on statecraft. The chapter also reviews common analytical flaws that impede progress toward thinking systematically about statecraft practiced among highly interdependent states.

Chapter 2 introduces an alternative understanding of statecraft rooted in "strategic manipulation." I argue that states can influence rivals other than by persuading or altering their expected utility via direct threats or rewards. Instead, states can manipulate foreign target choices by structuring the opportunity costs and risks of compliance. The chapter begins by assessing the role of risk in statecraft. Drawing on prospect theory, soft power, and neo-institutionalism, I then posit a theoretical framework to grasp the international and domestic conditions under which a state can manipulate a target's decision-making domain and the relative riskiness of compliance. In addition to generating propositions about the success of statecraft, the chapter outlines specific techniques of strategic manipulation and a method for operationalizing and testing the argument.

Part II consists of case studies that explore the variable success of Russia's energy diplomacy at securing compliance on bilateral, regional, and strategic energy security policies over time, across different sectors, and vis-à-vis rival Eurasian supplier states in the decade following the Soviet collapse. Chapter 3 specifies the core explanatory variables: Russia's international market standing and domestic regulatory institutions in the natural gas, oil, and commercial nuclear energy sectors. The ensuing chapters analyze the strategic effects on four states. Each begins with a brief description of Russia's approach to leveraging advantages in the respective energy sector and then reviews the central puzzles related to the cases at hand. Next is an in-depth narrative that describes the sequence of events and the relevant causal connections between Russia's diplomacy and shifts in the decisional domains, relative riskiness of options, and strategic decisions made by Eurasian targets. Chapter 4 illustrates Russia's uniform success at manipulating compliance from Turkmenistan and Kazakhstan on strategic natural gas issues. In contrast, chapter 5 examines how deficiencies in market power and institutional authority in the oil sector undermined Moscow's efforts at prescribing the terms of ownership, development, and pipeline politics with Azerbaijan and Kazakhstan. Chapter 6 traces the argument for strategic manipulation against the record of Russia's mixed success at securing commercial but non-imperial arrangements in the nuclear energy sector with Kyrgyzstan and Kazakhstan. The concluding chapter summarizes the findings and the implications of Russia's strategic manipulation for stretching theories of statecraft, as well as illuminates practical opportunities and challenges to engaging Moscow in future energy security cooperation.

Part I

Statecraft and Strategic Manipulation

1

❖

Theories of Statecraft and the Enigma of Russia's Energy Leverage

The primary goals of this book are to refine our understanding of statecraft and to explain how it works in the Eurasian energy context. The first section of this chapter specifies the central puzzle: Russia's mixed success at dominating energy security in Eurasia. The next section reviews single-factor explanations for statecraft, teasing out fundamental empirical challenges posed by Russia's variable record. The third section identifies crucial analytical flaws common to conventional formulations of coercion and inducements that limit our understanding of statecraft in highly interdependent settings. It also extends the traditional focus on relational forms of "hard" power to encompass the opportunities and challenges for exploiting "soft" power advantages for strategic effect. Drawing links between soft power and statecraft, this chapter lays the groundwork for crafting a theory of strategic manipulation developed and tested in the ensuing chapters.

Russia's Energy Puzzle in Eurasia

Energy statecraft has been a storied element of Russian diplomacy. Discretionary energy deliveries and prices were integral to the Soviet

Union's strategy for managing political control, mitigating instability, and shoring up support within the Eastern bloc.[1] Different interpretations of Europe's growing energy dependency on Moscow also fueled transatlantic discord over construction of the trans-Siberian natural gas pipeline in the early 1980s. Following the Soviet collapse, energy diplomacy animated Russia's early attempts to revitalize influence throughout Eurasia. While the Soviet demise bequeathed to Russia large energy endowments, it simultaneously decapitated centralized fiscal and political control over former republics and created a power vacuum along its immediate periphery. Accordingly, Moscow had strong incentives to rely on its energy dominance to reclaim lost assets and advance Russia's strategic ambitions in the region. As early as 1992, a rare consensus emerged across the Russian political spectrum that viewed control over the vast energy resources as a vital national interest and the linchpin to upholding Moscow's "special" security interests in the former Soviet space. By the close of the first decade, even pro-Western reform-minded Russian politicians looked to energy diplomacy as the crutch for forcibly reintegrating the former Soviet space under the aegis of a "liberal Russian empire."[2]

Notwithstanding these strategic aspirations and residual energy advantages, Moscow both succeeded remarkably and failed miserably at securing compliance across sectors and states throughout Eurasia. Russia had mixed success at wrangling concessions on critical energy security issues, reconstituting regional energy dependencies, and stemming newly independent states (NIS) energy security relations with extraregional actors. As depicted in Figure 1.1, at least three distinct patterns emerged: compliance, defiance, and mutual accommodation.

The first pattern was one of consistent NIS target state compliance with Moscow's maximum political demands. These states repeatedly succumbed to Moscow's intimidation and inducements by deferring to Russia's preferred policies for controlling regional gas supply and transit. These states reluctantly acceded to Moscow's preferred terms for bilateral transactions, as well as bowed to the Kremlin's attempts at reconstituting a regionally integrated gas network and limiting diversification of strategic gas ties. In the process, Moscow also succeeded at converting Russian corporate behavior into an effective instrument of foreign policy. Though unable to dictate firm behavior, the Kremlin adroitly shaped the strategic preferences of the huge gas monopoly, Gazprom, so that the company's foreign operations advanced Moscow's diplomatic agenda vis-à-vis specific target states. Turk-

FIGURE 1.1
Russia's Regional Energy Shadow, 1991–2002

Sector of Russia's Energy Statecraft	NIS Target	Strategic Outcome
Natural Gas	Turkmenistan, Kazakhstan Ukraine, Belarus	Compliance
Oil	Azerbaijan, Georgia Kazakhstan, Baltic states	Defiance
Nuclear (Uranium, fuel element)	Kyrgyzstan, Kazakhstan Ukraine, Armenia	Mutual Accommodation

menistan constituted the classic compliant state, but Kazakhstan, Belarus, and Ukraine too fell consistently into this camp when Russia played the gas card.

A second pattern saw Russia's oil diplomacy meet outright defiance. Target states flatly thwarted Moscow's regional preponderance and rejected preferred energy security initiatives, resisting outright or effectively renegotiating bilateral oil relations on more favorable terms. They also successfully solicited new energy investors, suppliers, markets, and transit networks beyond the post-Soviet space. This was compounded by the penchant among Russian oil firms for staking out positions that contravened the Kremlin's diplomatic objectives. These problems came into sharp relief in Russia's petroleum diplomacy toward Azerbaijan, the Baltic States, Georgia, and Kazakhstan that failed to prevent the latter from establishing independent foreign energy relations at Moscow's expense.

A third pattern was depicted by the mixed responses to Russia's commercial nuclear energy statecraft. In these cases, Russia failed to secure compliance with its most favored aims, but nonetheless averted outright defiance to attain narrow commercial objectives. Kyrgyzstan, Kazakhstan, and Ukraine consistently fell into this pattern. These states were susceptible enough to the lure of reconstituting stages of the former Soviet nuclear fuel cycle, conceding shares to Russian interests in joint uranium enrichment and nuclear fuel delivery projects, notwithstanding other competitive options. Yet, each state contained Russia's heavy-handed political influence by embedding respective deals within a multilateral framework of mutually beneficial commercial interests. To the extent that Moscow was able to rely on domestic agents to champion national interests, it did so to augment the appeal of doing business with Russia in the sector.

Surprisingly, the enigma of Russia's mixed success at energy leverage has not received much scholarly attention. The literature that exists is schizophrenic, as it treats Russia's aggregate energy power as either a diplomatic asset or liability.[3] Widely disparate expectations are premised on presumptions of Russia's energy firms as either mere instruments of state power or completely independent of Moscow and determined to undercut Moscow's diplomacy when it conflicts with their commercial strategies.[4] This overlooks, however, the variation in Russia's energy prowess across different sectors and issue areas. Those studies that disaggregate Russia's energy diplomacy tend to examine the consequences for economic and political issues. The analysis is target-centric, with emphasis placed on describing how domestic agents within the NIS perceived and managed energy dependence on Russia.[5] Yet, overlooked is how NIS policy choices and hedging strategies have been systematically circumscribed and even shaped by Moscow's posture in different energy sectors.

By contrast, the range of political economy relations in the post-Soviet space has recently received scholarly attention. Comparative studies tend to attribute causal weight to different target state perceptions of threat and national identities. These, however, are inherently difficult to disaggregate and measure, as well as fail to account for different responses to Moscow's policy initiatives across sectors and time. Kazakhstan, for example, a predicted "loyalist" on foreign economic issues on the basis of its weak national identity, became conspicuously opportunistic at diversifying strategic relations in the regional oil sector that both gained momentum as the decade elapsed and contrasted with growing dependence in the gas sector. Alternatively, Azerbaijan with its moderately strong national identity was initially prone to outright defiance at independence but became increasingly adept at working with Moscow to secure favorable oil deals by the end of the first decade of independence.[6] The target-centric focus of this literature lacks a theoretical framework to explain how, when, and why Russia succeeded at affecting these different strategic responses. Little attention is devoted to discerning links between Russia's energy diplomacy, the interpretations of strategic energy choices by NIS states, and the different trends in NIS compliance across sectors. The first step toward filling this void is to assess the empirical validity of alternative causal claims concerning the conditions under which Russia's statecraft is expected to work best.

Competing Explanations of Statecraft

Statecraft entails the deliberate use of specific policy instruments to influence the strategic choices and foreign policies of another state. It constitutes a unilateral attempt by a government to affect the decisions of another that would otherwise behave differently.[7] It is also different from concepts that treat power as an instrument of policy or a balance of capabilities.[8] Typically, the practice of statecraft involves the use of diverse policy instruments. Economic statecraft, for example, relies primarily on applying resources that have discernable market prices. The most widely analyzed forms are sanctions and inducements that entail the actual or threatened withdrawal/extension of economic resources to prompt policy change. Similarly, energy statecraft involves increasing or decreasing access to a resource, as well as to related property rights, pipelines, investment capital, prices and tariffs that are extended to deter, contain, or coerce a target. These tools of statecraft contrast with the value of military and diplomatic techniques that are generally stipulated in terms of violence, symbols, or negotiation.[9]

Influence attempts can be discrete, directed at realizing a specific political objective; or general, tied to attaining a broadly defined goal. Similarly, statecraft can be used to affect a specific foreign policy change by a target, or to realize secondary goals by signaling intentions to third parties. Unlike money, which is highly fungible, the conversion of a policy instrument into influence is determined by contextual factors related to the specific domain of application, scope or issue-specific objective, quantity of the resource applied, and costs relative to other techniques.[10] In the Eurasian context, this includes the effects of Russia's energy statecraft on an NIS target's policies regarding ownership, development, and export in respective energy sectors; loyalty towards Moscow's approach to regional energy security; and diversification of energy ties with Russia's great power rivals.

Two basic issues must be addressed to uncover the causal mechanisms for effective statecraft. First, how and under which conditions can states make compliance advantageous for a target?[11] Second, why in some cases can a state secure preferred policy responses from a target, while under other conditions it can only prevent least-preferred options or is altogether impotent to affect the policies of a target? Scholars typically attribute causation to four factors: relative power advantages, asymmetric interdependence, structural power, and domestic institutional fortitude. Although the purpose of this study is not to invalidate these claims, the empirical puzzle of Russia's mixed energy leverage exposes the limitation to each of these stand-alone theories.

Relative Power and Issue-Specific Dominance

Prominent theories of international relations attribute successful state-craft to relative power advantages. The most basic ascribes leverage to material power differentials and to concerns about inferiority.[12] Power is assumed to be generic and fungible across issue areas, vesting the dominant with both incentive and strength to punish noncompliance and extract uniform political concessions across issue areas. This is consistent with the theory of hegemonic stability that attributes a target state's compliance to the lead of an economically dominant power. A hegemon is inclined to supply public goods, such as the enforcement of coercive threats and the delivery of inducements that are otherwise undersupplied because of free-riding among reluctant allies. It also is poised to leverage raw material advantages in pursuit of less benign policy goals.[13]

Other scholars contend that relative advantage alone is too blunt to confer leverage. Rather, predictions of successful statecraft depend on "who is trying to get whom to do what." A hegemon's leverage varies consistently with the scope, domain, and magnitude of its relative power advantages.[14] This is especially apropos to the energy sector, where the sunk costs and appropriability of asset-specific investments should incline a hegemon to exercise imperial control. Because energy plants and pipelines are fixed assets and associated rents can be easily seized, a hegemon with colonial ambitions is well positioned to credibly brandish coercive threats to impose favorable property rights.[15]

Alternative "power" explanations predict that Moscow's leverage throughout Eurasia should be uniform and impressive. Moscow's regional dominance widened across all indices of material power over the years since the Soviet collapse, notwithstanding protracted economic, political, demographic, and military problems that marred Russia's transition.[16] The gap was conspicuous in the oil, gas, and nuclear sectors, as Russia's annual production in each industry exceeded respective demand across the entire post-Soviet space and consistently dwarfed the output of rival NIS suppliers. In these asset-specific extractive industries with relatively high sunk costs and few substitutes, Russian leaders should have faced especially strong incentives to reclaim control as a springboard for reintegrating the NIS.[17]

Yet, preoccupation with relative, issue, and asset-specific power offers little insight into to why Russia's leverage over the NIS varied despite its persistent regional dominance. Lost is appreciation for the reciprocal costs of exercising power, as there were many ways that Eurasian targets were able to temper, offset, redirect, or manage the pressure imposed by Russia. That Russia's interests and general preponderance could not be converted into uniform influence, and that Russia had to make concessions precisely on energy security contingencies where it remained dominant, suggest striking limitations to relative power arguments for leverage.[18]

Asymmetric Interdependence

The rising tide of globalization—marked by the dramatic expansion of commercial and information exchange, deregulation of financial markets, privatization of capital, growing importance of foreign direct investment, and diffusion of technological innovation—intensified interdependence that can be exploited by states. As distilled from the works of Albert Hirschman, Robert Keohane, and Joseph Nye, "sensitivity," measured in terms of the volume and distribution of specific resources exchanged, refers to the extent to which a country is affected by the actions of another.[19] If one state's effort to change a bilateral trading relationship disrupts a greater percentage of another state's overall trade, that asymmetry is predicted to be a source of leverage for the first state. Asymmetries of "vulnerability," which refers to the value each actor assigns to a specific relationship, speak directly to the issue of dependency. Where a target can readily adjust to and insulate itself from the unilateral efforts by another state, it is less susceptible to the other's statecraft. "Even though a hegemon may possess the world's largest consumer market or be the world's largest supplier of a particular set of goods, offering or denying access to these goods may have little effect on the behavior of foreign actors if alternate markets or suppliers can be found, and if the cost of shifting to the alternate source is lower than the costs of the sanctions."[20]

This set of arguments expects Russia's leverage to be significant but varied across the NIS. As part of the legacy of centralization and specialization in the Soviet planned economy, Russia remained the principal trade partner for the southern NIS. Kazakhstan's trade, in particular, was dominated by bilateral exchange with Russia following the Soviet collapse, despite concerted attempts by Astana to diversify economic ties with the outside world.[21] In the energy sector, Astana became increasingly sensitive to imports of Russian oil and gas for domestic consumption, while Moscow steadily reduced its already paltry levels of hydrocarbon imports (for domestic consumption) from Kazakhstan.[22] These arguments also predict that Russia should wield less energy leverage over Azerbaijan and Turkmenistan, as both states traded less with Russia as a share of respective total trade.[23] Yet, Russia remained an integral trade partner for both states, especially in the oil and gas sectors. Therefore, asymmetric interdependence models expect that Azerbaijan and Turkmenistan would be less sensitive than Kazakhstan to fluctuations in trade with Moscow, but that Russia should be able to block their competing strategic energy initiatives nonetheless.

Russia's mixed success challenges classic sensitivity and dependence explanations. Kazakhstan, the most dependent state on trade with Russia, exercised considerable autonomy to search out alternate oil export routes that circumvented Russian territory altogether. However, Turkmenistan, the least dependent of the southern NIS in terms of its aggregate and energy trade balance, was the most deferential to Moscow's preferred energy policies in the region. This pattern was especially puzzling, given the intrusion of outside trading states and availability of alternative energy options. By 1994, the percentage of southern NIS trade with industrialized countries began to displace Russian-NIS trade in all but two southern NIS (Kazakhstan and Kyrgyzstan).[24] Yet Russia's leverage over Turkmenistan's energy security policies continued to exceed its hold over Kazakhstan throughout the first post-Soviet decade.

Globalization and Structural Power

A problem with both traditional "power" analyses and asymmetric interdependence arguments is that they clutch to a narrow, relational perspective of statecraft. Structural power theorists, in particular, question the extent to which markets rule and draw attention to the indirect mechanisms through which states can exert power over global outcomes.[25] However, some states rely on the "second face of power," controlling not only what other states do but what they want.[26] By its sheer weight in "deterritorialized" markets and institutions, a preponderant state can skew material incentives and trigger policy adjustments from foreign targets merely by taking action at home.[27] Alternatively, power can come from the allure of beliefs, practices, and identities that are constructed through political, cultural, and social interactions, independent of balances of power or trade. Leverage is exercised in the spirit of emulation and consent, with the hegemon setting normative standards for domestic and regional targets alike. Conversely, the more independent a target's national identity, the less susceptible it is to another's example.[28]

Nowhere was Russia's structural presence greater than in the Eurasian energy sector. The residual Soviet pipeline network constituted a "steel umbilical cord" that bound Eurasian energy suppliers and customers to Russia.[29] Because most of Kazakhstan's oil and Turkmenistan's gas exports tapped directly into Russian pipelines, both states were at the mercy of Moscow's practices for regulating access to national export terminals and pipelines.[30] This plausibly stood to increase with the participation of Russian oil and gas firms in international energy consortia, and proliferation of debt-equity swaps for residual Soviet energy assets. Accordingly, action taken by Russia to reduce national subsidies, adjust domestic prices, reallocate pipeline access, and reorient national energy production toward international markets should have inflicted energy shocks across the SCCA. Because Eurasian supplier and transit states tied national security and welfare, as well as near-term political legitimacy, to the projected payoffs of extracting and exporting regional energy, they should have been increasingly responsive to shifts in Russia's domestic energy policies.[31]

Yet Moscow's structural power did not produce uniform deference on regional energy security issues. The monopoly over the residual main oil export pipelines and the growing presence of Russian oil firms in international consortia did not fundamentally alter the strategic calculus in Baku or Astana. Conversely, the insertion of Russian gas interests into projects underway in Turkmenistan strengthened Moscow's extraterritorial leverage. This variation suggests that market power alone did not equate with bargaining power or strategic influence.[32]

To the extent that Central Asian states strived to emulate political traditions and identities of other states, they did so with little regard for Russia. Turkmenistan fancied itself as the "Kuwait of Central Asia" and cultivated its own brand of pan-Turkic authoritarianism as the springboard for nationalist resurgence. Since independence, Azerbaijan too sought to create a new normative niche in the region, deriving little sustenance from Russian institutions and practices.[33] Ironically, compliance with Russia's energy security policies was most impressive among those states that harbored strong and autonomous national identities, such as Turkmenistan, and relatively weak in the case of

Kazakhstan, a state with limited statehood experience and only a nascent political identity.[34]

Finally, structural power arguments fail to capture the interplay between Russia's regional and global standing. Following the Soviet collapse, outside actors became increasing involved in mediating ethnic and separatist conflicts, conducting military training exercises, and investing in energy development and export projects in Eurasia.[35] Yet the discussion of Russia's structural dominance typically takes place in a geostrategic vacuum, with little regard for the presence of the People's Republic of China, European Union, United States, Turkey, Iran, and multinational firms and the opportunities that they introduced for diversifying structural relations throughout Eurasia.

Institutional Strength

A collective problem with arguments rooted in power and interdependence is that they assume that potential domestic capabilities, including those strategic resources under private ownership or control, can be readily marshaled for geostrategic objectives. Yet in states where corporate decision making on energy does not privilege noneconomic objectives, great power advantages do not readily convert into credible forms of coercion. Rather, this suggests the importance of the domestic institutional components to statecraft.

One hypothesis attributes international leverage to a regime's institutional "strength" relative to domestic legislatures and interest groups.[36] Policy makers that stand above the parochial interests and competition among societal forces—to extract, mobilize, and employ national resources toward a specific foreign commitment—can wield significant international leverage. Enjoying autonomy to impose unpopular policies, authoritarian regimes should be more effective at channeling national capabilities directly toward foreign objectives. Alternatively, "democratic statecraft" is tempered by the related difficulties of restricting legislative dissent and building supportive coalitions among competing interest groups that ultimately render the efficacy of leverage contingent on which group prevails on a specific policy issue.[37] As a quasi-democratic state, Russia's ability to practice energy statecraft is plausibly predicted to mirror the ascendancy of alternative domestic coalitions comprised of concentrated energy interest groups, financial oligarchs, and independent-minded regional leaders, on the one hand; and environmental protection lobbies, advocates of *realpolitik,* and "Great Russian" nationalists, on the other hand.[38] Periods of intense competition between these domestic lobbies, however, should match inconsistencies in Moscow's energy leverage.

An alternative institutional argument stresses the importance of a central leadership's capacity to monitor and enforce domestic compliance. The degrees to which political authority is centralized and policy makers can observe the actions of functionaries and firms determine a regime's competence at securing extraterritorial compliance. Governments that can detect and reverse the opportunism of self-interested administrators and interest groups to promulgate and implement coherent foreign policies are taken more seriously by target states. Conversely, gaps in administrative oversight encourage bureaucracies and firms to exploit advantages in information and expertise by pursuing policies to satisfy their narrow interests, thus crippling the coherence and credibility of a hegemon's statecraft.[39] Suffering from rampant shirking among competing federal branches, agencies, and local authorities, the Russian government should be expected to exert only marginal and ad hoc leverage, marred by contradictory policies toward the ownership, extraction, and export of the region's energy.[40]

Arguments related to regime type and institutional capacity apply to target states as well. Those regimes that are institutionally insulated from disenfranchised political opponents or interest groups, and that are free to redistribute the costs or benefits attendant to changing international conditions, are well positioned to rally domestic support to thwart foreign pressure. However, regimes that must accommodate organized or concentrated interest groups that benefit from international pressure, are more likely to comply with another country's statecraft.[41] Some scholars suggest that the effectiveness of statecraft ultimately hinges on the interaction of special interest-groups within the influencing and target states that are able to exploit weak state structures to pressure home governments to impose and respond to sanctions, respectively.[42] Other studies find that issues related to institutional stability and capacity in a target are inversely related to compliance. Those regimes that are economically or politically unstable are more responsive to foreign pressure than are those with stronger capacity.[43]

Neither institutional strain systematically captures the variation in Moscow's energy leverage. Although the Central Asian regimes notably differed in leadership skill and in degrees of authoritarianism, they shared common defining attributes of neopatrimonialism, weak infrastructural and societal institutions, and arbitrary rule during the first decade of independence.[44] Furthermore, interests in the Russian energy sectors were not monolithic, as respective lobbies were comprised of numerous private and semiprivate companies with divergent strategies for tapping domestic and international resources and markets that at times compromised Russia's energy statecraft.[45] These models also understate the coercive potential of weak state structures. It is precisely because democratic governments are restrained by domestic dissent, that they are less likely to engage in strategic bluffing and that the credibility of the their threats are enhanced when selectively issued.[46] Neither can it be assumed that a democratic government is a mere transmission belt for parochial commercial interests. Unable to prevent its own firms from collaborating directly with foreign states and companies for the exploration and export of Eurasian energy resources, the Russian government nonetheless was able on occasion to induce corporate and regional compliance by amplifying the political risks of independent behavior. As summed up by one specialist, Moscow was free to act "opportunistically and negatively" in the region, with sufficient strength to disrupt, stall, and guide the private interests of the national gas industry.[47]

Similarly, explanations that turn on weak governing institutions typically understate the consistency and effectiveness of a state's extraterritorial reach. While such arguments account for Moscow's inability to prevent provincial leaders from contracting separately with foreign energy consortia, they slight the federal government's capacity to exploit subnational rivalries to induce compliance.[48] For the most part, Moscow's heavy-handed gas policies were not subverted by the "involuntary defection" of Russia's powerful gas lobby, despite the government's weak enforcement powers and heated battles with corporate executives over regulation of the industry.[49] Moreover, that the Russian government's capacity to do so remained constant, despite the increasing consolidation of vertical power under the Putin regime, speaks to the nuanced effects of domestic institutions.

Moving Beyond the Coercion-Inducement Dichotomy

Each of the above single-factor explanation identifies conditions conducive for two basic forms of statecraft: coercion and positive inducement. The most ubiquitous and heavily studied is coercion, where the focal point for diplomacy lies with issuing "a demand on an adversary with the threat of punishment for noncompliance that is credible and potent enough to persuade him that it is in his interests to comply with the demand."[50] Coercion is distinguished from "brute force," as it involves latent pain and holding power in reserve.[51] It can be used either to "blackmail" another state into conceding something of value, or to encourage an opponent to cease a specific action.

The logic of coercive diplomacy rests on a straightforward utility maximizing model of state decision making. Success occurs, according to Robert Pape, when a state can alter the substantive meaning for an adversary of accepting or resisting specific demands. It obtains "when the benefits that would be lost by concessions and the probability of attaining these benefits by continued resistance are exceeded by the costs of resistance and the probability of suffering these costs."[52] Alternative strategies of coercion focus on changing different components of this decision calculus to ensure that the expected costs of noncompliance exceed the anticipated advantages of resistance. For example, punishment strategies inflict pain on an adversary's population, or attack key economic choke points to overwhelm the political will to resist. Risk strategies, however, raise the probability that an adversary will suffer unacceptable costs with noncompliance by gradually increasing damage until demands are met. Denial strategies thwart a target's noncompliance, reducing the probabilities that it will be able to realize alternatives to compliance or to extract meaningful value from resistance.[53] Success for each of these strategies requires that the initiator communicate explicit threats that can be credibly carried out if resisted.

The second form of statecraft is comprised of positive inducements. Inducements change the focus from raising the probabilities and direct costs of noncompliance, to improving the expected utility of accommodation for a target state. This can be accomplished by creating mutually beneficial exchanges, or by granting better terms of trade on one issue in return for compliance on another.[54] Inducements can be used defensively to resolve a crisis, satisfy the appetite of a greedy opponent, or reassure a nervous adversary; as well as proactively to socialize or coax another state into accepting a policy defeat or changing political course.[55]

Positive inducements operate in substantively different ways to sanctions. Carrots can signal sympathy to a target and generate constructive spillover effects onto other issues that ease the pain of compliance. They can promote mutually beneficial payoffs of compliance, especially when they involve exchanging goods with increasing marginal utilities.[56] In contrast to coercive diplomacy, inducements also can stimulate domestic political support in both the inducer and target states that reinforce the impetus for compliance. As noted by William Long, proffering incentives creates new constituencies that gain increasingly from exchange. As potential "win-win" exchanges, incentive strategies are less prone to inciting a "rally around the flag" effect within the target, and can create a domestic lobby with a stake in supporting the political concessions that are demanded. Such strategies also are unlikely to spark interest in exploring alternative responses or to encourage third parties to disrupt an exchange.[57]

However, there are higher bargaining costs to proffering inducements. Threats are cheap to issue and only costly if they fail. When extending carrots, however, a state will have to ante up with the target's subsequent compliance, requiring follow-through that is unnecessary for coercion.[58] Under conditions of international anarchy, contracting between states to extend and sustain an inducement is especially difficult to observe and enforce. In the absence of international institutions or domestic checks that can bolster the credibility of compliance, targets have strong incentives to cheat on and extort an inducer. Concessions are more apt to degenerate into appeasement that, in turn, increases both the anticipated domestic and international costs to the inducer of being played the sucker. Furthermore, extending diplomatic carrots is a second best option for dealing with future rivals. Because states are presumably preoccupied with maintaining a reputation for tough bargaining and are loath to risk conferring material advantages on an expected rival, they are inclined to forego proffering inducements today that could make coercion more difficult tomorrow.[59]

The end of the Cold War and gathering trends of globalization confound these distinctions. Great powers are increasingly called on to influence the choices of smaller states across a wide array of security concerns that include WMD nonproliferation, energy security, terrorism, organized crime, and ethnic and civil strife. These issues are distinguished by asymmetries of interests that offset obvious regional and global power advantages, complicate the process of mobilizing domestic support, and confound strategic bargaining. They create new vulnerabilities while simultaneously expanding the scope and altering the potency of different nonmilitary policy tools. The situation is compounded by the broadening and deepening of transnational ties among subnational actors that operate beyond the purview of respective home states. Together these trends create a new strategic context that exposes conceptual limitations common to nearly all work on statecraft.[60]

First, traditional assessments of statecraft narrowly focus on "hard" material instruments of statecraft. Often overlooked, however, are the strategic dimensions to "soft power" that involve "getting others to want what you want" by shaping the situational context and opportunities surrounding a target's choices. As elucidated by Joseph Nye, states can control policy outcomes not only by exerting direct pressure, but by setting the political agenda and framing the terms of debate.[61] With expansion of global markets and deepening penetration of transnational pressures, a state can shape a target's options, values, and domestic incentives associated with exchange that under certain circumstances can be exploited for political effect. Interdependent commercial supply and distribution networks, in particular, empower states to alter the incentives for third parties that, in turn, can affect the range and substance of choices available to a target. Although less able to command multinational firms and subnational actors to do its foreign bidding, a state can guide private incentives by altering the business climate and public support at home to ensure a coincidence between diplomatic objectives and subnational practices.[62]

An ancillary problem relates to the narrow conception of a target's situation and decision calculus. Because the purpose of statecraft is to get another state to undertake action that it does not otherwise value, the challenge is not only to induce a specific policy course, but to discourage subsequent defection. Attention is generally confined to altering a target's cost calculus because, as Pape claims, "the benefits of resistance are not usually manipulatable."[63] This, however, neglects situations where targets can be coaxed either to adjust their bargaining strategies or to pursue different agendas that they find ultimately more rewarding than the alternatives. Consequently, a target may comply with the wishes of another state not because it is being explicitly coopted or is under duress, but because its decision-making circumstances have changed, rendering compliance not only tolerable but desirable.[64]

Several empirical episodes illustrate the point. Germany in the 1930s, for example, exploited the monetary dependence of the small states of southern and eastern Europe to manipulate the appeal of accepting Berlin's domination. By altering the value and availability of the home currency, Berlin was able to affect indirectly the value of respective national currencies to extract wealth, and to harmonize the incentives of firms and government agencies within a target state with Germany's needs for war and expansion.[65] In the case of Rhodesia, although the white minority government initially exploited British and UN sanctions to alter domestic markets to reward supporters and punish dissent, the respective adjustment ultimately institutionalized

incentives that fostered international compliance and eventual acceptance of majority rule with Zimbabwean independence.[66] Even in contests over territory, where differences between states can be indivisible, a
target's preferences over actions are not fixed and can evolve in
response to new opportunities or the reconstruction of national identities. During the 1970s, for example, shifts in the external and internal
settings dramatically altered the value that the Egyptian leadership
placed on the "occupied territories," paving the way for President
Anwar Sadat's "peace initiative" with Israel.[67] Given the increasing den-
sity of global interdependence and reduced reliance on territorial bases
of power, national attachments to strategic assets are becoming more
diffuse and malleable than in the past. Although transformation of a
target state's preferences might not occur at the outset of a dispute,
there is no logical reason to preclude adjustment over time.

A second flaw of "hard" power myopia relates to the restrictive
characterization of positive inducements. Skeptical of the relevance of
economic power for security concerns altogether, pessimists indiscrimi-
nately associate inducements with appeasement—the "original sin" of
diplomacy. As epitomized by the Munich crisis of 1938, the costs of
inducements are not controllable and are fundamentally detrimental to
subsequent attempts at coercion.[68] Appeasement ceded strategic advantages to Germany, while simultaneously whetted Berlin's appetite for
future concessions and undermined the credibility of Anglo-French
threats to stand firm in the ensuing Polish crisis.[69] Although inducement
optimists counter by pointing to the strategic and domestic advantages,
they too blur the line between constructive incentives and counterproductive appeasement. This is due largely to a fixation on concessions as
the bulwark of inducements. Yet, it is possible to induce without conceding, as states can undertake self-interested action on one issue that
renders political compliance on another more rewarding to an adversary. A state can induce an adversary, not only by acquiescing as part of
a political quid pro quo, but by taking action at home or abroad that
effectively alters a target's options so that the preferred outcome
becomes mutually desirable. Even at the extreme, appeasement is not
uniformly self-defeating or irrational. Not only are there cases of successful appeasement, including successive British concessions to the
United States from 1895 to 1905, but systematic studies of crisis behavior have uncovered that reputation or prior conciliatory acts do not fundamentally discredit a defender's threats. In short, the efficacy of
appeasement varies with the nature of the adversary's ambition, number
of potential challengers, character of the inducement proffered, and

availability of other reinforcing incentives.[70] Therefore, thinking about inducements must be distinguished from explicit policy concessions and from unique cases of misguided or poorly executed diplomacy.

A third problem relates to the treatment of inducements as pure substitutes for coercive diplomacy, notwithstanding differences in substantive logic. In practice, policy makers alternate their use of carrots and sticks. Russia, for example, simultaneously threatened and enticed Azerbaijan's compliance with its Caspian energy security policies. The same held for Moscow's gas diplomacy toward Kazakhstan and Turkmenistan, but with dramatically different results. If positive and negative incentives are substitutes, why did they produce different outcomes in comparable circumstances? By omitting comparative analysis, many studies fail to gain purchase on the practical utility of this dynamic relationship or on the relative effectiveness of these logically distinct forms of statecraft.[71] A full understanding of the relative effectiveness of positive and negative inducements requires examination of cases where they are applied together but produce different results in target behavior.

A fourth shortcoming with the traditional conception of statecraft is that influence attempts are assessed as static, one-sided contests. Yet, the presumption of a linear relationship between threat (promise) and response does not comport with the strategic dynamics of statecraft. Targets do not stand idle in the face of international pressure but actively work to offset an initiator's efforts either by taking steps to redirect or reduce the potential consequences of coercive threats or by imposing greater costs on the coercer and its domestic allies.[72] At the same time, initiating states are not oblivious to a target's potential countermoves but factor them into their initial and follow-up actions. The dynamics of this strategic interaction presumably have intensified with the growing complexity of multidimensional interdependence. This warrants moving beyond strict target-centric analysis, with greater appreciation for how initiators can alter the availability and value of countermoves, as well as the expected utility of compliance for target states.

A more accurate assessment of statecraft also needs to take into account that influence can be garnered without explicit quid pro quos. Although there are obvious advantages to analyzing issues where matters of statecraft are publicly stated and clearly brought to a head, these cases represent a limited set of interactions that have not been resolved at other levels. The phenomena of "Finlandization" and "Yanqui-ization," for example, reflected the implicit influence of the Soviet Union and the U.S. in respective strategic orbits. Given respective situational backdrops, Moscow and Washington rarely had to issue explicit threats or proffer inducements to influence respective target states. The implied consequences of noncompliance usually sufficed, except in notable episodes where diplomatic activity escalated into a public crisis or failed outright.[73] The fixation on crisis diplomacy constitutes a potentially serious selection bias that neglects the many successful applications of statecraft that never reach a fever pitch.

The preoccupation with assessing crisis diplomacy also neglects the subtle practice of statecraft over an extended period. A state can preempt another's behavior by shaping the context within which a target makes its choices. Accordingly, the value of a specific instrument of statecraft "must be measured for its contributions to the background level of pressure as well as to the spikes in the level of threat."[74] Focus on noncrisis influence attempts can preserve the heuristic value of the cost-benefit framework for understanding statecraft, notwithstanding valid critiques that underscore deviations from rational decision making that can distort crisis behavior and make it difficult to predict state responses.[75]

Finally, that statecraft is typically assessed as either working or not presents both empirical and conceptual problems. While the effectiveness of compellence is blatant, as states either do or do not change their policies in response to threats, successful deterrence is almost impossible to discern from an adversary's overt behavior. Statecraft also almost never produces clear-cut outcomes, as targets rarely surrender outright or modify behavior without pursuing their own policy objectives. Because states pursue diverse objectives, the absence of compliance along one front also does not necessarily imply the failure to realize other objectives. The British, for example, pursued a range of objectives with the embargo on Rhodesia, including signals to Black Africa that it was committed to preventing the breakup of the Commonwealth and avoiding sanctions imposed by South Africa. It is too simplistic to code the collapse of the embargo and early problems with realizing the stated objectives of changing to majority rule as an outright failure.[76] Conversely, successful statecraft does not necessarily require the full satisfaction of every demand, as an adversary can modify is behavior to satisfy an initiating state's minimally acceptable stipulations. As David Baldwin cautions, success defined as compliance with explicit threats raises the bar too high as "(t)hird parties, secondary goals, implicit and unstated goals are likely to be significant components of such undertakings."[77]

An important step toward conceptual clarity, therefore, is to think of success in terms of altering the decision calculus of an adversary rela-

FIGURE 1.2
Target Decision Tree

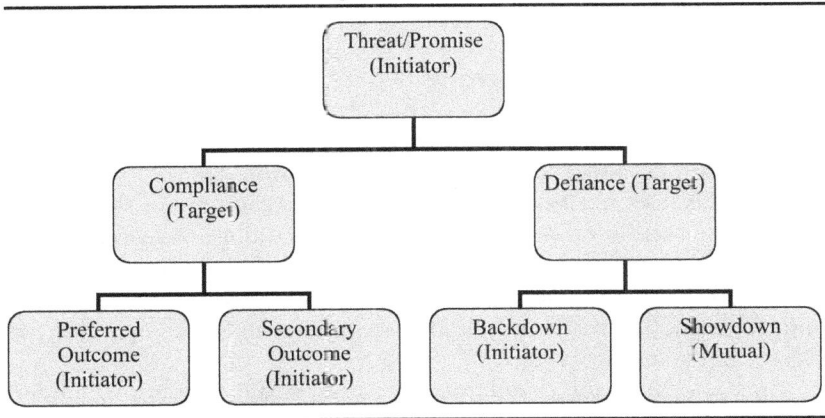

```
                    ┌─────────────────┐
                    │  Threat/Promise │
                    │   (Initiator)   │
                    └─────────────────┘
                   ┌──────────┴──────────┐
        ┌──────────────────┐      ┌──────────────────┐
        │   Compliance     │      │ Defiance (Target) │
        │    (Target)      │      │                   │
        └──────────────────┘      └──────────────────┘
          ┌──────┴──────┐           ┌──────┴──────┐
┌───────────┐ ┌───────────┐ ┌───────────┐ ┌───────────┐
│ Preferred │ │ Secondary │ │ Backdown  │ │ Showdown  │
│  Outcome  │ │  Outcome  │ │(Initiator)│ │ (Mutual)  │
│(Initiator)│ │(Initiator)│ │           │ │           │
└───────────┘ └───────────┘ └───────────┘ └───────────┘
```

tive to the priorities pursued by an initiator and options available to a target, as diagramed in Figure 1.2. Initially, a target can choose to resist or comply to a credible threat or promise. If the target resists, the initiator may respond either by backing down before a target accedes or by forcing a showdown. Both outcomes reflect the failure of statecraft, given the initiator's initial preferences and targets choices. Alternatively, if a target chooses to acquiesce, then the episode can terminate in one of two ways. First, a target can comply with the initiator's preferred outcome. Second, a target can concede by meeting the initiator's minimally acceptable demands, including mutually beneficial accommodation. While this outcome constitutes a relative success, it is less preferable as the target opts not to pursue the most offensive policies to the initiator while continuing to privilege its own preferences.[78]

Expanding the Diplomatic Arsenal

Theoretical examination of statecraft requires attention to the circumstances under which one state can marshal its hard and soft power resources to convince another to make the desired political changes. The efficacy of statecraft depends not only on the promise of rewards or pain, but on the corresponding ability of an initiator to structure a target's decision-making context. I refer to this as "strategic manipulation" that, unlike strategies of coercion and inducement, does not prejudice either threat-based or threat-reducing elements of statecraft. Rather, manipulation encompasses the application of both forms in a single concept.

This broader conception of statecraft places the onus of success on the initiator. In contrast to recent studies of sanctions and coercion that analyze the domestic conditions surrounding target state responses, the premium here is placed on assessing how a sanctioning state can shape these conditions and the utility of compliance relative to roads not taken.[79] This does not neglect the political costs for a target state or the dynamic interaction between initiator and target, but recognizes that an initiator must invariably anticipate an adversary's countermoves. It is precisely because the practice of statecraft is a dynamic contest that an initiator must build into its diplomacy provisions to compensate for likely target state responses.

Similarly, attention to manipulation is especially relevant for understanding the strategic implications of soft power. The globalization of commercial, financial, production, and distribution relations increase at once the breadth and depth of interdependence. States both lose and gain new dimensions of sovereign control amid rising trends of globalization. They lose control over firms and subnational actors that are increasingly beholden to the dictates of networks and markets. Yet states that are increasingly ensconced in mutual dealings have at their disposal multiple avenues for shaping respective interactions. As noted by George Shambaugh, "when one state controls the activities of firms within another state's territory, it not only decreases the physical resources at that state's disposal, but also undermines is political integrity by challenging its ability to control actors and activities within its territory."[80] As depicted in Figure 1.3, this creates new and more intimate opportunities for states to skew the decision making of target states without resort to strictly bilateral forms of threats or punishment. To the extent that states can determine the behavior of third parties or actors operating within another state, they can expand the scope of their influence by shaping the situational context and structuring the choices presented to a target. Viewed in this light, globalization creates new opportunities for practicing more indirect, nuanced, and intimate forms of soft diplomacy.

Identifying the conditions apropos to strategic manipulation also is important for redressing the "ex ante-ex post" dilemma that otherwise complicates international strategies of coercion and inducement. Because confrontations are costly to all sides, states have incentive to reach peaceful resolutions beforehand that allow them to realize their

FIGURE 1.3
Framework for Analyzing Statecraft

Relative Power

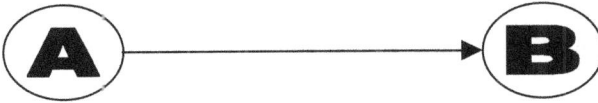

Interdependence

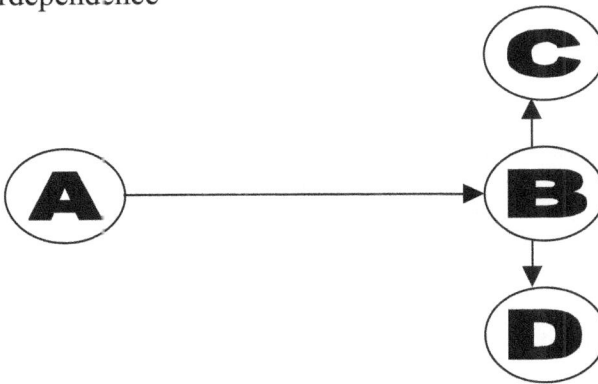

Manipulation

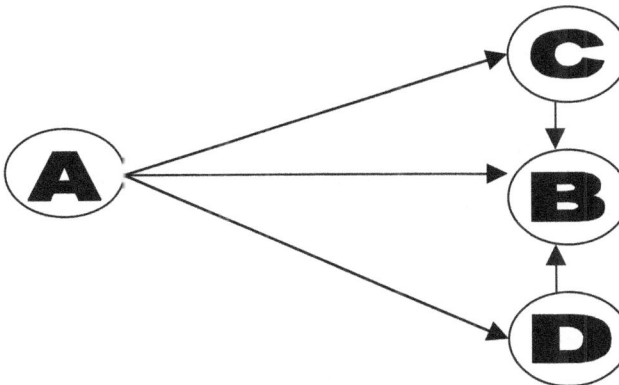

interests without paying hefty prices. If rivals share complete information on how each evaluates the outcomes and costs of a showdown, then it is relatively easy for them to identify a mutually acceptable solution. Under conditions of international anarchy, however, states must bargain with incomplete and asymmetric information, as they know more about their own willingness and ability to follow through on threats and promises than can be observed by respective adversaries. This adds confusion to each state's expectations about the others' preferences and responses. Strategic interaction is driven by efforts to demonstrate resolve, as each state seeks to marshal its capabilities and send diplomatic signals to convince the other that it is determined to back up its threats or promises.

The rub for coercive diplomacy is that the credibility to follow through on threats is intrinsically dubious. This is because the costs of carrying out the threat often are greater than the potential benefits derived from an adversary's compliance.[81] Except under conditions where the stakes are immense and the costs of confrontation are negligible, it is often more beneficial not to carry out threats or promises. Because states have incentives to exploit information asymmetries and to exaggerate their commitments, not all promises or threats are taken seriously by an adversary. Although an initiator can improve the credibility of diplomacy by trading on a reputation to honor threats and increasing the domestic "audience costs" for failing to uphold them, uncertainty regarding an initiator's willingness to pay costs ex post is never fully removed from a target's ex ante decision calculus.[82]

It is here that the focus on soft power and strategic manipulation differs significantly from classic forms of coercive diplomacy. Unlike threat- or reward-based statecraft that relies on pledges that are inherently suspect, manipulation privileges preemptive diplomacy that alters opportunity costs and benefits of compliance. This entails undertaking specific actions up front to affect the relative appeal of different courses of noncompliance, as opposed to threatening to alter the probabilities and utility of compliance. Because opportunity costs and benefits of compliance are presented ex ante, whereby targets make choices between policies that carry certain gains and losses, the uncertainty and credibility of ex post actions becomes less salient.

Conclusion

By including the dynamics of soft power and strategic manipulation, this study broadens the scope of statecraft. The purpose is to highlight not only different forms of statecraft, but to understand when and why different strategies are successful. Instead of examining how states are able to directly punish and reward a target state, this approach emphasizes the incentives and causal links that translate soft power into strategic manipulation. To build theories about how states are able to manipulate indirect and dispositional forms of statecraft necessitates adopting a microfoundational approach to understanding the incentives confronting public and private actors to line up behind a nation's influence attempt. This, in turn, raises several basic questions: What are the dimensions to soft power and how can they be manipulated? Is manipulation sufficient to incite meaningful changes in a target's strategic behavior? Under which conditions is strategic manipulation most likely to produce preferred outcomes?

2

❖

Strategic Manipulation

This chapter presents an explanation for how states can manipulate interdependence to secure strategic aims short of war. The key challenge here for the initiating state is to marshal both "hard" and "soft" power in a manner that alters a target's decision-making context. Success turns on exploiting a target's vulnerability by framing the range and substantive content of choices so that it draws comparable conclusions about the value of compliance.

The chapter begins by introducing "risk" into the traditional formulation of statecraft and the common critiques. The second section addresses these shortcomings by developing a theory of strategic manipulation. Building on insights from prospect theory, soft power, and neo-institutionalism, I generate propositions regarding the specific techniques and conditions under which policy makers can manipulate the decision frames and relative riskiness of compliance for successful statecraft. I deduce that when a state enjoys global market power in a specific energy sector, as well as operates within a clearly delineated regulatory system at home, it will be able to shape a target's decision-making context so that strategic compliance holds out more favorable prospects than noncompliance. The final sections outline the research methods and design used to illustrate the argument for explaining Russia's variable energy leverage that follows in part II.

Manipulating Risk Versus Uncertainty

Coercive diplomacy and traditional forms of statecraft are typically assessed in terms of persuasive power; the ability to pressure policy makers in a target state to change preferences over specific bargaining strategies. The causal link between policy tools and strategic effects hinges on how the specified variables influence the exchange of information and, in turn, the cost-benefit calculations of target states. The presumption is that decision makers choose policy courses that maximize net expected utility, given a set of fixed preferences over outcomes. Targets weigh the payoffs of each possible outcome of a given course by the probability of its occurrence, summing over all possible outcomes for each option, and selecting that option with the highest expected utility. The utility of compliance is assessed relative to that of available forms of noncompliance; there is no difference between actions that either increase the costs or decrease the benefits of compliance by commensurate values. This approach has come under attack from those that question the validity of the rational actor assumption altogether. Individual decision makers, it is argued, confront emotional, normative, motivational, and cold cognitive constraints that impair rational maximization of choice.[1] Yet, even without jettisoning the rational choice perspective, strict adherence to a simple expected utility model is problematic.

First, there is considerable evidence, both experimental and historical, that policy makers evaluate costs and benefits differently depending on their predicament. Prospect theory reveals that individuals are more sensitive to changes in values than to changes in the level of net expected utility. As applied to the political arena, decision makers tend to be more sensitive to gains and losses from a specific reference point than they are to changes in aggregate levels of wealth and welfare.[2] One implication is that decision makers value what they have more than what they stand to gain from making a good situation better. Another is that policy makers tend to inflate outcomes that are certain compared to those that are probable, while giving more weight to possible outcomes than to those that are either moderately or highly probable. Because losses loom larger than gains and probabilities are weighed differently, policy makers will prefer to take greater risks in order to avert even certain losses, while becoming risk averse in the face of sure gains.[3]

Given loss aversion and the variable risk-taking propensities associated with distinct decision frames, a target state can be expected to respond differently to threats and promises of equivalent values, as well as to variations in the presentation of alternatives. Situated in the domain of gains, for example, targets will more likely be conservative and prone to comply with an initiator's coercive measures in order to avert a possible loss. Alternatively, a target with a baseline in the domain of losses will be acutely sensitive to incurring even an improbable loss with compliance, thus prone to taking further gambles in defiance of an initiator's threats. Only by conspicuously reducing the measure of loss that a target will incur by complying can an initiator overcome a target's desperation to avoid loss and induce compliance. This suggests that a target's frame can bear directly on the efficacy of an initiator's statecraft, with fundamentally different consequences for compliance than predicted from an expected utility perspective.[4]

A second fundamental shortcoming with the simple expected utility principle is that future utilities are discounted and conflated with present utilities into a single value function.[5] However, a national leadership's assessment of the value of compliance is affected not only by the costs and benefits at a particular moment in time, but by the overall expected value of peace or conflict for the foreseeable future.[6] For example, if a target has positive expectations that the other state will sustain free and open trade (cooperation) over the long term, then the expected value of trade (cooperation) will be close to the value of the benefits of trade (cooperation), thus making compliance more acceptable in the short-term. Conversely, for any given expected value of conflict, the lower expectations of future trade, the lower the expected value of trade, and therefore the less likely a target is to comply with the security demands of an initiator at that moment. Because trade expectations differ across targets and circumstances, responses to specific forms of statecraft are likely to vary significantly.

A third problem with the traditional expected utility formulation is that it is preoccupied with the role of uncertainty in competitive bargaining situations. Typically, a premium is placed on demonstrating resolve and credibly signaling information about following through on threats and promises. An initiator must possess the bargaining skill to communicate and exploit private information to convince a target that it is not bluffing and will stand firm behind its commitments.[7] This formulation, however, neglects the role of risk in bargaining situations. Unlike uncertainty, which typically concerns incomplete information and the probabilities assigned to net expected utilities, risk pertains to the stakes at issue for decision makers. Risk is about the values assigned to a potential outcome, and thus tied to a decision maker's assessment of losing an important value or failing to obtain a desired goal. Accordingly, risk-taking propensities vary consistently with differences between prospective positive and negative outcomes tied to a specific choice.[8] Risk-acceptant behavior involves selecting an option that holds prospects for more extreme outcomes (positive and negative) than the range of other available options. Even in a situation where a mutually acceptable deal is conceivable and an initiator is able to extend a credible threat, target policy-makers could still be driven to resist if compliance is tied to a likely and significant loss relative to unlikely but significant gains associated with other options. Alternatively, if a target is satisfied, lingering questions about the credibility of an initiator's threats may not outweigh the overall penchant for risk-aversion in selecting an option that holds the smallest difference between possible positive and negative outcomes. In short, risk-taking propensities reflect the magnitude of prospective positive and negative outcome values for an option that vary distinctly from information asymmetries and the probabilities associated with specific choices.

That states must contend with risk as well as with information asymmetries has direct consequences for effective statecraft. Because a target's baseline assessments, or decision frames, are dynamic they create opportunities for influence.[9] On gauging how an adversary perceives its predicament (either in the domain of losses or gains), an initiator can alter information accordingly. It can tailor specific policy instruments to amend the target's view of the status quo or to clarify a likely future situation in order to elicit a favorable response. The former can be achieved by reassuring a target that is otherwise convinced that it has a lot to lose from compliance.[10] Furthermore, an initiator's diplomacy can clarify a target's calculus by sending signals that effectively manipulate expectations of future conflict.[11]

Although exposing descriptive limitations of simple expected utility formulations, these insights do not by themselves present a theory of statecraft. The extant literature on prospect theory, for example, tends to assume that an initiator can readily marshal generic forms of power to shape the context within which target states make decisions. Which dimensions of hard or soft power can be exploited at the discretion of a national leadership to produce the desired policy from a target, and which instruments are relevant for doing so on a particular issue, however, are left unexplained. Similarly, without specifying a priori issue saliency, researchers may inadvertently associate a target's risk-taking behavior in one issue area with its desire to avoid loss on a completely unrelated issue.[12]

Another problem concerns the difficulty of altering a target's decision frame. Targets that are predisposed to view their predicament in the domain of losses are inclined to dismiss small bargaining gestures as insignificant, regarding direct threats and promises that are inconsistent with prior beliefs with suspicion. In negotiating Cold War arms control agreements, for example, both superpowers were more sensitive to losses in security derived from reducing their own arsenal than from the security produced by the other's downsizing.[13] Though clear signals can mitigate asymmetries of information, they are not necessarily effective at reconciling divergent predispositions. The bombing pauses in 1965 that were meant by the United States to demonstrate restraint were dismissed by North Vietnamese leaders as signs of weakness that, in turn, stiffened Hanoi's resolve at waging war.[14] These episodes suggest that it is very difficult for an initiator to tailor statecraft toward reordering the relative weights of different policy options and revising the substantive preferences within a target's specific decisional context.

Third, prospect theory is silent on the causal link between a target's frame and substantive foreign policy orientation. Making peace, as well as war, is a risky proposition that can expose a target to foreign and domestic exploitation.[15] For NIS targets situated in the domain of losses, under which conditions is it more advantageous to assume greater risks by acquiescing to Russia's energy statecraft, banking on Moscow's reciprocal goodwill and endorsement by domestic audiences, than by throwing caution to the wind in challenging Russia? Both may be high-risk strategies for securing something of value but carry radically different implications for compliance. Thus, an initiator must influence a target's risk-taking orientation, as well as the riskiness of compliance (i.e., gap between prospective positive and negative outcomes) relative to other feasible options to manipulate behavior.

Another typical limitation of prospect theory stems from an unrealistic view of national decision making. The discussion of frames and expectations of gains and losses tends to aggregate individual assessments to decision-making groups, assuming away the institutions that delimit the governing capacity of a national leadership. But to use strategic goods in pursuit of foreign policy objectives, policy makers must not only frame a target's choices and manipulate reference point calculations; they must also be able to pay the domestic political price of overcoming obstacles to mobilizing resources and implementing respective policies. Even within extremely hierarchical state structures, a central executive must contend with agency costs, as government bureaucracies and private actors enjoy residual control over policy resources that allow them to pursue agendas at odds with national interests. These administrative problems are especially relevant to energy statecraft, as resources and infrastructure are typically owned by both state and private actors, and domestic regulatory responsibilities are usually delegated to multiple government agencies and quasi-state actors. The global dispersion of markets and corporate transactions render multinational firms increasingly beyond the grasp of national security policy-makers who might be determined to exploit commercial relations for purposes of statecraft.[16] Shirking by these domestic agents can effectively create substitute exchange relations either at home or abroad to alter the value of compliance for a target state, further complicating the initiator's statecraft. Because neither administrative nor corporate compliance can be taken for granted, a state's capacity to alter a target's frames and the relative riskiness of compliance via energy diplomacy can vary with the institutions that regulate exchange within the respective national energy sector.

To recap, missing from the application of prospect theory to statecraft is focused exploration of how an initiator can effectively employ policy instruments to amend a target's decision making, changing its prevailing frames and affinity for specific policy options. The challenge for an initiator, therefore, is to take steps that unambiguously but innocuously alter available choices and expectation levels to compensate for a target's heightened sensitivities to future losses or gains. This raises two central questions for a theory of strategic manipulation: Which specific strategies are appropriate for a target's risk orientation? Under which conditions can an initiator's statecraft have the desired effect?

Manipulating for Strategic Effect

Assumptions

There are four suppositions that underlie the microfoundations of strategic manipulation. These factors specify which actors matter most for practicing statecraft, what they want, and the setting within which they interact. Clarity on these basic issues facilitates systematic understanding of how changes in specific conditions can affect the success of statecraft.[17]

The first assumption is that foreign policy leaders practice statecraft to advance relative influence. The traditional realist preoccupation with international relative position does not suffice, as there are objective circumstances that can turn the pursuit of power by security-seeking states into an intense and destabilizing security dilemma.[18] Notwithstanding how statesmen may define national interests on a specific issue, they are likely to want "more rather than less external influence, and pursue such influence to the extent that they are able to do so."[19] Accordingly, when confronted with a set of unfavorable options, they may care less about maximizing potential gains and more about limiting damage, as either could advance the state's relative influence depending on the strategic context.

A second assumption is that a state's international influence turns on the leadership's capacity to balance the ends and means of statecraft.[20] Because sanctions and inducements involve more than rhetoric and impose adjustment costs on domestic actors, decision makers must have the capacity to make the policy trade-offs needed to uphold international commitments. To engage in energy diplomacy, statesmen must be able to devise a strategy and secure the cooperation of domestic actors (both government and private) that possess the critical expertise and control over respective policy resources. The link between statecraft and internal politics rests with the institutional mechanisms that lay down the ground rules for extracting and directing public and private resources in support of diplomatic priorities.[21]

Third, I assume that formal authority to oversee the formulation and implementation of foreign policy is hierarchical. Although the specific division of responsibility varies across regime type—with authoritarian rulers wielding near complete authority and democratic leaders generally more responsive to institutional and societal checks and balances—all foreign policy systems are supervised by a central executive. Yet, the specific delegation of authority and allocation of implementing authority among agents vary across regimes. Accordingly, foreign policy decisions are the product of interaction between principals—central executives empowered to devise and oversee policy—and agents—administrative actors tasked with carrying them out. To influence the behavior of another state, a central executive must convince foreign targets to sacrifice preferred policies or to take a risk on compliance, as well as secure the dutiful implementation of statecraft by functionaries and interest groups at home.[22]

Finally, policy makers (in both the initiator and target state) are presumed to adhere to the general principle of "situational rationality."[23] As purposeful actors, statesmen reach decisions based on assessments of the values and probabilities of different options in light of subjective baselines, goals, risks, and prevailing conditions. They weigh otherwise commensurate values and probabilities with due regard to reference points and other conditions that lead them to vary in their risk-taking propensities. In this regard, situational rationality lies between both "outcome rationality," which judges decisions strictly in terms of goals and environmental conditions without consideration of the decision-making process, and "procedural rationality," which places emphasis squarely on the intervening reasoning processes that can corrupt the optimization of specified goals under given conditions.[24] It also offers a bridge between prospect theory's powerful empirical finding of loss aversion and broad interpretations of rational decision making. That decicion makers respond differently to choices with comparable net utilities under different situations, violates the basic conditions of invariance and linear fusion of probabilities and utilities embedded in expected utility theories.[25] Yet, because rational choice can broadly subsume the pursuit of self-interest by decision makers as they understand it, this challenge is not necessarily a general indictment of conservative rationality. Accordingly, policy makers can be treated as wholly rational as long as they adhere to the principles of transitivity and completeness in making their cost-benefit calculations *from* a specified reference point. This implies that within each frame a decision maker maintains a hierarchy of preferences whereby if option A is preferred to option B, and option B to option C, then option A is also preferred to option C. If one option from a given reference point is preferable in one aspect and is at least as good in other aspects, it will be preferred to lesser options.[26]

Furthermore, frames are shaped by situational factors. Exogenous pressures, such as the actions of other states, can amend the sequence and availability of alternatives confronting a decision maker. Because situational factors are partly responsible for "motivating" common cognitive responses, they exert an autonomous influence on frames.[27] Of course, this does not discount that frames also can be the product of cold cognitive processes. Rather, it draws focus to the external features that contribute to clarifying whether a situation is generally considered to be one of gains or losses prior to a state or individual's choice. An explanation for common patterns of strategic manipulation need not depend on knowing how options are precisely interpreted; once the general circumstances are clear in a specific issue area, predictions of the framing effects and riskiness of compliance become possible, irrespective of the peculiar cognitive makeup of a target decision maker.

Accordingly, focus on issue specific frames can ameliorate the aggregation problem of applying individual attributes derived from prospect
theory to group decision making.[28]

To sum up, national decison makers presumably are committed to
advancing the relative influence of their states in the international
system. They are attuned to the opportunity gains and losses attendant
to strategic interaction with another state. Statesmen are rational in that
they seek to optimize values given specific goals, strategic frames and
domestic institutional constraints. In their efforts to influence the
behavior of other states, they are sensitive to how targets frame and
assess different options, as well as are constrained by the domestic
agency costs of securing cooperation from nonstate actors. Thus, successful strategic manipulation rests on both what targets will acquiesce
to and what domestic agents and interest groups will accept.

From these assumptions, I deduce a theory of strategic manipulation that attributes success to an initiator's capacity to control the value
and riskiness of compliance for a target. Insights from prospect theory
suggest that this requires leverage over both the "editing" and "evaluative" dimensions to a target's decision making. The former consists of
the procedural power to affect the framing of a given choice problem
for a target, shaping trade-offs between risks and losses and the values
of different options. By prescribing the range and order of choices presented, an initiator can bias a target's risk-taking propensities and
policy inclinations in favored directions. This can be achieved, for
example, by juxtaposing favored outcomes that require deep concessions to other chancy prospects, or by comparing outcomes that
demand a target to play it safe against other potentially positive outcomes.[29] In addition, an initiator must be able to influence the substantive agenda of a target's decision choice. This evaluative dimension
involves shaping the content of specific policy options considered by a
target to affect the risk of compliance relative to other choices.

It follows that all else equal, energy statecraft will be more successful when a central executive can affect the domain and value that a
target assigns to an exchange of the strategic good, and can ensure that
domestic agents will pursue complementary policies. Effective statecraft
depends less on the raw power to coerce or enforce compliance, than on
the indirect power to define issues, initiate and order options, and
enhance the substantive appeal of compliance relative to other policy
options that are available to foreign and domestic targets. The key to
manipulating compliance, therefore, is the ability of an initiator to
reframe the choice set available to a target so that expected losses from

compliance are either pitted favorably against other risky choices or redefined as forsaken gains, and that a target's respective risk-acceptance is biased toward compliance.[30]

The logic of this argument raises two issues concerning the conditions propitious for effective statecraft. First, under which circumstances and by which methods can an initiator formulate the options available to accentuate or amend a target's decision-making frame? Second, how is an initiator able to manipulate the riskiness of a target's choice to accede to specific political demands relative to other policy courses?

Framing Effects

The first element of strategic manipulation involves framing a target's specific choice problem. This entails structuring the set of available options—by including or excluding specific options and presenting choices in a certain manner—to accentuate or alter a target's expected value and risk-taking propensity.[31] The ability to do so is largely a function of the target's vulnerability, or the net value that it assigns to an exchange with the initiator, plus the costs of adjustment if it is altered.[32] The less value a target places on the specific terms of an exchange, the less it stands to lose with a change in that relationship. Adjustment away from an initiator here is viable and relatively painless, notwithstanding modest transaction costs of substituting for the previous level of exchange. Conversely, the value of an established relationship can be exceedingly high if the target cannot easily replace the benefit if lost. The more difficult it is for a target to secure access to new energy markets or supply, the more vulnerable it is to disruptions in the existing set of relations. Thus, the lower the opportunity costs to a particular exchange, the more valuable it is for a target to preserve that relationship.[33]

It follows that the capacity to manipulate a target's domain rests on the net vulnerability of that target. The more vulnerable a target to an established exchange, the greater the value that it places on the choice set presented by the initiator. Conversely, the less value that a target assigns to a specific exchange, the less control an initiator has over the range of potential outcomes, and the less input it has on a target's expectations of gains or losses. In this case, the target can make choices with little regard for the initiator's statecraft.

Initiators can seize on a target's vulnerability to manipulate the probabilities and values of different outcomes in an issue area. One technique involves framing how a target codes outcomes in terms of gains and losses.[34] An initiator can exploit a target's vulnerability by changing the number of options available at any given time. The trick is to add or subtract options in a way that improves the appeal of favored outcomes. For example, an initiator can take action to foreclose or saturate certain energy markets or export routes, presenting a target with a choice between alternatives that offer equally undesirable outcomes. It also can reorder options to frame a favored outcome in the most positive light. This entails positioning a target's choice in the domain of gains by framing concessions in terms of the prospects of forgoing rewards as opposed to incurring certain losses. Framing a political concession as a missed opportunity rather than a direct loss, reduces the incentive for a target to gamble for high value outcomes and can exploit the inclination to sacrifice unrealized benefits. Alternatively, an initiator can manipulate a target into taking risks in the domain of losses by situating a preferred outcome among other more undesirable outcomes. The key here is to order options so that political compliance does not stand out either as the least attractive risk under consideration or as inferior to others with limited prospects of yielding high value.

A second framing technique involves sequencing a target's choice. As discussed by Zeev Maoz, it is possible to manipulate a target into undertaking "myopic," path-dependent decision making by disaggregating issues and framing each choice in terms of possible foregone gains.[35] Instead of demanding a painful policy adjustment, an initiator can present a target with a series of choices that pit modest concessions against future gains. These "salami tactics" speak to the conservatism in the domain of gains by setting up a chain of decisions. By presenting it with the choice of either continuing along a path of least resistance with modest concessions, or making a stark change in policy course that involves taking a large risk on an outcome inconsistent with past choices, an initiator can entrap a target into settling on an outcome that it would otherwise not chose if presented with a single decision.[36]

A third framing procedure is to order options in a way that discounts similar outcomes and discriminates against factors that do not appear directly relevant to an immediate problem.[37] An initiator can structure a target's choices so that it must decide between one pipeline option (preferred by the initiator) that has a very high probability of being constructed but offers only modest potential payoffs; and another option that has only a low probability of being constructed, yet offers potentially equivalent payoffs. Because modest payoffs associated with both options cancel out, the decision becomes one between a pipeline that has a high probability of undergoing construction and another that is unlikely to be constructed. In addition, an initiator can structure a decision to exclude factors that do not bear directly on the outcome of a specific choice. An initiator can frame a choice between alternative pipeline projects in terms of the relative cost-effectiveness of each route, while excluding from consideration the overall probability of constructing each route.[38]

From this we can deduce conditions pertaining to an initiator's ability to manipulate a target's decision frames. In the energy sector, vulnerability can be measured in terms of market power. Market power is a function of the percentages of global imports/exports accounted for by a state's consumption/supply of a specific resource.[39] The greater the percentage, the greater the concentration of exchange, the lower the probability that alternate trade partners will be available, the higher the costs of adjustment, and the greater the capacity of a state to orchestrate the framing effects for foreign targets in that sector. Conversely, the weaker the market power, the lower the concentration of exchange, the more likely opportunities exist to diversify relations, the lower the costs of changing the terms of an existing relationship, and the more difficult it will be for a state to manipulate a target's decision choices. All things being equal, the lower the opportunity costs of compliance for other states, the more likely a state will be able to exploit strategic energy advantages to shape a target's reference point, decision domain, and risk-taking propensity.

Containing Agency Costs

An initiator must not only recast a target's frame; it must ensure that a target's risk-taking propensity is orientated toward compliance. This entails manipulating the substantive appeal of specific options so that compliance is assessed as either the least risky or most risky choice, depending on the target's domain. Faced with structuring a target's decision within the domain of losses, for example, an initiator must ensure that its risk proneness is oriented toward compliance. Alternatively, once casting the target in the domain of gains, the initiator must ensure that the corresponding risk-aversion favors the preferred outcome.

Market power alone cannot determine the relative appeal of specific options. In his critique of vulnerability analysis, R. Harrison Wagner demonstrated that mere asymmetries in economic relations are not sufficient to cover the costs to an initiator of sacrificing potential gains from trade for political concessions. Influence is possible only if an initiator can exploit favorable terms of trade. This requires the capacity to mobilize available resources either to create new gains from trade or to forego some market options for the explicit purpose of securing political concessions.[40]

The challenge is compounded by the discretionary actions of domestic agents and firms. For example, energy firms can exploit market power in supplying a strategic good either by offering favorable terms of trade to customers that cannot be matched by rival suppliers, or by restricting the target's access to that market altogether. Although both strategies can produce commercial profits, they carry very different implications for relations with the target. Moreover, under certain circumstances firms act more as "risk minimizers" than as global "profit maximizers." They prefer stable access to domestic markets or financing arrangements to gambling on the commercial payoffs (while incurring penalties from a home state) associated with uncertain international market ventures.[41]

This suggests that for market power to translate into a potent tool of strategic manipulation, an initiator must be able to ensure that domestic agents and firms line up behind its diplomacy. Domestic actors that wield direct responsibility over the use of national resources must undertake policy initiatives to reinforce the central executive's desired framing effects and bolster the substantive appeal of compliance consistent with a target's specific reference point. Domestic actors must be encouraged to pursue policies toward a target that either maximize or minimize the variance between positive and negative outcomes associated with compliance (relative to available options).[42] If a target is situated in the domain of losses, the initiator must be able to count on domestic agents and firms to present substantive choices in a manner that either entices a target to take a risk on compliance or that diminishes the value of gambling on defiance. Given the tendency to discount moderate to high probabilities associated with negative gambles, they must present options that make it very painful for a target to resist.[43] At the same time, because small probabilities tend to be inflated, domestic agents need to propose policies that lower the probabilities of a target incurring negative outcomes by complying. This necessitates that domestic agents present a target with an option set consisting of carrots and sticks that threaten significant pain for defiance but that offer prospects for maximizing gains from compliance relative to negative values associated with gambling on resistance.

For a target situated in the domain of gains, an initiator must encourage domestic agents and private actors to offer substantive policy options that render compliance as the best choice among other possible gains. The proclivity to discount moderate to high probabilities of positively valued outcomes relative to other certain outcomes, requires that domestic agents present the compliance option as the least risky choice. What matters most is bolstering the credibility of the initiator's commitment to assuring even small gains associated with compliance.[44] In practice, this requires that domestic agents present the compliance option (i.e., a home state's preferred energy supply or pipeline option) as the one that offers the minimum variance between positive and negative outcomes relative to missed opportunities for the target.

The above analysis reveals that the containment of agency costs is integral for determining a state's capacity to exploit market power. Because government agencies or interest groups that enjoy comparative advantages, depend on foreign investments, and stand to profit from deeper engagement with target states possess different strategic outlooks than do those with little or adverse foreign exposure, a central executive cannot assume domestic compliance.[45] With different stakes in extraterritorial behavior, domestic actors may choose to respond differently to available opportunities, imposing additional political costs on the framing of a choice problem and altering the substantive meaning of compliance for a target. Firms can pursue alternative market strategies either that are commensurate with a home country's preferences or that present foreign targets with additional options. Private companies can opt to fund a new pipeline that mitigates the coding, cancellation, or sequencing effects of the home state's procedural control. Similarly, energy firms can advance new options that offer a target either more attractive outcomes if they work or less costly results if they fail, thereby distorting the relative appeal of the home country's preferred outcome. Such domestic "intrusion" into the international exchange relationship can undermine the coherence of a central executive's statecraft.[46] Therefore, a central executive must search out administrative mechanisms for securing the compliance of domestic winners and losers of foreign exchange if it is to succeed at leveraging market power for strategic effect.

It follows that there is a domestic dimension to strategic manipulation that involves increasing the "win-sets" for diplomacy that bring divergent domestic interests into line.[47] One low-cost approach derives from the procedural capacity to alter the opportunity costs of compliance for domestic groups that could otherwise subvert the formulation and implementation of a central executive's statecraft. This concerns the administrative authority to "stack the deck" and improve the substantive appeal of preferred options for those groups directly affected by the exercise of statecraft. This contrasts with the reliance on costly mechanisms for monitoring, sanctioning, or overcoming domestic opportunism by resting on the discretion to modify the relative appeal of policy alternatives so that compliance with national objectives becomes a rewarding strategy for potential challengers.[48] In the energy sector, for example, national firms can be induced to reinforce energy diplomacy by mandating lower energy prices at home that, in turn, create private incentives for national energy suppliers to be more aggressive at protecting foreign market share. In this case, the energy company would have strong commercial reasons to comply with the state's competitive foreign policy, irrespective of the threat of extraterritorial sanctions on deviant behavior.

The institutions put in place to delegate responsibilities for negotiating, approving, and implementing national policies are critical to this administrative dimension to strategic manipulation. Theories of property rights and agency suggest that the most effective institutions in this regard are those that concentrate decision-making authority.[49] A single actor that has complete discretion over the "bundle of rights" within a regulatory process can more easily derive the full benefits and costs of its decisions and limit the costs of delegation, than is otherwise possible with the division of decision-making authority. Where regulatory rights are concentrated there are fewer preferences that need to be brought into line, reducing significantly the need for costly oversight and enforcement.

Yet, to preempt defection it is not necessary that decision responsibilities be fully concentrated. At issue is not the ability to overturn a legislative veto or reverse the opportunism of domestic agents; what matters most for strategic manipulation is the ability to lower the opportunity costs of internal compliance. The premium here is placed on a central executive's capacity to anticipate the probability of a veto and to initiate policies that alter the substantive appeal of compliance, effectively discouraging, not preventing, defection. This depends above all on how clear decision-making authority is delineated within a policy-making process.

The beauty of domestic institutions that clearly delineate decision-making authority lies in their subtlety. Possessing discrete responsibilities, domestic administrators and interests groups must bear the explicit costs and benefits of respective policy behavior. Because policy preferences are transparent, policy makers can readily identify potential allies and adversaries, assess the implications of slight differences in the cost of alternative policy options, and target policies accordingly. This obviates the need to bludgeon societal interest groups into strict conformity via costly sanctioning mechanisms, and opens up avenues for indirectly inducing domestic compliance. A central executive, for example, does not need to exercise direct control over the decision making of domestic energy companies that might otherwise have strong preferences for dealing independently with foreign governments. Instead, a leadership can modify the appeal of specific policy options so that companies and bureaucrats favor compliance with the state's diplomacy. By tapping discrete authority to alter energy supplies, financing arrangements, prices, export duties, or pipeline access at home, a central executive can affect the appeal of engaging in energy markets and projects abroad for the same domestic firms. Vested with clear and exclusive discretion to initiate policies on a specific issue, a central executive can manipulate proposals to discourage domestic noncompliance.[50] From this we can deduce that the more transparent and exclusive a central executive's regulatory authority is in a domestic energy sector, the more effective its capacity to shape the framing effects and riskiness of compliance for a foreign target.

Conversely, the more imprecise the allocation of policy-making authority, the more difficult it should be to discern discrete policy responsibilities or observe the behavior of state administrators and private interest groups. The cost of bargaining and the potential for domestic deviation increase as multiple domestic actors are assigned overlapping authority to formulate, amend, and implement national policies. Redundant responsibilities shield rival claimants from accountability, and insulate decision making on separate issues within a regulatory stream. This encourages opportunism by self-interested administrators and private actors, as well as makes it hard to distinguish potential allies from adversaries on specific policy issues. Procedural advantages to initiate policy change in one area are offset by the difficulties of identifying potential challengers and curbing their incentives to act with impunity. This fosters the shirking of national policies, raising the costs of organizing and channeling the commercial responses of domestic agents consistent with a state's external ambitions.[51] With respect to regulation in an energy sector, policy-making opacity complicates problems of oversight and divorces the consequences of decisions taken in one realm, such as domestic pricing, from those involving access to foreign markets and supply. Accordingly, the more opaque the regulatory process, the weaker the domestic capacity to orchestrate the framing effects and riskiness of political concessions for a target.

FIGURE 2.1
Strategic Manipulation Matrix

	+ Market Power −	
Clear	Target Compliance	Secondary Outcome (mutual accommodation)
Institutional Clarity	Secondary Outcome (mutual accommodation)	Target Defiance
Unclear		

Combining these international and domestic dimensions yields three hypotheses regarding the scope of a state's strategic manipulation, as captured by Figure 2.1. The upper-left cell demonstrates that the more market and regulatory power a state enjoys in a particular energy sector, the more successful it should be at securing strategic concessions from a target. Possessing both elements, an initiator can structure the framing effects and guide a target's risk-taking propensity toward compliance. In the Eurasian energy context, Moscow should be able to dictate the terms of NIS energy development and export in those sectors where Russia enjoys significant global market power and the government has clearly defined preemptive discretion to coopt key potential domestic challengers.

The lower-right cell reflects the obverse. Here, a state should be deprived of strategic leverage in those sectors where it lacks both market and domestic institutional power. Unable to manipulate either the range or substantive appeal of a target's policy choice, an initiator will find it especially difficult to extract meaningful political concessions from asymmetrical energy relations or to blunt countermoves by targets to evade coercive pressure. Accordingly, Russia's attempts at manipulating regional energy security developments should fall short precisely in those sectors where rival suppliers are presented with more attractive options and where the Kremlin lacks discrete authority to offset the opportunity costs of domestic compliance. In these cases, Moscow's diplomacy can be readily subverted by rival investors or low cost options for transiting energy exports, and by unrestricted competition among domestic agents vested with ambiguous authority and divergent incentives to pursue separate energy policies.

The middle cells represent conditions propitious for extracting modest concessions from a target state. In those scenarios where the initiator enjoys *either* global market power in a specific sector *or* the clear allocation of regulatory authority, leverage should be confined to obstructing only the most offensive policies pursued by a target state. In these cases, the initiator can constrain a target's choice either by shaping its domain or affecting the relative value of compliance in order to secure compliance with its secondary or mutual interests. In the Eurasian context, this suggests that Russia should be able to secure mutual accommodation from a target state by manipulating the latter's energy trajectory or risks of doing business with Moscow. Yet, Russia should face considerable problems extracting strategic concessions from otherwise independent-minded targets. Consequently, the strategic interaction that takes place is prone to produce commercial energy deals that are less susceptible to overbearing political pressure.

Operationalization

The argument presented here is that the success of energy statecraft rests on an initiator's ability to alter a target's decision-making context. The explanatory variables are related to the initiator's market power and domestic institutional capacity. The combination of these two factors determines the capacity of an initiator to affect a target's assessment of opportunity costs and risks assigned to available choices. The dependent variable, compliance, refers to the size of the concession made by the target. If the theory holds, the initiator will secure compliance with its statecraft to the extent that it can marshal its market power and domestic institutional authority in the respective energy sector to structure a target's choices so that the latter derives greater prospects either from taking a risk on compliance relative to other unattractive options, or from settling on compliance to avoid the costs of pursuing other uncertain opportunities for gain. Alternatively, an initiator's demands will be least effective if it lacks both the market and institutional power to manipulate the feasibility and substantive value of the options considered by a target.

The usefulness of applying this argument for strategic manipulation rests on several conditions. First and foremost, there must be evidence that a target's compliance is volitional. Not only should it be costly to comply, but a target should evince at least an honest attempt to explore alternative policy courses. Second, the validity of this approach depends on ruling out straightforward expected utility explanations.[52] The crux of this argument is that targets assess differently certain gains and losses of compliance relative to other options that hold out similar expected utilities. For this to be plausible there must be evidence of consideration by a target of other potentially more advantageous options that were nonetheless rejected in favor of the seemingly more or less risky path chosen. Finally, there must be evidence to demonstrate that a target's risks-taking for compliance mirrors expectations of the respective decision frame. Those targets confronting losing prospects are expected to take great risks on compliance. Conversely, a risk-averse target's choice to comply must constitute less of a gamble relative to vague positive options.

Measurement of the causal chain linking the independent and dependent variables requires an assessment of a target's domain and risk-taking in the specific energy sector at issue. Although it is often difficult to disentangle subjective and objective criteria, I discern a target's preexisting frame using the following a priori indicators: (1) global significance of the national energy sector (present and projected percentages of global market); (2) availability of international markets; (3) condition of the domestic infrastructure and access to investment capital; (4) level of diversification of the economy; and (5) domestic political strength and legitimacy of the regime. If these frame indicators have positive values—that is, the national sector represents a large share of global energy supply and export; there are many available customers; the domestic energy infrastructure is well maintained and there are realistic plans for development and investment; the national economy is diversified; and the political leadership is stable and derives its legitimacy from sources other than energy rents—then the target is considered to be in the domain of gains. Conversely, negative values suggest that the target state is in the domain of losses. The extent to which subjective interpretations diverge from objective criteria, as reflected in the public commentary by target leaderships, will be noted and highlighted.

The capacity of an initiator to manipulate a target's decision choice and the relative riskiness of compliance is a function of how clearly the national regulatory authority is specified, as well as the extent of a state's market power in the specific energy sector. I investigate the former by examining the formal regulatory structure of an initiator's domestic energy sector. I identify the specific division of authority for determining ownership and use of energy resources and infrastructure, pricing and tariff policies, and the construction and access to distribution and export facilities among state and nonstate agencies. This is supplemented by insights into the informal administrative structure gleaned from personal interviews and international commercial analysis. Market power is measured in terms of both aggregate percentages of supply and competitive advantages at delivering energy to international markets. A state is traditionally considered to wield significant influence over foreign markets if it controls nearly half the supply of the good.[53] In the case of strategic goods, such as energy, however, the relevant percentages are typically much lower. This is because energy is essential to all aspects of a state's military, industrial, and consumer sectors, and that even marginal fluctuations in supply have potentially severe implications for the breadth of a target's national activities.[54] Moreover, market power in the energy sector is not determined solely by raw supply, as states must be able to deliver energy to foreign markets. A robust assessment must incorporate parameters critical for determining the commercial cost-effectiveness of pipelines. This includes factors related to distance, economies of scale, ramp-up capacity, intensity of incremental competition presented by alternative pipelines, wellhead economics, and creditworthiness of prospective customers.[55] Accordingly, I regard an initiator as wielding significant market power in the oil and gas sectors if it controls roughly 30 percent of supply and export to foreign markets, as well as possesses competitive advantages at reliably delivering low-cost energy via shorter-distance and wider-diameter pipelines than available through other routes.[56]

As noted by other scholars, risk is difficult to operationalize.[57] Although it derives directly from domain and influences a target's choice, it is independent of both. Decision makers rarely formally rank different options, or provide public access to calculations that extend beyond preferences for one option over another. In keeping with the extant literature, therefore, I measure risk in terms of relative variance in outcomes.[58] For example, in choosing between option A and B, if the best outcome for option A is better than the best for option B, and the worst outcome for A is worse than the worst outcome for B, then option A is considered to be more risky than B. Alternatively, if both the best and worst outcomes for option A are better than the corresponding values for option B, then option A is defined as less risky.

Finally both the magnitude of the demands issued by the initiator and degree to which the target state acquiesces to them are used to indicate the level of strategic compliance.[59] As noted above, initiators seek to influence a range of target state behavior; from modest demands for changes to the terms of specific energy security policies and transactions, to more robust demands for altering a target's strategic energy orientation and trade partners. Similarly, target states have a range of responses at their discretion. Accordingly, I code the relative effectiveness of an initiator's statecraft in terms of the level of compliance with the initiator's demands placed on both the specific terms of energy deals and the target's overall energy security orientation in the sector. A target is considered to be compliant if it concedes to maximum demands for changing both specific policies and strategic trade partners. It is coded as defiant, if it does not accommodate an initiator's minimum desires for either. Alternatively, mutual accommodation obtains when a target meets the minimum commercial interests of an initiator on common energy security policies.

Method of Inquiry

The comparative case method provides a preliminary test of the argument and propositions advanced in this study. By employing a structured, focused, comparison of a small number of cases, I intend to demonstrate how changes in market power and the clarity of domestic regulatory authority in an energy sector influence the efficacy of energy statecraft, and thus a state's ability to manipulate a target's compliance on critical foreign policy issues. Each case study addresses the following questions: What were the main strategic objectives of energy statecraft? To what extent did the target modify its foreign policy course in response to the other's energy diplomacy? Was the target's response consistent with the principle of expected utility maximizing? How did a target perceive its energy trajectory and available strategic choices? What role (if any) did market power and efforts to enlist domestic firms and subnational actors play in affecting target state calculations of the probabilities and value of compliance? How did a target interpret direct versus indirect forms of an initiator's statecraft? Finally, how (if at all) did a state's energy policies change the opportunity costs and risks of compliance for domestic firms and foreign targets?

I rely on "process-tracing" to investigate these questions.[60] This method, which places the premium on qualitative analysis of the historical record, allows for identifying the microfoundations of effective statecraft. Attention is devoted to specifying causal mechanisms, zooming in on not only the interests and calculations of policy makers, but the processes through which market and regulatory factors shape (or do not shape) the formulation and implementation of effective energy statecraft. This is complemented by the use of "congruence procedures" to evaluate the consistency between the evolution of independent and dependent variables within each case. We would expect that if a target's decision frame changed over time but an initiator's market power and institutional capacity remained the same, that the success of the latter's statecraft would be determined by changes in how it applied its market and institutional power to alter the opportunity costs and risks of available options.[61] To flesh out the complementarity between the independent variables and their causal links with the foreign policy decisions of target states, I also employ counterfactual analysis. This technique is used sparingly to underscore the role of contingency in key decisions by initiators and targets, and to evaluate the viability of plausible courses of action that were available and contemplated by respective decision makers but were not pursued.[62]

I deliberately eschew a "large-n" statistical study in favor of the process-tracing method of select cases for several reasons. First, the literature on economic statecraft is dominated by empirical studies that privilege debate over the efficacy of alternative definitional and measurement techniques over exploration of causal mechanisms. Because issues related to soft power are inherently difficult to quantify, statistical methods are not well suited at this point to advance this new agenda. Although statistical analyses can demonstrate the plausibility of strategic manipulation—identifying correlations between market power, institutional authority, and compliance—they cannot uncover the agencies through which these variables are converted into leverage. Quantitative studies cannot tell us if strategic manipulation is instrumentally practiced at the behest of concentrated private interests or is a state-initiated and -sponsored policy. Because firms and subnational entities are increasingly responsive to global commercial pressures, their extraterritorial behavior may reflect narrow self-interests rather than the prodding of home governments. Although this difference may not affect a target's behavior, is has dramatic implications for evaluating the efficacy of statecraft. It is only by examining the process of policy formation that we are able to discern "who is doing what to whom and why," concerning the "softer" dimensions to statecraft that do not involve explicit threats or culminate in dramatic policy showdowns. Furthermore, process tracing individual cases allows for testing the causal connection between factors related to strategic manipulation and energy security against the performance of rival explanations. This will help root out potentially spurious correlations in the extant debate on statecraft, as well as illuminate the relative merits of new variables that could warrant future testing in larger statistical studies.

Ultimately, the success of the comparative case method depends on the quality of the case selection. For this, I have turned to a structured, focused comparison of Russia's energy statecraft toward southern NIS suppliers. Eurasia provides both a "strong test" and natural laboratory for examining competing explanations for energy security for several reasons.[63] First, the small states of the former Soviet south were primary targets of Russia's energy diplomacy following the collapse of the Soviet Union. Accordingly, the region is rich with data on Russia's energy diplomacy that vary across sectors—including the oil, natural gas, and commercial nuclear industries; across targets—involving all eight states in the South Caucasus and Central Asia; and across time—including the evolution of respective policy courses over the first decade following the Soviet collapse. I select the three energy sectors due to the variation in respective international market and domestic regulatory structures. The oil sector, for example, is distinguished by a high level of global integration, as Moscow retained only marginal market power, and by Russia's fragmented domestic regulatory authority. In contrast, Russia's standing in foreign gas markets was significant and growing, while its governing authority was attenuated but clearly delineated. The inclusion of different stages of the nuclear fuel cycle provides for alternative combinations of weak market power and transparent regulatory authority that rounds out the testing. Breaking down the cases along sectoral lines reveals the mixed effects of energy statecraft both within and across the southern NIS. Moreover, contrasting different national cases with the Kazakh case for each industry highlights the sectoral dimension to energy statecraft, as well as the limits to target state hedging strategies vis-à-vis Russia. Because these cases provide significant variation (across and within each case) on both the independent (market power and institutional delineation) and dependent (compliance) variables, they present strong tests for assessing the generalizablility of the argument.

Second, Russia's Eurasian energy diplomacy from 1992–2002 unfolded against the backdrop of relatively constant international and domestic conditions that allow for the systematic testing of the propositions advanced in this study against those made by rival theories. Both Russia as the initiator state and the southern NIS suppliers as targets constitute least likely cases for the argument. Given Russia's multifaceted regional preponderance, its efforts to coerce or induce favorable energy policies from these small states should constitute "easy cases" for relative power and asymmetrical interdependence explanations.[64] Yet, it is difficult to reconcile Moscow's variable leverage with these arguments, especially in the energy sector where Russia possessed both the incentive and muscle to impose imperial control. Similarly, the combination of strong, autonomous southern NIS leaderships and a conspicuously "weak" Russian government offers "easy cases" for traditional domestic institutional explanations of regional leverage. However, the uniform political insulation of southern NIS regimes cannot explain why some regularly conceded to Russia while others did not. "State strength" arguments also are hard pressed to explain how a highly penetrated and divided Russian government managed to secure the compliance of the concentrated national gas lobby on critical issues of energy statecraft, but was unable to wrangle the same deference from the more diffuse domestic oil industry. At the same time, Russia represents a "hard case" for strategic manipulation, given that Russia had little sway over the processes of globalization and only modest influence over the behavior of its own footloose private and quasi-private energy entities. If the argument can hold in this case, it is likely to have significant implications for states, such as the United States, that enjoy tremendous soft power across multiple dimensions.

In addition, the southern NIS suppliers stand out as strong cases relative to other targets of Russia's energy diplomacy. The focus on statecraft among rival Eurasian energy supplier states mitigates some of the peculiar political, strategic, and psychological dimensions of dependency that complicate assessment of Russian-NIS consumer state relations.[65] Not only are the supplier states the subject of contemporary policy concerns, but analysis of Russia's respective energy statecraft is not confounded by variables unique to NIS consumers, such as intimate association with the EU (Baltic states), flirtation with establishing a confederation with Russia (Belarus), or intense intrastate divisions over Eurasian and Western strategic orientations (Ukraine). Because the southern NIS suppliers were reluctantly ejected from the Soviet Union and subsequently maintained some of the most asymmetrically dependent energy relations with Russia, they are unlikely candidates for demonstrating the relative utility of indirect methods of control. Despite this legacy, since independence these states pegged sovereign control, regime survival, and economic welfare to stable and diversified foreign energy ties. They became increasingly loath to take cues from Moscow and sensitive to any advances from Russia that could be construed as forceful reintegration. That any of these states would not only comply but willingly forego policy options that would decrease dependence on Moscow is curious.

Finally, these cases offer an empirical corrective to several contemporary studies of Russia's statecraft. As noted by others, the focus on Eurasian case studies is important for redressing a conspicuous American-centric bias in the extant literature on statecraft.[66] But for the findings to offer broader implications it is imperative that they not be grounded in stylized snapshots and that they remain duly sensitive to the idiosyncrasies of regional politics. The purpose of this study is not strictly to present a rigorous challenge to specific arguments: it is also to add necessary richness to inform more constructive policies toward the region. This is especially warranted in these cases, as policy makers across the globe have been befuddled by what on the surface appear as irrational and inconsistent Eurasian strategic energy choices. This concern should not be dismissed as a penchant for marginal quibbling or correcting historical footnotes that otherwise distract from the broader intellectual agenda of advancing deductive reasoning. Rather, it is precisely because energy security sits at the apex of strategic concerns in Eurasia, and because those who study these specific issues both as policy makers and as area specialists are not inclined to think deductively about them, that it is critical that we get the empirical record right in our efforts to produce generalizeable arguments. Thus, it is my hope that this attempt at bridging the "area-theory" gap in the study of international affairs will provide a constructive springboard for stimulating and communicating strategic analyses of Eurasian security issues to a wide audience of experts and practitioners of statecraft.

Part II

Case Studies

Part II

3

❖

Russia's Strategic Energy Predicament

This chapter reviews Russia's varied market and institutional profiles in the natural gas, oil, and nuclear sectors during 1992–2002. In the gas sector, Russia enjoyed near-superpower status within regionally operated markets. This was complemented by a clearly delineated domestic regulatory structure that vested the state with limited but discrete decision authority. By contrast, the Russian oil industry was a marginal player in the global market, reactive to the strategic maneuvers by international swing producers. Notwithstanding conventional depictions of an emerging oil juggernaut by the end of the decade, Russia's oil industry also remained hobbled by diffuse and opaque property rights. By comparison, Russia's stature in the commercial nuclear sector was mixed. On the one hand, it was a secondary supplier in front- and back-end nuclear fuel cycle markets, constrained by competitive practices beyond its control. On the other hand, domestic regulatory authority was concentrated and clearly delineated under state auspices. This cross section of Russia's energy footprint illuminates critical differences among key political interests and players not captured by traditional measures of Moscow's relative, relevant, and structural power advantages over the southern NIS, offering a baseline to assess Moscow's capacity to manipulate strategic outcomes analyzed in the chapters that follow.

The Shadow of Russian Gas

Russia emerged from the rubble of the Soviet collapse with unrivaled proven and probable reserves, as well as comparative advantages at landing gas in established and emerging foreign markets. Natural gas constituted the mainstay of the economy, and was dominated by the world's biggest gas company, Gazprom. Ironically, as Russia presided over the world's largest reserves, the government lacked institutional capacity to compel compliance from the national gas lobby. Accordingly, Moscow was left to exert extraterritorial influence via the industry's "strength" in foreign markets and the government's "weakness" at home.

Global Giant

From 1991 to 2002, Russia was to the global natural gas industry what Saudi Arabia was to oil, controlling over 32 percent of the world's proven gas reserves and approximately 25 percent of international gas production.[1] Moscow's share of global total proven reserves dwarfed its nearest competitors Iran (15 percent), Qatar (7 percent), Saudi Arabia and the UAE (4 percent), and the United States and Algeria (3 percent), and was virtually ten times the size of proven reserves in the Caspian region. In spite of the fluctuation in domestic gas production—that peaked in 1991 at approximately 23 tcf, and slipped to a low of 20.2 tcf in 1997 before recovering to 20.6 tcf from 1998 to 2002—Russia remained the largest exporter of natural gas, controlling as much as 50 percent of the world's gas pipeline exports during the period.[2]

The regional structure of the gas sector accentuated Russia's stature.[3] Russia was Europe's leading gas supplier, as it inherited a 25 percent stake in the established hard currency markets of Western Europe. Over the course of the decade, Russia's deliveries increased to cover 42 percent of the European Union's expanding demand. In addition, Russia dominated gas export markets in Eastern Europe and the former Soviet Union, initially controlling over 80 percent of the supply to both regions.[4]

Russia's huge strategic footprint was augmented by the significant economies of scale and unique opportunities for expanding shares in key foreign markets afforded by the country's vast export pipeline infrastructure. This presented Moscow with competitive advantages versus suppliers in Central Asia, North Africa, and the Middle East for meeting the booming demand not only in the established European market, but in the world's fastest-growing gas markets in Northeast Asia. Advantages in geographic proximity and reserve base rendered Russia a cost-competitive supplier to these markets over both the short and long terms. Because of the "tyranny of distance" of gas pipelines—as well as the prohibitively high capital costs of constructing and operating new pipelines and developing alternative methods for monetizing gas (such as transporting liquefied natural gas [LNG])—Russia was well positioned to reap the benefits of the noncommercial decisions made by Soviet authorities to build long- distance pipelines from Siberia to Europe. These legacy pipelines enabled Russia to export huge volumes with low marginal costs and lucrative economies of scale that afforded Moscow advantages at incremental competition in European markets. By contrast, regional competitors had either to start virtually from scratch by constructing large-scale pipelines, or to develop new ways for ramping up throughput of existing pipelines that effectively put them at a disadvantage for meeting the expanding European demand for gas. Moscow's sterling reputation among European consumers as a proven supplier reinforced these technical advantages.[5]

Russia's shadow also loomed large over the emerging Northeast Asian market that was projected to constitute 20 percent of global demand by 2010. The substantial size of the reserves in the Kovykta, Sakhalin, and Sakha fields conferred a potential competitive advantage at piping gas over rivals in the Middle East, Southeast Asia, and Central Asia. By contrast, the volumes available, transit distances, and rugged terrain made the economic and technological feasibility of delivering Caspian gas to meet long-term demand in Northeast Asia a highly uncertain prospect. In addition, the cost-effectiveness of an alternative LNG supply option was rendered moot for Central Asian exporters, given their landlocked status and the exorbitant capital investments in ports and liquefaction that were required. Although questions lingered about the accessibility and production of Russia's East Siberian and Far East fields—as well as over Moscow's ability to attract the necessary financing to construct regional pipelines vis-à-vis smaller Southeast Asian suppliers—the magnitude of the projected rise in Asian demand coupled with the size and proximity of Russia's reserves provided Moscow with a potential advantage at reaching competitive contracts with China, Japan, and South Korea. The tremendous uncertainty concerning future demand and commitment to diversified energy consumption, viability of national trunk pipelines, and political costs of piping gas to emerging Northeast Asian markets were on average higher for all large-scale non-Russian suppliers and pipeline options.[6]

A State within a State

Russia' global enormity was mirrored at the domestic level, as it inherited approximately 86 percent of the Soviet natural gas reserves and preserved the consolidated structure of the industry. Throughout the 1992–2002 period, approximately 65 percent of Russia's total proven gas reserves and 90–95 percent of Russia's gas production was carried out by 37 production associations solely owned by a single joint-stock company, Gazprom. The company alone controlled 25 percent of global proven gas reserves, and accounted for 20–25 federal budget revenues, nearly 20 percent of foreign exchange earnings, and up to 8 percent of Russia's GDP.[7] In addition, the firm's monopoly extended across the entire chain of gas production, processing, and transportation. This included sole ownership of Russia's high-pressure natural gas transit infrastructure and export pipelines. Although the gas distribution network was privatized in the mid-1990s, Gazprom steadily reacquired approximately 10 percent of the low-pressure pipelines by 2000, due to the rising indebtedness of regional and local distribution companies.[8]

Gazprom possessed near exclusive authority to administer access, production, and distribution of Russia's natural gas that complemented its dominant ownership stake. Unlike other natural monopolies that shared licenses to underground minerals with subsidiaries and regional authorities, Gazprom held exclusive rights to exploit roughly 70 percent of Russia's established gas fields and many others under development.[9] Gazprom also possessed full discretion to determine access and set prices along the domestic processing and transit chain, as well as exercised exclusive authority to set tariffs for interregional transmissions that deprived federal and regional coffers of additional tax revenues. In 1997, the Federal Energy Commission (FEC) was assigned to regulate Gazprom's monopoly over the internal transmission system, including tariffs on access to the pipeline system. The FEC also supervised 15 percent of the transmission capacity that was devoted to third-party access. In practice, however, Gazprom's prerogative to set transportation charges remained intact, as it could independently determine the spare capacity available for independent suppliers. Furthermore, Gazprom set transfer prices that governed transactions between the company's production, sales, and transportation subsidiaries.[10]

Gazprom's foreign trade subsidiary, Gazexport, was charged with determining spare capacity and access to export pipelines.[11] Although Gazprom coveted this monopoly, by 1994 it began to shed nonpaying obligations in the NIS and grant large-scale access to Eurasian pipelines to the private gas trader, Itera. Itera cleared noncash settlements via reciprocal multistage barter deals, offsets, and unusual payment schemes with insolvent NIS customers in return for collecting hard currency rents on access to the company's pipeline system. Although the preferential treatment enabled the trading company to increase NIS gas deliveries and fostered a de facto division of labor over regional gas transmission by the end of the decade, Gazprom retained full control over the Russian pipeline network. Gazprom also hoarded ownership and access to the pipelines for delivering gas beyond the former Soviet Union, especially to the prized European market.[12]

Privatization and corporatization of the gas sector effectively divested the state of its historical ownership of the industry. Although the state retained a 40 percent (and later 37.5 percent) stake in Gazprom, the company's chief executive officer was officially delegated a proxy to exercise the government's shares. Successive attempts at asserting the government's control over Gazprom's internal operations, increasing tax revenues collected from the gas monopoly, and imposing open access to the domestic pipeline system were significantly attenuated by the management's own initiatives. As concluded by prominent Russian financial experts, "whenever the government announced that it wanted change, Gazprom responded with its own restructuring plan, effectively coopting the (government) reformers."[13] It was not until 2001 that the Russian government rescinded the proxy and appointed officials to manage the state's shares. But even with the majority of the company's supervisory board directly selected by the state and the appointment of a new chairman with close Kremlin ties, the government still lacked sufficient expertise and clout to compel reform of Gazprom's corporate practices. Instead, the government settled for specifying the long-term guidelines for corporate restructuring, while deferring to the decisions by the firm's opaque management regarding strategic priorities for investment, production, sales, transmissions, and exports.[14]

Notwithstanding Gazprom's leading position, the Russian government maintained partial but clearly defined rights to administer wholesale and retail prices for domestic consumers. Although charges along the gas chain were essentially treated as transfer prices internal to Gazprom, sales to distribution companies and end-users were strictly regulated by the state. Prices were set initially by an inter-government committee (later the FEC), and approved by the prime minister.[15] The federal government used this authority to keep prices artificially low (and at times barely covering the cost of delivery) for domestic industry, electricity generators, and private consumers, relative to prices for both exports and other fuels. For example, the power generation and industrial sectors, the largest domestic consumers of gas, were charged at some points less than 3 percent of the export price. Prior to 2000, Gazprom annually received payment for less than 40 percent of the gas it delivered at home, of which less than 20 percent comprised prompt cash payments. Yet the gas monopoly earned roughly 50–70 percent of its revenues from delivering much lower volumes of gas to hard currency markets in Europe.[16] The political commitment to suppressing domestic prices and the physical danger of disconnecting distribution effectively insulated domestic consumers from changes in price or supply in wholesale industrial, power generation and residential heating markets, notwithstanding the dramatic contraction of the economy or considerations of marginal costs incurred by Gazprom.[17] Russian industrial and household consumers effectively lacked structural incentives to conserve or to pay for internal gas deliveries altogether. Thus, despite the tremendous fluctuation in industrial gas prices between 1993 and 2000, which peaked at 60 percent of international prices in 1995, Gazprom experienced a swelling nonpayments crisis.[18]

The Russian government also retained exclusive mandates to levy royalty fees and excise taxes, and to impose export tariffs and quotas. Following the Soviet collapse, Gazprom was exempted from paying almost all taxes and duties. This was reversed in 1993, as the government came under pressure from international lending institutions to impose excise taxes on Gazprom as percentages of wholesale prices to bolster revenues. This was followed by the introduction of geology and royalty fees on production. The state also was vested with authority to expropriate a significant proportion of gas export revenues that reached as much as 55 percent of total intake in 1993. Furthermore, the federal government legally claimed an annual percentage of the export pipeline capacity under the "state needs" provision. This authority was available to assist the state at meeting political obligations to deliver gas across Eastern Europe and NIS, and to pay for foodstuffs and equipment imported from Western Europe.[19]

In a nutshell, the regulatory structure of the Russian gas industry, though centralized and opaque at the firm level, was relatively straightforward in terms of administering production, pricing, transmission, and exports. Although the gas monopoly effectively enjoyed unsupervised discretion to manage internal operations and dictate the quantity of gas produced, distributed, and exported, ownership and regulatory responsibilities were clearly divided between Gazprom and the state, with the latter retaining limited but critical authority to set domestic prices. In this structure, domestic pricing policies directly affected the opportunity costs (monetary and nonmonetary) of doing business at home and abroad for Gazprom, even though the demand among industrial and household end users was inelastic. The bifurcation of regulatory authority generated distinct state and firm interests in the natural gas sector that, as discussed in chapter 4, provided the government with a potent lever for guiding Gazprom's profit-making activities consistent with Moscow's statecraft.

Russia's Troubled Oil Sector

Unlike the gas sector, Moscow's grip over the oil industry seemed to weaken, despite the impressive recovery in production and exports from the deep depression that immediately followed the Soviet collapse. Russia never came close to becoming an influential swing producer capable of independently altering the incremental commercial value of Eurasian oil development. In addition, the regulatory environment in the domestic sector remained opaque, notwithstanding progress toward liberalization and reform of corporate governance practices.

Russia's Recovery and the Global Market

The Russian oil sector was hit hard by the Soviet collapse. After successive years of decline, production bottomed out by 1996, contracting roughly 47 percent from its peak in 1987 when the Soviet Union led the world in national output. This was primarily due to the virtual collapse of investment that curtailed new drilling and the industry's capacity to increase recovery from depleted fields. In contrast to the Soviet heyday, when the republic accounted for nearly 90 percent of the Union's 5 million barrel per day (bbl/d), independent Russia's net exports plunged to 3.2 million bbl/d from 1993 to 1995. Similarly, crude oil shipments via the Russian pipeline system in 1996 represented only 56 percent of the 1990 throughput.[20]

By the end of the decade, the Russian oil industry seemed to bounce back. Domestic production and exports soared during 1999 to 2003 in response to rising world oil prices, devaluation of the ruble, and growing confidence in the domestic investment climate. Oil companies pumped out 7.59 million bbl/d in 2002—more than a 25 percent increase over the 1998 level—that positioned Russia as the world's leading producer in 2002 for the first time since the fall of the Soviet Union. This dramatically outpaced the combined rate of oil development in Azerbaijan, Kazakhstan, and Uzbekistan that peaked at roughly 530,000 bbl/d in 2002. Exports recovered as well, climbing to 4.2–5.17 million bbl/d, making Russia the world's second leading crude oil exporter behind Saudi Arabia between 2000 to 2002. The perpetuation of steep international oil prices and drop in domestic consumption firmly ensconced Russia as the world's third largest producer and second largest exporter of crude oil for the coming decade.[21] This outlook was reinforced by efforts to extend or develop six pipeline systems. Among the achievements were: completion of the first stage and operational development of the Baltic Pipeline System with an export capacity of 240,000 bbl/d in December 2001; delivery in 2001 of the first oil shipments via the Caspian Pipeline Consortium's (CPC) project that was slated for an export capacity of 1.34 million bbl/d by 2015; and exploration of an oil terminal at Murmansk that was intended to boost the potential for Russian "big oil" to service up to 10 percent of the large American market.[22]

Despite regaining stature as a premier producer and exporter, Russia's comparative standing was limited. Russia's commercially viable oil reserves totaled less than 10 percent (5–6 billion tons) of those "technically" suitable for extraction under ideal conditions. This represented only 4.5 to 5 percent of global proven oil reserves (and at most 15 percent of total reserves). Although this was significantly larger than the estimated 15–33 billion barrels of proven oil reserves in the Caspian Basin (2–3 percent of the global total), Russia's reserves were dwarfed by the 25 percent and 77 percent of global proven reserves controlled by Saudi Arabia and OPEC, respectively.[23] By 2000, over 70 percent of the oil reserves in operation yielded low flow rates that rendered development "only marginally commercial." The enthusiasm for Russia's post–1999 surge in production, therefore, was tempered by concerns that the growth in output could not keep pace with rates of discovery or depletion of reserves. Unless redressed by greater investment and more efficient exploitation, Russia was expected to experience a dramatic decline in output over the following decade.[24]

The tightly oligopolistic structure of the world oil market in the 1990s further complicated Russia's commercial woes. In the integrated market, oil delivered to one part of the world directly affected the price of deliveries to another region.[25] The geographic concentration and poor quality of crude exports placed the Russian oil industry at a distinct disadvantage in searching out new markets in this regard. The Urals-blended export crude, for example, was suitable for refineries restricted primarily to northern Europe and the Mediterranean. Yet within these markets—the destination of 98 percent of such exports—Russian suppliers accounted for only 10–16 percent of aggregate consumption, and were especially vulnerable to incremental competition by suppliers from the North Sea and the Middle East that were more proximate and that incurred lower marine transportation costs.[26]

Although Russia possessed a competitive edge at piping crude to established inland markets in Western and Central Europe relative to competitors from Europe and the Middle East, there were real limits to how much additional Russian oil could be consumed by those markets. The West European market, for example, was saturated by relatively cheap deliveries from the North Sea. This trend was expected to continue despite the projected steep decline in output from the North Sea by 2010, as rates of growth in demand were anticipated to flatten in conjunction with shifts in regional demographics, fuel efficiency, and the substitution of natural gas. Moreover, the two main existing western transit routes—the Baltic Pipeline System and Odessa-Brody Pipeline—could not handle the projected growing volumes available for export to these destinations.[27] The potential for expanding Russia's footprint in its traditional Central European market also became increasingly suspect due to the downturn in consumption sparked by economic transition and the expected rise in long-term demand for higher quality crude. The same held for markets in Mediterranean Europe, as the annual demand for Russian oil was projected to climb incrementally due to gradual rates of economic growth, rising substitution rates of natural gas, and stiff competition from Middle East suppliers.[28]

In addition, Russia's underdeveloped regional oil infrastructure handicapped the prospects for breaking into emerging Asian markets. Although the rising Chinese, Japanese, and Korean demand for oil accounted for over two-thirds of the expansion in world consumption by 2000, these economies were overwhelmingly dependent on supplies from the Middle East and OPEC. Even with large-scale foreign investment in oil fields in the Russian Far East, the future cost-competitiveness of Russian supplies was uncertain owing to extreme winter temperatures, rough terrain, high level of seismic activity and environmental risk to regional fisheries. The limited proven reserves in East Siberia also precluded plans for constructing a robust pipeline infrastructure in the region.[29] While Saudi Arabia serviced nearly 13 percent of global demand, Russia even under favorable investment, production, consumption, and transportation scenarios maintained the potential for capturing at most 2.5–3 percent of this demand, and for sustaining only modest growth in niche markets.

The constraints were compounded by the short- and medium-term dynamics of the international oil market. Incremental market power and the ability to mold the behavior of rival suppliers is defined as "the ability of a single, or group of buyers or sellers to influence the price of the product or service in which it is trading."[30] In light of the integrated structure of the global market and the inelastic demand for crude oil over the short run, those suppliers that enjoy the most favorable ratio between accessible reserves and production wield the greatest leverage over other oil suppliers via price adjustments. What matters most for credibly influencing immediate changes in international oil prices is the spare capacity to quickly and cheaply swing a few million barrels of oil per day; increasing production and deliveries in some instances to reduce prices, or idling production to prop them up in alternative scenarios. For example, by maintaining surplus production capacity and tight national ownership over production decision making, the Saudi government deftly marshaled its vast oil potential to exercise market leadership both within and outside of OPEC.[31] The capacity to expand production vested Saudi Arabia with the equivalent of a "nuclear deterrent" in the global market; Riyadh could credibly induce rival suppliers to restrict supplies and enhance its medium-term market shares as compensation for falling prices by threatening to boost production.[32] Yet, even Saudi Arabia's market power over short-term prices was structurally limited by the high demand elasticity of crude oil over the medium- to long-terms. The substitution effects of alternative resources, combined with the tax policies of consumer states, technological innovation, increased production by OPEC and non-OPEC suppliers, and evolution of multiple trading mechanisms insulated both consumers and rival suppliers from the sustained predatory rent-seeking practices of any single producer.[33]

By comparison, Russia lacked significant swing leverage over global market prices. Russia's impact on the international market was most pronounced during the initial post-Soviet period when production declined precipitously, due to the even greater drop in domestic consumption rates. Over the course of the decade, however, Russian oil firms were in no position to rapidly shift prices on spot or contract markets, as they operated at near full capacity and were dependent on transit states for piping exports. They were ill suited to go "head-to-head" with rival suppliers for market share or to alter significantly the output value of specific foreign oil development projects by dramatically deflating contract prices, let alone challenging Saudi Arabia's deterrent over the medium term. To the extent that the Russian oil industry exerted any independent influence on the global oil market, it was by broadening the production base to dampen the effects of short-term cutbacks, restore price stability, and discipline aggressive pricing policy by rival suppliers. The impact, however, was modest as it depended on acting with other non-OPEC suppliers, transit states, and consumers to temper price aggression by any single producer. This was primarily a defensive posture that deprived Russian companies the clout either to orchestrate cooperation or to impose price restraints directly on rival suppliers by unilaterally withholding production.[34] In search of foreign investment, the Russian oil industry also lacked available capital to set the terms for oil field and pipeline development in other countries, or to provide the critical technologies needed to access new, small, hard to recover reserves from remote East Siberian fields. Thus, Russia was a price-taker in the international oil market, and given its limited import capacity, represented a small and declining potential market for other suppliers, including those in the Caspian Basin.[35]

Opaque Ownership and Control

In addition to the market constraints and deteriorating structure of the reserve base, Moscow's shadow in the oil sector was circumscribed by the domestic regulatory system. In sharp contrast to the gas sector and to the nationalized oil industries of the Middle East, the Russian government did not directly manage the development and export agendas for the national petroleum industry. Rather, as one of the early targets of liberalization and commercialization, the oil industry was rapidly divested of the state's monopoly ownership and direct control over production, refining, and distribution facilities. Replacing the 38 production associations that constituted the backbone of the Soviet oil industry were more than 110 independent oil producing companies in Russia, with over 50 firms involved in crude and condensate processing and approximately 150 firms that exported Russian crude outside of the post-Soviet space. By 2000, there were approximately 28 refineries, a dozen oil processing facilities, and 132 independent oil producing firms in Russia.[36]

The dramatic withdrawal of state ownership and decentralization of administrative control of the Russian oil industry was neither straightforward nor complete. Though ownership changed hands from the state to a mélange of private firms, banks, and quasi-state entities, it remained highly concentrated, consistent with the dual aims of preserving hierarchical guidance and promoting independence. After an initial period of "spontaneous privatization," when enterprises and regional authorities took independent control of specific assets, the federal government spearheaded a program to manage the distribution of production enterprises. The first stage was characterized by the government's efforts to reduce and then consolidate residual ownership stakes in the oil industry. With the 1992 Presidential Decree on Privatization, the lion's share of the Russian oil producing and refining complex was reorganized into roughly a dozen independent, vertically integrated holding companies. Initially, the government retained majority stakes in the holding companies, which in turn, held majority interests in the subsidiaries that controlled specific production, processing, and refining facilities.[37] The second wave of privatization in 1995 introduced the "loans for shares" arrangement that enabled banks and financial institutions to acquire at first temporary control and then outright ownership rights to retain or auction the state's shares in the principal oil holding companies in exchange for financing the government's budget deficit. In 1997, the government abolished all restrictions on foreign ownership of oil assets, creating opportunities for foreign investors to participate legally in the oil privatization process.[38]

The evolving privatization process transferred ownership of the Russian oil industry unevenly to private hands. The core element consisted of the ten to fourteen independent financial groups that as a result of the vertically integrated structure of the oil industry collectively accounted for almost 89 percent of production and 79 percent of refining capacity in the country by 2002.[39] In addition, there were small independent production companies, as well as private extraction firms, export companies, and wholesale and retail distribution enterprises. The state, however, retained 100 percent of the holding shares in the pipeline network operated by the joint-stock company, Transneft. Although the government permitted the construction and operation of the CPC pipeline that was independent of the Transneft system, and subsequently reduced its holdings in the company to 75 percent, the state retained exclusive voting shares and operational control of Transneft throughout the period.[40]

The successive waves of "privatization" dispersed ownership and control in the Russian oil sector. In principle, this should have rendered the national oil industry both less influential and more vulnerable to manipulation by the government. The state not only owned the pipeline network and possessed limited ownership stakes in the vertically integrated companies, but was well positioned to exercise political leadership by redefining issues and altering private preferences that mirrored policy making on international oil issues in other "weak" states.[41] Yet the transfer of ownership did not translate into clear regulatory authority. Instead, what emerged was a "recombinant" property rights structure in the Russian oil sector characterized by "a multiplicity of different kinds of owners, a blurred distinction between private and public forms, and mixed relations of ownership which (implied) participation by the banks." Both shareholders and administrative agents were fragmented among a diverse set of employees, managers, banks, and foreign investors, as well as multiple federal, regional, and municipal authorities. This effectively obfuscated the distribution of decision making and economic power, enhanced the practical autonomy of firm managers, and reduced responsiveness of the industry to specific political and financial pressures.[42]

The government also lacked formal authority to prescribe the commercial preferences of the fractured domestic oil industry. Partial licensing and tax responsibilities were spread among multiple federal agencies, regional authorities, and semiprivate entities. Successive "Laws on Oil and Gas" were conspicuously vague about stipulating "joint jurisdiction" (dva klucha) between federal and regional governments over the issuance of licenses to lease oil exploration and production rights from the state. Consequently, administrators at different levels wielded independent and conflicting mandates to grant/withhold concessions to oil reserves across the country.[43]

Similarly, the Russian oil taxation regime was marred by complexity and redundancy. For most of the decade there were over 100 federal, regional, and local taxes and special payments levied on oil firms that were assigned primarily on the basis of production not profit. The government's "tax take" averaged between 50 to 60 percent of gross revenues per annum, but during certain periods exceeded 100 percent of net revenues for domestic sales and exports for the typical Russian oil producer. The punishing tax regime was insensitive to fluctuations in world prices or that the industry operated at a loss, despite accounting for as much as 8 percent of Russia's annual GDP and 35 percent of its foreign trade earnings in 2000.[44] Different federal agencies shared authority with regional and municipal governments to assess various corporate, VAT, special, excise, export, and royalty taxes on oil revenues. This haphazard approach to taxation undermined the inducement effects of federal tax exemptions, as regions and localities were under no obligation to reciprocate and were free to levy separate taxes at discretionary rates. In 2001, the tax code was reformed to place emphasis on revenues, reduce the tax burden to levels that firms could pay, and to simplify the scheme to three payments—a royalty tax, normal profit tax, and excess profit tax. A year later, these taxes were replaced by a single production tariff projected at 16.5 percent of oil prices by 2005. Yet, Russian oil firms were deprived of exemptions from corporate taxes and had to pay higher export duties with rising oil prices, as well as had to cover special regional tariffs and hand over payments to the federal government that in some cases exceeded 90 percent of net revenues.[45] As only one of several claimants to regulatory authority, the federal government lacked the political gravitas to offset the significant opportunity costs associated with expanding operations at home for national oil companies and foreign investors alike.

The division of regulatory authority among the executive and legislative branches of government also significantly raised the costs of oversight. Transit fees, for example, had to be approved by both houses of the Russian parliament and endorsed by respective provincial administrators that pursued separate political and economic agendas. The same applied to production sharing agreements (PSAs), the industry standard for clarifying long-term tax rates and rules for private investors while allowing host governments to retain title to oil resources. The Duma (lower house of parliament), in particular, exploited its legal mandate to specify the fields and deposits eligible for development on PSA terms. By holding up ratification of the 1995 PSA legislation, the Duma effectively confused responsibilities and watered down the law by stipulating the government's right to renegotiate arbitrarily the contract in the event of a "substantial change in circumstances." Similarly, regional administrations concluded specific PSA agreements without coordinating with Moscow. Due to the cumbersome and politically charged process for passing agreements, the Russian government was unable to reach a single PSA agreement from 1995 to 2002 and was handcuffed in implementing prior agreements.[46] This exacerbated uncertainty about property rights in the sector that discouraged foreign investors, despite the industry's immediate financial needs and available reserve base.[47]

Moscow also possessed only modest authority to set the agenda for petroleum pricing, transport, and exports. Although domestic oil prices were partially liberalized in September 1992, there were five federal agencies, one Duma committee, as well as multiple regional authorities, consumers, producers, and international financial lenders vested with formal rights to adjust domestic oil prices.[48] The regulatory scheme became significantly more transparent in 1995, as Russian domestic oil prices were mostly liberalized, and export quotas and duties were significantly curtailed. Yet national prices for crude were constrained by periodic gluts in the world supply, as well as by the deterioration of the national refining industry, the limited capacity of the Russian export infrastructure, and the introduction in 1998 of government mandated domestic delivery requirements for crude oil and refined products. As a result, Russian oil production was virtually "shut in" and significantly underpriced at home, ranging from less than 1 percent to as much as 60 percent of international levels during the period.[49] This effectively reduced the value of domestic sales and reinforced incentives among Russian oil firms to diversify upstream and downstream profiles abroad, irrespective of shifts in Moscow's regulatory policies.

In addition, there were multiple government agencies charged with conflicting mandates to allocate access and set throughput prices for the Russian pipeline system. From 1992 to 1995, the government controlled exports via a quota system and by designating authority to trading companies to act as exclusive export agents. This system was rescinded in 1995 when export rights were liberalized and pipeline access was assigned in proportion to a firm's production.[50] This empowered multiple actors—the Ministry of Fuel and Energy, Federal Energy Commission, an Inter-Departmental Commission on Access to Export Pipelines, the state transport monopoly, Transneft, the Union of Oil Exporters, and all of the Russian oil companies—with overlapping authority to supervise mandated domestic deliveries, negotiate export prices and contracts, and monitor firm auctions for export quotas. In practice, access to the Russian pipeline system was granted via an opaque network of ad hoc favors among rival coordinating agencies. Transneft, in particular, exploited the administrative confusion to adjust delivery schedules and tariffs with little concern for international prices and total transport costs. This scheme discouraged production for the domestic market and imposed stifling export bottlenecks. It also generated an incentive among state agencies to maintain a gap between internal and export prices as a source of rent from auctioning export pipeline access. Thus, opaque access to the Russian pipeline system afforded opportunities for parochial enrichment, producing conflicting mandates among rival regulatory agencies to exploit oil shipments either to maximize short-term rents or advance the industry's long-term international competitiveness.[51]

The confused ownership and administrative mechanisms spawned competing production interests among Russia's new "petroarchy" beyond the grasp of the state. Companies, such as YuKOS and the regionally owned Norsi and Bashneft companies, concentrated initially on domestic upstream and downstream projects; others, such as the state-owned company Rosneft and Zarubezhneft pursued joint projects in the Middle East. By the end of the decade, however, YuKOS embraced an expansionist global profile and courted American partners with transparent corporate governance practices and offers to deliver oil directly to the U.S. market. The largest private Russian oil major during the period, LUKoil, solicited strategic collaboration with foreign companies and retail outlets (filling stations) in the West, while pursuing an aggressive upstream strategy in Iraq and the Caspian Basin. Commercially motivated to reduce production costs, attract foreign investment, modernize technology, and search out more solvent customers, LUKoil expanded participation in exploration projects with Azerbaijan, Kazakhstan, and other international oil companies to secure stakes in the development of onshore and offshore Caspian deposits.[52]

Added to the confusion was the intrusion of regional interests.[53] Vested with autonomous jurisdiction over ownership and revenues, and with variable rights to distribute federal tax proceeds, approve production-sharing agreements, and to direct oil investment, subfederal level political bodies pursued independent energy agendas in respective localities. Municipal administrations exercised these rights to favor specific development and pipeline projects, and to assign additional fees and taxes to fund local infrastructure projects. Although the decentralization of regulatory authority did not lead regional actors to pursue energy strategies strictly at odds with Moscow, it severely circumscribed federal authority to control commercial and political opportunities in the oil sector across the country.[54]

The administrative disarray in the regulatory process also generated competing incentives for oil exports. As both the state's regulatory agent and *de facto* owner (via subsidiaries) of 90 percent of the national pipeline system, Transneft possessed distinct commercial interests and thrived on the administrative confusion. Because tariffs and contract prices were determined largely by the volume of throughput shipped across specific segments, Transneft had an incentive to preserve the state's pipeline monopoly as a means to recoup budgetary shortfalls generated by delivering oil to low-priced domestic markets. In closed "intergovernmental" deliberations, the company decided the terms of access and service contracts, as well as championed the construction of numerous costly projects that displaced construction of private export pipelines and terminals that offered potentially greater access and lower transit fees to Western and Far Eastern markets.[55] This directly challenged the commercial interests of the oil producers that lobbied either for long-term and predictable pipeline contracts with Transneft, or for opening up ownership of the export infrastructure altogether to private competition. The tension came to a head at the end 2002, when the Russian oil majors collectively complained to the prime minister that Transneft's predatory practices and export strategy were having a "negative influence on the possibility for growth in production, and as a result, [were] causing significant losses to the federal budget."[56] Transneft's authority also conflicted with the empowerment of the FEC to oversee nondiscriminatory access to the pipeline system, as well as with mandates accorded the Ministry of Energy and various regional administrations to promote greater transparency in the pipeline system to attract foreign investment in the sector.

Finally, the peculiarities of the 2001 state system for auctioning export quotas paradoxically imposed an administrative barrier between domestic and international prices. The program offered to bolster competitive access to export pipelines that would only benefit a few larger producers without offering attractive alternatives in the home market. At the same time, the plan constrained the government's capacity to dangle sales on the domestic market as a lucrative substitute for participating in international ventures. As summed up in a comprehensive IEA study:

> The auction scheme institutionalized official administrative limits on exports, as well as a large gap between international and domestic prices. Because the amount bid for export access was driven by the size of the price gap, the government had the incentive to keep the gap as wide as possible. In effect, the additional cost for export access would function like an export tax, as a wedge between domestic and international prices.[57]

Russia's Nuclear Legacy: A Mixed Bag

The breakup of the Soviet Union also dramatically altered Russia's stature in the commercial nuclear industry. Although it inherited much of the expansive Soviet nuclear infrastructure, Russia's global standing significantly constricted. Almost overnight the country was relegated to a marginal swing player in front-end and back-end nuclear fuel service markets, as well as was no longer self-reliant at critical stages of the fuel cycle. The nuclear industry also accounted for only a small portion of the domestic wholesale energy market, operating at a loss and competing at a distinct disadvantage against conventional power generation technologies. Yet, domestic regulatory authority was concentrated and clearly delineated. Ownership and control were predominantly in state hands, consolidated under the Ministry of Atomic Energy (Minatom). Although oversight within the nuclear complex was divided internally among the ministry's executive and regional offices, Minatom nonetheless possessed exclusive authority to establish production targets and supervise the industry's international ventures, as well as retained discrete authority to set prices for foreign sales of commercial nuclear services. The combination of limited market standing at home and abroad, with streamlined state regulatory mechanisms distinguished the nuclear complex from the gas and oil sectors, creating both commercial constraints and political opportunities for using nuclear energy as an instrument of Russian state policy.

Front-end and Back-end Supply

At independence, Moscow retained ownership of over 80 and 90 percent of the Soviet nuclear industrial and R&D capacity, respectively. This included supervision of over 150 research institutes, employment of nearly one million people, and operation of 29 of the 43 residual Soviet nuclear power plants (NPPs). The complex was nearly self-sufficient at all stages of the nuclear fuel cycle: extraction of uranium, preparation of uranium concentrate, production of uranium hexoflouride, uranium enrichment, production of fuel assemblies, handling of spent nuclear fuel, and disposal of radioactive waste. Russia also inherited the only spent fuel storage facility, as well as supplied all of the front-end (uranium conversion, fabrication, enrichment) and back-end (spent fuel storage and reprocessing) nuclear fuel services for the 75

Soviet-designed reactors located throughout Eastern Europe and the former Soviet space.[58]

The nuclear sector seemed to fare better than other traditional power producers throughout the first decade of transition. While overall electricity production dropped 22 percent from 1990 to 1998, nuclear power generation declined only 12.5 percent. It then jumped 14 percent after 1998, and was positioned to cover a larger portion of Russia's future electricity demand.[59] By the end of 2002, there were 30 NPPs in operation at greater fuel efficiency than at anytime during the Soviet period, with plans in place for running 38 to 44 plants by 2020.[60] With signing of the "Megatons to Megawatts" agreement with the United States in 1993, Russia also was expected to earn $12 billion by 2013 from foreign sales of 15,000 metric tons of low enriched uranium (LEU) blended-down from highly enriched uranium (HEU) that was extracted from dismantled nuclear warheads and military stockpiles. Coupled with new enrichment service contracts for reactors in Western Europe, the Russian nuclear sector was poised to perpetuate the Soviet legacy as a powerful force in the global nuclear fuel, power generation, and spent fuel reprocessing markets.

Though impressive, Russia's footprint in the global nuclear industry was far from decisive. With respect to uranium supply, Russia occupied a secondary position that was projected to erode over the medium term. Because the Soviet Union accounted for only 15 percent of the world's reserves of low-cost uranium, with its collapse, Moscow's share dropped precipitously by 70 percent, leaving only 5 percent of the world's natural uranium reserves on Russian territory. This constituted less than 20 percent of the reasonably assured low-cost resources inherited by Kazakhstan.[61] Furthermore, Russian uranium production declined nearly 40 percent from 1992 to 1996, after which it held steady at 2,500 tons per annum. By the end of 2002, Russia had sunk from the fifth- to eighth-largest supplier of natural uranium, producing as little as 30 percent of that supplied by Kazakhstan and Uzbekistan and representing only 4.6 percent of global output.[62]

Russia's natural uranium production also could not meet more than 40–50 percent of the projected demand from Soviet-type reactors (both at home and abroad) through 2015. This placed a premium on securing alternative sources via international joint ventures and secondary supply methods that included reprocessing, constructing fast breeder reactors, accessing uranium stockpiles, and reenriching depleted uranium.[63] Although uranium extracted from the military stockpile immediately augmented Russia's delivery capacity for the ensuing twelve to fifteen years, these supplies drove down prices and retarded rates of primary production. Moreover, the eventual depletion of these nontraditional stockpiles would effectively accentuate Russia's long-term disadvantages at competing against major low-cost suppliers of reasonably assured natural uranium, such as Australia, Canada, Kazakhstan, South Africa, Brazil, and Namibia.[64]

Russia's capacity to service new uranium markets was further restricted by trade barriers imposed by important customers. Although Russia supplied nearly 25 percent of the natural and enriched uranium for Europe (earning $150 million annually), its trajectory of growth was subject to EU restrictions. In 2000, for example, the EU Commission proposed a long-term energy agenda that called on member states to harmonize nuclear safety standards, certify that Russian imports conformed to western safety standards, shut down Soviet-built first generation plants, and diversify sources of fresh fuel. This was incorporated into an EU legislative package in 2002 that obliged existing and prospective member states to protect the internal enrichment industry and limit the share of Russia's future sales on the common market.[65] Throughout the decade, the United States too imposed trade barriers to protect the American industry from the dumping of "cheap" Russian uranium. Under the 1992 "Suspension Agreement," Russia could sell uranium in the United States as long as there was a matching sale from an American producer.[66] Contractual delays also resulted in only $2 billion (38 percent) of the projected $12 billion in LEU shipments to the U.S. under the "Megatons to Megawatts" plan by end of 2001. Coupled with efforts by private investors beginning in 1998 to renegotiate the commercial terms and the rates of releasing uranium feed supplies on the market, Russia faced significant restrictions on growing its long-term profile in Western markets.[67]

Russia's precarious position extended to the uranium conversion, enrichment, and fuel fabrication markets.[68] Though Moscow serviced nearly 25 percent of the global conversion market, its primary customers were confined mostly to traditional customers in Eastern Europe and the NIS. By contrast, seven of Russia's sixteen international customers, including Germany, Spain, Sweden, Switzerland, and the United States, transacted with other foreign providers. Although additional services that came from Russian weapons uranium were large enough to compensate for immediate shortfalls, because Russia's conversion plants were tied closely to their enrichment facilities for technical and logistical purposes, they were ill-suited to fill the supply gap once nontraditional stockpiles were depleted.[69] Similarly, while Russia possessed the largest enrichment capacity for low-cost gas centrifuge plants among the top five suppliers, its commercial power was hamstrung by oversupplied and regionally segmented foreign markets. In this highly competitive setting, Russia was the main supplier to East European and NIS markets that were both significantly smaller than the United States and West European markets, and were growing at a much slower pace than those in Asia. Moreover, Russia was subject to mounting pressure from traditional customers that were increasingly interested in expanding purchases from extraregional enrichment companies to obtain lower costs and more flexible contracts.[70]

The situation was more tenuous with respect to fuel fabrication. These markets were annually oversupplied by 50 percent, with high barriers to entry due to the close connection between reactor purchases, fuel fabrication, and fuel services. While Russia covered approximately 17 percent of the global market (14 countries), foreign competition stiffened. By the end of the decade, Moscow had lost ground to rival suppliers in long-standing Czech and Finish markets, as well as faced curbs on new inroads to European customers that were under pressure by the EU to diversify long-term nuclear fuel supply. Furthermore, prior to 2000, Russia was critically dependent on Kazakh supply of pellets for fuel assemblies used in Soviet-designed RMBK and VVER-1000 reactors operating both at home and abroad. Although Russia developed a new line of pellet production at two of its own facilities, it could not fully compensate for Kazakh imports to meet the expected demand for its fuel assemblies.[71]

A similar constellation of commercial pressures obstructed Russia's growth in power generating markets. With independence, Russia supplied less than 10 percent of the world's total nuclear power generation. Expansion into new markets far afield was especially constrained by the residual transmission infrastructure. The capacity of the Russian nuclear sector to steal market shares from either foreign competitors or from other Russian conventional power generators was significantly limited due to "the lack of relationships abroad, general entry barriers for Russian electricity, and UES' [the Russian grid operator's] political connections both domestically and abroad."[72] Likewise, there were tight margins for substituting nuclear energy deliveries in traditional export markets. By the end of the decade, 90 percent of foreign electricity sales were to indebted CIS members that generated export revenues for the Russian electricity monopoly significantly below the costs of producing the power. Even had Minatom been able to break into the

Russian electricity monopoly's export markets, the increased revenues would not have been sufficient either to support the sales or to secure a niche in primary power supply for these foreign customers.[73] At the same time, traditional customers in Finland, the Czech Republic, and Ukraine solicited foreign contracts for new plant construction that effectively broke Russia's long-standing power generation monopoly in these established markets. The pressure intensified with decisions to close Kazakhstan's Soviet-designed fast breeder reactor in December 2002, and Lithuania's two RBMK units by 2017.[74]

The stringency in foreign markets was compounded by the squeeze on nuclear power generation at home. Although the share of total electricity production from the nuclear sector grew from 10 to 15 percent from 1990 to 2002 (with plans to increase to 20 percent by 2010 and 40 percent by 2030), nuclear energy competed at a disadvantage against conventional power generation technologies for additional market shares. Factoring in investment, operating, and fuel cycle costs, the per unit expense of generating electricity from new nuclear plants was higher than power produced by new natural gas turbines under virtually any commercial scenario. In spite of the fact that domestic electricity sales comprised one-third of Minatom's budget, nuclear power generation failed to "generate enough money into the future even to properly maintain nuclear plants much less offer support to any of the other ventures of Minatom."[75]

Finally, Russia's potential commercial power was restricted at the back-end of the nuclear fuel cycle market. As the inheritor of the only reprocessing and long-term storage facility for Soviet origin spent nuclear fuel, Moscow was uniquely poised to exploit foreign NPP operators supplied with Russian fresh fuel. Given limited temporary storage capacity at plant sites and the technical and financial problems posed by alternative methods for disposing of spent fuel, Russia potentially was well positioned to dictate the terms of waste processing reduction for either once-through or plutonium fuel cycles for Soviet/Russian-designed NPPs. By expanding the service capacity to 20,000 tons to accommodate imports of foreign-origin spent fuel, Russian nuclear industry officials projected that the country could capture at least 10 percent of the burgeoning global market, earning an estimated $21 billion over the ensuing decade.[76] Yet, there were technical, political, and commercial barriers beyond Russia's direct control. First, Russia's storage capacity was already pushed to the limit handling 15,000 tons of its own spent nuclear waste, leading domestic critics to castigate the plan as not only "impermissible but criminal."[77] Second, excess reprocessing capacity at plants in Britain and France offset Russia's competitive advantages for reprocessing services that did not include long-term storage of spent fuel stocks. Given that many of the potential new customers had established ties with West European reprocessing services, there were added political and legal benefits to sticking with them. Third, Kazakhstan expressed growing interest in challenging Russia for at least $1.2 billion in the international spent fuel market.[78]

Fourth, the United States and EU essentially controlled the destination of the most lucrative supply of spent fuel. Approximately 80–90 percent of the world's fresh and spent nuclear fuel was of U.S. origin that by law and international agreement required Washington to approve every detail regarding the transfer and disposition of spent fuel. This imposed significant political obstacles to transactions between Russia and interested foreign customers, such as Taiwan, South Korea, and Japan. According to U.S. State Department officials, a primary consideration for granting consent to the retransfer of U.S.-origin spent fuel was the assurance that Russia would fully comply with appropriate standards for safe and secure disposal. This purportedly included full disclosure of the terms for transport, storage, and disposition of each foreign contract.[79] Because these rights could be extended at Washington's discretion, the United States also was well positioned to attach additional political conditions to tie up future spent fuel storage contracts, including demands for a two-decade halt to fuel reprocessing and abrogation of Russia's contracts to construct light-water reactors in Iran. Although Russia balked at both conditions, senior Minatom officials openly acknowledged that Washington's approval would be decisive to the ultimate success of any spent fuel import program.[80]

By the same token, Russia's prospective back-end fuel cycle market niche was hemmed in by the EU, which controlled the existing market for non-U.S. origin spent fuel. Competitive commercial interests and concerns for avoiding accidents that could shake global confidence in the nuclear sector compelled the European nuclear lobby to keep close tabs on Russia's nuclear waste processing. This included passage of an EU Commission directive that required all member states to limit exports of spent fuel to those "foreign countries that have the legal, regulatory, and technical capability to manage it safely and that have formally agreed to the import." Consistent with Brussels' preference for disposing of radioactive waste and skepticism of Minatom's environmental credentials, the EU Commission issued a 2002 directive that further circumscribed the residual demand for Russia's spent fuel services by requiring member states to establish national burial sites by 2018.[81]

Minatom in Control

Although Russia's market power was severely constrained across the nuclear industry, the government's institutional clout remained firmly intact throughout the first decade of independence. The implosion of the super-centralized Soviet system notwithstanding, ownership of Russia's commercial nuclear fuel cycle complex defaulted primarily to the state, as initially all of the nondefense assets remained 100 percent government-owned. In 1993, however, the state formally laid down guidelines for creating three ownership profiles for the commercial sector.[82] The first group consisted of 171 exclusively state-owned facilities, including federally owned financial entities and research institutes, as well as commercial fuel cycle and power generation facilities under the proprietary domain of federal, regional, or municipal governments. The second group comprised approximately 157 open joint-stock companies that the federal government owned at least 49 percent of the voting shares, as well as retained exclusive authority to license operations, appoint federal representatives to the respective governing boards, and orchestrate privatization schemes. The three most prominent joint-stock companies were Rosenergoatom, TVEL, and TENEX that supervised construction and operation of Russia's nuclear power plants, managed nuclear fuel development and production, and administered Russia's trade in fresh fuel and enrichment services, respectively. Each joint-stock company was solely government-owned, and organized as a vertically integrated corporation guided largely by commercial considerations. In an effort to streamline nuclear commerce in Fall 2001, the government reorganized Rosenergoatom as a single state nuclear company, as well as folded TENEX into TVEL with the new shares purchased directly by the State Property Ministry. According to senior officials, the main purpose of the latter was to integrate all stages of nuclear fuel supply into a single entity that could at once boost the efficiency of state regulation and international competitiveness of Russia's nuclear fuel cycle industry.[83] The third set of nuclear entities consisted of several hundred limited liability companies that were privately owned and operated as subsidiaries of state-owned nuclear facilities, performing a variety of nontechnical services for commercial profit.[84]

Formal regulatory authority over the commercial nuclear sector was consolidated among a small group of government agencies. Beginning in 1992, legal control over roughly 98 percent of Russia's nuclear program came under the purview of Minatom, including supervision of all policies related to fuel cycle services, commercial power generation, international sales and scientific and technical cooperation, export control, and waste management.[85] This included exclusive authority to set targets for nuclear power construction and generation and fuel cycle production. It also entailed oversight of foreign reactor and fuel sales, as well as authority to act as the state's primary contracting agent for the "Megatons to Megawatts" deal with the United States.[86]

Minatom inherited several direct and indirect instruments to secure compliance from subordinated entities. It was delegated authority to administer sales of nuclear material from the national stockpile, certify nuclear operators, set standards for monitoring and enforcing quality control at nuclear facilities, provide technical approval for nuclear exports, monitor internal compliance with national export control laws, and approve the (de)classification of nuclear information and travel clearances for top-level managers in the nuclear complex.[87] Minatom also supervised corporate governance practices at both state-owned and open joint-stock companies, as well as was authorized to adjust priorities, coordination, and allocation of federal financing and foreign assistance for the nuclear industry. With authority to request or deny privileges for subordinated nuclear facilities, the ministry capitalized on the residual corporate culture of centralization pervasive within the nuclear complex, its own strong representation within interdepartmental government commissions, and regular access to senior echelons within the executive branch.[88]

Although the lion's share of administrative responsibility rested primarily with Minatom, the authority to set prices for the industry was bifurcated. Minatom set export prices for Russian fuel cycle and secondary services. By Summer 2001, the ministry, via Rosenergoatom, acquired the exclusive right to sell nuclear-generated power directly to Georgia, Finland, and Ukraine. This empowered Minatom to break the monopoly of the national grid operator, UES, and to modify the export tariffs for nuclear-produced electricity.[89] However, the situation was the reverse for domestic power generation. UES was authorized as the national grid operator with direct access to wholesale customers nationwide. It exercised discrete authority to purchase power from Minatom's nuclear plants, transact directly with wholesale customers, set wholesale prices, and control national tariffs managed through the Federal Wholesale Market for Electricity (FOREM). This division of authority provided Minatom with both regulatory muscle and incentives to advance narrow commercial ambitions abroad to offset institutional barriers to breaking into the domestic power generation market.[90]

Minatom's dominance over the nuclear complex was not tantamount to monolithic control. In 1995, for example, a presidential decree relaxed the federal government's exclusive jurisdiction over the nuclear complex, stipulating both "joint" and "independent" supervisory responsibilities to regional and municipal administrations.[91] Protracted federal budgetary shortfalls also constrained the effectiveness of Minatom's financial control mechanisms. The ministry faced regular difficulties covering operating expenses and salaries throughout the complex, and, at times, received less than 50 percent of the annual federal outlays necessary to support defense activities. Even as federal financing for the sector began to pick up in 2000, Minatom's fiscal control had clearly slipped, as salary payments remained delayed and unadjusted for inflation, and unpaid leaves, reduced work schedules, downsized production, and difficult social conditions remained the norm. Accordingly, institute directors had de facto discretion to look beyond traditional vertical channels for extrabudgetary relief.[92] In addition, political infighting within the ministry compromised capacity of the federal center to guide responses to the social and economic crisis afflicting the nuclear complex. Conflicts over funding priorities for civilian power generation, defense production, and reorganization and control of profit-making activities in the nuclear industry sent mixed signals throughout the bureaucracy.[93] Consequently, there was a stark contrast between Minatom's decisive administrative control over federal-level policy making on vital issues of trade and commercial strategy across the fuel cycle and power generating complex, and its fragmented oversight of fiscal matters and implementation of operations at local facilities.[94]

Conclusion

How much power did Russia broker as an international energy player? As is clear from the discussion above, Russia's post-Soviet market and institutional prowess varied, even as it inherited vast reserves that dwarfed that of its immediate neighbors, introduced new ownership structures, and experienced recovery in all sectors by the end of the first decade of independence. Concerning its market power, Russia regained superpower status in the gas sector while it was only a secondary player in world oil and nuclear fuel cycle and power generation markets. Accordingly, Russia had potentially more resources that it could bring to bear for diplomatic purposes in the gas industry than in the other sectors. At the same time, the domestic regulatory institutions varied across the three sectors. Although ownership was concentrated in starkly different state versus private hands in the gas and commercial nuclear sectors, both industries were distinguished from the oil sector in that federal administrative control was clearly delineated. Although the state did not have exclusive authority to regulate the gas industry (and it had no control over management at the firm level), it retained key rights to affect the decision-making calculus for both governmental and nongovernmental actors within the sector. By contrast, the confusion and opacity of the domestic oil industry mitigated oversight and enforcement of state policy. As a result, the state was well poised to exploit Russia's respective market potential in both the gas and nuclear sectors, while its grip over the rising and increasingly internationally oriented oil sector remained conspicuously constrained. Moscow's effectiveness at manipulating these various market and institutional conditions for strategic ends in the former Soviet south is the subject of the next three chapters.

4

❖

Russia's Gas Diplomacy

Manipulating Compliance from Turkmenistan and Kazakhstan

In January 2002, Russian President Vladimir Putin launched a bold initiative to form a Eurasian Gas Alliance with former-Soviet producer states, Turkmenistan, Kazakhstan, and Uzbekistan. Notwithstanding benign claims of creating a "common system" of management, Moscow was poised to dominate the cartel, using it to: (1) impose favorable terms for regional gas production and export; (2) arrest the influence of outside energy players; (3) pave the way for regional reintegration; and (4) strengthen Russia's competitive edge at landing gas in lucrative international markets. The initiative marked the culmination of a decade of success at reclaiming residual natural gas assets in the region, as Turkmenistan and Kazakhstan not only agreed to coordinate policies, but conceded to arrangements that tied security in the sector to Russia for the foreseeable future.

The success of Russia's gas diplomacy raises basic questions about energy statecraft. Why did rival Eurasian suppliers choose to increase reliance on Russia; action that defied national objectives and respective geostrategic and commercial opportunities? Why was Russia more effective at securing compliance by avoiding strategic showdowns than by issuing direct coercive threats? How was Moscow able to accomplish these feats in the natural gas sector, where the state was too weak to impose political control?

The chapter addresses these issues in three parts. The first section analyzes the market and institutional components to Russia's gas diplomacy from 1991 to 2002. As discussed in chapter 3, though Russia was the world's natural gas superpower, key actors within the government retained limited but clear authority to regulate the industry. These factors empowered Moscow to alter the prospects for regional producers and align domestic political and commercial interests, giving teeth to a competitive gas strategy. The second section reviews how Moscow manipulated Turkmenistan's deteriorating energy trajectory. By reframing Ashgabat's strategic energy choices from the domain of gains to the domain of losses and inducing the Russian gas lobby to make concessions on narrow substantive issues, Moscow managed to guide Ashgabat into gambling on compliance among other risky prospects. The third section traces the success of Moscow's gas diplomacy at tapping its global and institutional advantages to adapt to Kazakhstan's transition to the domain of gains, and to coax the domestic gas lobby to reinforce its firm diplomacy. Here Moscow deftly manipulated Astana's options so that compliance was a safe but suboptimal bet relative to the uncertainty of diversifying relations with China or the West.

Russia's Gas Strategy

Though Russia presided over the world's largest gas reserves, the government lacked authority to impose energy policies on the highly concentrated national industry. Accordingly, the state was saddled with balancing a broad national strategy of domestic energy subsidization and regional reintegration, with the gas lobby's commercial interests in expanding competitive advantages in key foreign markets. Although there was convergence on the general contours of regional statecraft, to the extent that the Russian gas monopoly, Gazprom, considered other viable options Moscow could not bludgeon compliance.

The bifurcated regulatory structure in the Russian gas industry gave shape to distinct private and state interests. First, as the largest Russian company and energy producer in the world, Gazprom had strong commercial incentives to cover marginal costs for servicing cash-strapped domestic consumers; extract itself from the chain of debts in the economy by increasing the share of prompt cash and noncash receivables; and collect monopoly pipeline rents from potential domestic competitors.[1] In addition, it shouldered responsibility prescribed by the state to provide gas to price-insensitive domestic markets and to fulfill its tax obligations. Consequently, Gazprom faced the challenge of turning a profit, while providing cheap energy to consumers that had little capacity or incentive to pay, let alone respond to price and supply shifts.

Gazprom also faced constraints and opportunities in its foreign operations. On the one hand, the firm was obliged to provide nonmonetized exports to customers in the NIS, both as part of the state needs program and as payment for transiting Russian gas to lucrative extra-regional markets. On the other hand, Gazprom strived to protect its market share for lucrative deliveries to Europe from foreign and domestic competitors alike. Saddled with enormous reserves and a growing need for cash receipts and capital investment, Gazprom sought to enlarge its presence in established and emerging markets, becoming a "gas octopus" controlling the flow from China to Great Britain. Enhanced foreign sales and acquisition of equity in downstream transit, trading, distribution, and storage facilities across the NIS offered potential relief from domestic and regional nonpayments to the extent that commercial conditions favored construction of new pipelines and deliveries of larger volumes of Russia gas, and that receipts from domestic deliveries exceeded marginal costs.[2]

Given the specific combination of political and commercial concerns, there were several courses open to Gazprom in its dealings with Eurasian gas suppliers. One strictly predatory strategy involved protecting established markets from Eurasian competition. This entailed exploiting Gazprom's pipeline monopoly to restrict access altogether to Central Asian suppliers for both monetized gas exports and noncash sales to Russian consumers. Given Russia's vast reserves and low fixed costs at delivering gas, even barter deals by Eurasian rivals on the home market threatened to deprive Gazprom of profits, as long as the firm could cover marginal costs. Protecting the home market offered potentially greater profits in light of the difficulty collecting transit rents from foreign suppliers that, in turn, incurred their own nonpayments problem in the Russian market. This strategy would effectively strand potential Eurasian gas suppliers that were dependent on Gazprom's trunk lines for access to both the Russian market and cash paying customers in Europe.[3]

A second strategy featured dumping. As recounted by David Woodruff, Gazprom had strong commercial reasons to practice price discrimination at home.[4] The combination of segmented domestic and foreign markets, and the relatively low marginal costs of supplying gas to concentrated, large-scale customers in the power generation and industrial sectors via the existing pipeline network, significantly reduced the costs of supplying the cash-starved home market. Gazprom, therefore, had a powerful incentive to charge domestic customers what they were able to pay in either cash or offsets, as long as the firm's marginal costs were covered. This was an especially attractive policy, should the Federal Energy Commission (FEC) play along by setting domestic prices accordingly, given that domestic consumption accounted for 70–90 percent of aggregate gas production throughout the decade. Although domestic receipts were proportionately lower than export revenues, by all accounts they continued to generate profits, albeit at a lower rate.[5] Relative to the high fixed costs of constructing long distance pipelines to both established and emerging foreign markets, the opportunity to earn lower but certain profits from delivering gas to the home market was potentially appealing.

As a complement to this domestic posture and to conserve on the higher costs of exporting Russian gas, Gazprom retained the option to dump cheap Eurasian gas onto foreign markets. This entailed exploiting the company's regional pipeline monopoly to land large volumes of Central Asian gas in foreign markets, even below the marginal costs incurred by Eurasian suppliers. This strategy offered to maximize Gazprom's profit-making activities by embracing the government's regulatory obligations to supply low cost gas to domestic consumers, while capitalizing on the "Turkmen bubble" to extract monopoly rents for the initial delivery of Eurasian gas to foreign hard currency markets that, in turn, could be sacrificed later in favor of higher priced Russian exports.[6]

A third viable strategy for Gazprom was to expand global market presence, while exploiting its regional pipeline monopoly to deliver cheap Central Asian gas to the home market. The beauty of this strategy was that it provided relief from low profit sales without readjusting gas subsidies to domestic producers and households. Gazprom could export gas to profitable European markets, as well as expand into Northeast Asian markets to reduce exposure to domestic nonpayments problems.[7] The company also could pipe Central Asian gas onto the home market, collecting monopoly transit rents in the process. This two-pronged approach could free up production for export during periods of soaring foreign demand, while minimizing competition from regional suppliers. It also could enable Gazprom to earn significant commercial profit, as long as the firm expanded foreign deliveries on a competitive basis or found it cost effective to leave gas in ground. Gazprom could commit large-scale deliveries of Central Asian gas to the Russian market, reducing the amount available for exploring alternative extraregional pipeline projects and market niches that competed directly against the firm's interests.

In short, there was no obvious commercial imperative that followed from Gazprom's concentrated stature in markets at home and abroad. The appeal of each strategy was contingent upon the specific combination of prevailing international market conditions and the government's pricing policy at home, both beyond Gazprom's direct control. Not surprisingly, Gazprom was sensitive to fluctuations in both factors, and was prone to altering its commercial strategy accordingly.[8]

What was good for Gazprom, however, was not necessarily good for the state. An important concern for the government was to promote the survival of the industrial sector and stave off additional social costs of large-scale factory closures during the tumultuous early post-Soviet economic recession. This required maintaining stable energy deliveries to wholesale and household customers. Boosting efficiency in a gas market dominated by a national monopoly was a lesser concern. Accordingly, Moscow relied on the authority to set prices to perpetuate the circulation of cash and nonmonetary transactions in the economy. Although the wisdom of this course was hotly contested, the government nonetheless imposed artificial prices while extracting high tax revenues from Gazprom to sustain the "virtual economy." The challenge was to set domestic gas prices low enough to keep industrial and household customers afloat, via cash or nonmonetary payments, but not so low to reduce the capacity of Gazprom to meet its severe tax obligations.[9]

Over the long term, the state's domestic gas strategy was aimed at boosting production, competition, and administrative control over gas transit. Russia's *Energy Strategy through 2020* called for ramping up production in 2000 from 595 bcm to 700 bcm, and increasing exports from 200 bcm to 240–280 bcm by the end of the period. Related reforms included breaking up Gazprom's upstream operations, and establishing equal access to the internal transmission system. By employing various tax incentives, the government hoped to induce oil companies to process their natural gas reserves (instead of flaring them off), and to encourage independent gas companies to exploit the country's smaller- to medium-size reserves. The strategy also called for the state to acquire a controlling stake in the management of both the internal and export pipelines. Although the plan was for domestic prices to rise gradually, doubling as early as 2005, there were no provisions for the eventual liberalization of prices.[10]

Complementing the internal reforms were provisions for reorienting Russia's international gas footprint. Although no single agency was charged with formulating Moscow's gas diplomacy, there were several overarching objectives specified in the *Energy Strategy*. This included employing the natural gas infrastructure as an instrument to orchestrate economic and political reintegration of the post-Soviet space. In addition to harmonizing production and processing activities across the region, the *Energy Strategy* summoned state and private entities to exert pressure on regional states to ease Russia's access to international markets and facilitate "the realization of the export potential of Russian fuel and energy resources." NIS suppliers also were encouraged to orient energy sales to the Russian market, and to acquiesce to the "penetration of Russian capital in all forms" into their energy sectors in order to serve Russia's strategic interests.[11]

Under President Putin, the consolidation of national interests and influence over Eurasian energy security issues figured prominently in the effort to improve Russia's international competitiveness. The government acknowledged the limited capacity of the Russian gas sector to dominate specific regional markets, due to gathering local pressures for price liberalization, short-term contracting, and diversification of supply. The *Energy Strategy* called for reorienting the energy lobby's push from capturing specific markets to upgrading Russia's stature in various markets across the globe. Emphasis was placed on tapping the country's vast domestic resource potential and export infrastructure to "preserve Russia's exclusive role as the largest provider of energy raw materials in the international community (writ large)."[12] As Putin made explicit on several occasions, the focal point of strategic energy policy was to steer market mechanisms to uphold the country's role as *primes inter pares* in the Eurasian gas equation, and to exploit this dominance as a springboard for "achieving competitive advantages in global markets."[13]

Russia's strategy for global expansion also necessitated improving prospects for closer integration into established European and emerging Asian gas markets as a segue to smoother relations with the EU and membership in the WTO. The preferred course involved reaching accommodation with international transit norms embodied in the Energy Charter Treaty sponsored by the EU. The Ministry of Fuel and Energy, in particular, regarded ratification of the treaty as critical for attracting new loans to Russia's energy complex and paving the way for enhanced trade earnings over the coming decades.[14] Gazprom vehemently objected to relaxing control over the European market, however, complaining that the measures not only would deprive it of $4.5–5.4 billion per year in export earnings, but would render impotent Russia's capacity to produce and transport gas altogether. Company officials argued that granting equal access to Russian pipelines under the Energy Charter would effectively flood lucrative cash-paying foreign markets with cheap Turkmen and Kazakh gas that, in turn, would intensify competition, lower prices, and reduce the state's critical export earnings. In a blatant attempt to coopt the state's own rationale, they claimed that this would "considerably weaken Russia's influence in the world and in the European political process."[15]

In sum, the balance of authority in the gas sector institutionalized divergent state and private interests concerning the direction of Russia's energy diplomacy. While there was considerable overlap, it was clear that neither the state nor Gazprom was captured by the other. Though the most divisive issues concerned reform of the domestic transmission system, there were significant discrepancies over preferred strategies for controlling the Eurasian gas network. For Gazprom, the critical issue was how Central Asian suppliers could be most effectively folded into the firm's commercial strategy, given the constraints imposed by the state's pricing policies and regional market conditions. Where Eurasian gas suppliers fit into this equation varied, with the firm at times seeking to strand, dump, or exploit Central Asian gas. Alternatively, the state was determined to lock up Russia's energy influence in region, using it as an instrument for political and economic reintegration and reclaiming lost global stature. Although willing to cooperate with customers in the "far abroad" in the hopes of securing bigger political and economic payoffs, the government was more inclined to use natural gas as a lever to foil regional diversification. The challenge at home, therefore, was to enlist Gazprom's compliance in a manner that made commercial sense to the company.

Accentuating Risks for Turkmenistan

Turkmenistan's acquiescence to Moscow's natural gas diplomacy was both dramatic and incongruous with its energy security ambitions. In the immediate wake of the Soviet collapse, the regime in Ashgabat was bullish on diversifying energy and diplomatic relations. Yet, by the middle of the ensuing decade Turkmenistan's gas industry lapsed profoundly back into Russia's strategic orbit. Ashgabat watched the price of its exports plummet, and its huge reserves become virtually stranded from lucrative foreign cash markets. Increasingly dissatisfied with the sector's trajectory, Ashgabat stepped up efforts to woo risky export routes, markets, and foreign investment opportunities. Amid this shift in strategic outlook, Ashgabat took successive gambles on placing its energy and strategic future at Russia's discretion, capped by the concessionary terms of a twenty-five-year bilateral energy agreement signed at the beginning of 2003. By Ashgabat's own admission, the leap towards Russia was very risky, especially given Moscow's predatory practices and intolerance of competition in the natural gas market.[16]

Curiously, Turkmenistan's strategic deference did not track with the intensity of Russia's coercive diplomacy. Beginning in 1993, Russia restricted Turkmenistan's natural gas exports as part of a strategy for asserting regional influence. Moscow took particular exception to Ashgabat's handling of a 1992 price dispute with Ukraine that, in turn, lead Turkmenistan to halt supplies to Kiev that saddled Russia with increased gas subsidies to debt-ridden Ukraine as compensation. Russia responded by reducing Turkmenistan's gas export quota to Europe by 20 percent and relegating Ashgabat's deliveries to insolvent markets in Eurasia. The pressure intensified at the end of 1993, when Moscow terminated Ashgabat's access to lucrative markets in Europe. The President of Turkmenistan, Saparmurat Niyazov, subsequently cut a new deal with Russia that established a joint venture, Turkmenrosgaz, to exploit the country's gas reserves. This also obligated Ashgabat to sell Russia gas at prices well below world market rates, in return for piping Turkmen gas to European markets. Dissatisfied with restricted access to Europe and in an affront to Moscow's coercive energy diplomacy, however, Niyazov later canceled the deal, suspended the joint venture in 1997, and aggressively solicited pipeline options to circumvent Russia's stranglehold. Moscow then denied Turkmenistan access to the Russian pipeline system altogether, virtually depriving Ashgabat of all hard currency earnings from its gas exports until 1999.[17] But Turkmenistan again responded by opening a direct pipeline to Iran and attracting Western interest in exploring an alternative trans-Caspian pipeline to Turkey. It was not until 2000 that Turkmenistan resumed deliveries to Ukraine (and later to Russia), receiving only modest compromises from Moscow on the terms for sales to other NIS. By the close of the year, following the relaxation of Moscow's direct pressure, Ashgabat conceded to placing annual volumes and the long-term value of its gas exports increasingly under Moscow's sway.

The deference to Moscow was neither convenient nor cost effective, given Ashgabat's initial endowments and strategic predicament. Russia's decision to deny Turkmenistan's access to European markets in 1993 cost Ashgabat an estimated $2 billion per annum, equivalent to 50–80 percent of yearly export earnings through 1998.[18] By subsequently signing up for enhanced deliveries to Russia and Ukraine, Ashgabat not only sacrificed potential hard currency revenues but was forced to accept 40–60 percent of the annual payments in poor quality goods and services that were rarely delivered on time. According to Niyazov, Russia effectively "robbed" Turkmenistan, buying gas at rates as low as $18 per thousand cubic meters while selling it for as much as $120 in Europe.[19] By increasing Ashgabat's reliance on insolvent NIS markets, Russia severely restricted Turkmenistan's gas production and economic growth, as foreign customer indebtedness created severe cash flow problems for the government.[20]

Turkmenistan's deferential energy security disposition was not simply due to a lack of alternatives. Exploration of multiple options for plugging into new regional markets lied at the crux of Ashgabat's long-term gas strategy. Throughout the period, as depicted in Appendix 1, Ashgabat was courted as a potential partner in five prospective international pipeline projects: the U.S.-backed 1,300 mile Trans-Caspian Pipeline (TCP) project slated to transit Turkmenistan's gas through Azerbaijan and Georgia to Turkish and European markets; an alternative pipeline route to Europe via Iran, Turkey, and Bulgaria; a 765-mile pipeline to Pakistan (and possibly India) via Afghanistan; a route to China and Japan; and a shorter 175-mile pipeline to northern Iran. These routes presented Turkmenistan with viable options for bypassing Russian pipeline operators, as well as with gateways to expanding Turkish and European markets and to emerging outlets in southwest and Northeast Asia.[21]

Remarkably, the likelihood of success for each alternative project did not fundamentally worsen throughout the decade. Although each pipeline was marred by problems—due to uncertainty over sufficient volumes of Turkmen gas, financing, transportation, and political barriers that directly affected respective costs and probabilities of success—each remained promising, attracting foreign interest and prospective investors. Progress was made on all fronts, ranging from the opening of the small trunk line to Iran in 1997, to the "on-again-off-again" exploration of a pipeline to Pakistan. As the utility of exploration seem to rise, however, Ashgabat's interests in each pipeline seemed to wane. In the case of the TCP project, Ashgabat's misgivings heightened in early 2000, just as the U.S. stepped up efforts to boost the commercial viability of the pipeline. According to the U.S. Ambassador, Jon Wolf, the option remained firmly on the table, despite Niyazov's recalcitrance.

> It's not a question of finance—in fact, this project is readily financed. We (the U.S.) continue to support the project because we think that it is important for the countries of the region on the east side of the Caspian and the countries on the west side of the Caspian to have an opportunity to export their natural resources—gas and oil—to Western markets without having to go through some of the world's largest competitor nations. . . . We have always made clear that this project needs to be financed commercially, but that we would be able to help with U.S. government-backed Export-Import Bank credits and Overseas Private Investment Corporation (OPIC) insurance to support the project. OPIC insurance would be very important because it would reassure commercial investors that they would have a government guarantee for the project. And we have been very consistent in the position for two years.[22]

Moreover, as Azerbaijan was angling toward committing its own gas to the TCP project, which significantly bolstered the commercial appeal to international investors, Niyazov became conspicuously pessimistic, redefining the project as a threat to Turkmenistan's security. Ironically, Niyazov lost interest and realigned energy security ties toward Moscow when the situation seemed most propitious for reducing the country's vulnerability in the gas sector.[23]

Similarly, the forced removal of the Taliban regime in 2001 created conditions ripe for revisiting alternative eastern oriented gas pipelines previously championed by Ashgabat. In May 1992, Niyazov initiated discussions with Pakistan over the prospects for constructing a pipeline linking the two countries via Afghanistan. Over the ensuing decade, Ashgabat explored this option amid enormous international pressure that, according to insiders, nearly put Niyazov on a collision course with regional neighbors, including Russia.[24] Yet, Turkmenistan seemed to reverse direction by consummating the long-term gas export contract with Russia at the end of 2002 at the exact moment that the commercial prospects for a pipeline to Afghanistan received widespread international endorsement.[25] This was especially curious when compared to the long-standing uncertainty concerning the pipeline capacity available to Turkmen gas and the dubious profitability of noncash payments offered via the aging Russian transit system.

Finally, Ashgabat's willingness to gamble on deference to Russia marked a distinct break with Niyazov's satisfaction with the *status quo ante* and general preference for keeping open the country's pipeline options. In the immediate aftermath of the Soviet collapse, Turkmenistan was firmly situated in the domain of gains, as it inherited several benefits not available to other southern NIS. The country's natural gas potential was widely regarded as world-class at the time, estimated at holding up to 14 trillion cubic meters (tcm) of proven and unproven reserves (ranking fourth in the world), with a projected annual production capacity between 90 and 130 billion cubic meters (bcm) by 2010.[26] Due to low domestic consumption rates, net exports ranged between 85 and 95 percent of domestic gas production that, in turn, fueled optimism that the country was poised for economic success as the world's third largest supplier (behind Russian and Iran) to dynamic gas markets in Europe and Asia. The earning potential was expected to lure large-scale foreign investments, mitigating concerns about the economy's heavy reliance on natural gas export revenues.[27]

Added to the mix were geopolitical and geoeconomic benefits that accrued from the Soviet divorce. Independent Turkmenistan was geographically separated from Russia, contiguous with Iran, and proximate to Turkey. Although far from lucrative gas markets in Europe and Asia, the location afforded strategic opportunities both to distance Turkmenistan from "uncomfortable alliances" and to develop, albeit at considerable commercial cost, direct transit routes to Turkey, South Asia, and the Middle East that could circumvent Russia altogether.[28] From 1992 to 1996, Turkmenistan also registered the lowest turnover of trade with Russia among the southern NIS, at the same time that it increased trade with other industrialized states. The country's huge national gas reserves buoyed optimism for diversifying trade partners, affording a break in relations with its "oldest brother" without being treated as a "younger brother."[29] In addition, Turkmenistan's dependence on the Russian pipeline system, though considerable, was not complete. Ashgabat did not rely on it for domestic processing or delivery, and was actively engaged with Tehran to construct an integrated system for delivering gas from its eastern territories to northern Iran. Nor did Turkmenistan rely exclusively on Russian pipelines to export gas to customers in Kazakhstan or Uzbekistan.[30]

Turkmenistan's freedom to maneuver vis-à-vis Moscow was buttressed by early domestic conditions. Although a multinational state, the overwhelming majority of the population identified themselves as ethnically Turkmen (74 percent).[31] Moreover, the essentially "sultanistic" regime was insulated from external interference and the domestic reverberation of Russia's coercive diplomacy.[32] In the still centrally planned system, the president had both the political incentive and strength to pass down the short-term costs of deadlock with Russia, in exchange for acquiring potentially greater energy rents from new markets. Accordingly, Niyazov looked to the availability of prospective substitutes for Russia's pipelines and markets to sustain domestic subsidies and avert indefinitely the immediate political and economic difficulties of liberalization experienced by other post-Soviet states. As summed up by the first Foreign Minister of Turkmenistan, "Niyazov was counting on the country being able to survive on the proceeds of oil and gas, which would generate enough wealth that everyone would have a Mercedes."[33] Moreover, Niyazov did not look to Russia as a compelling political model. He rejected outright Moscow's identification with political liberalism, preferring instead to cultivate a unique brand of pan-Turkic "patrimonialism" as a springboard for constructing an independent identity as the "Kuwait of Central Asia."[34]

This strategic predicament gave rise to impressions in Ashgabat that with time, Turkmenistan would be able to establish favorable terms of trade with Russia and to diversify energy security relations. It spurred Niyazov's confidence at navigating a neutral, self-sufficient strategic course to becoming a "gas republic" that was not strictly beholden to Russia.[35] Although initially reluctant to exit the Soviet Union, when presented with a *fait accompli* of its demise, Niyazov embraced bilateral diplomatic and foreign economic relations that were "equidistant" from rival regional and global coalitions.[36] Codified in the concept of "positive neutrality," Turkmenistan undertook a series of steps aimed at detaching Ashgabat from the CIS, avoiding entangling alliances, and shielding the country from Russia's external influence. The strategy was heralded as "both the ends and means of entering the world community," and featured an "open door" policy to attract international investment and diplomatic support needed to develop the national energy infrastructure and explore additional gas export options to Europe, China, and southwest Asia.[37] Consistent with these ambitions, Ashgabat became a member of the Non-Aligned Movement, and designated Iran and Turkey as fraternal partners for constructing alternative pipeline routes to European markets. At the same time, Niyazov took care not to alienate Russia. Although resistant to joining the ruble zone and signing a free-trade agreement with Moscow, Ashgabat granted ethnic Russians dual citizenship and established bilateral military and border security arrangements. Similarly, Ashgabat pursued cost-effective access to the existing Russian-dominated gas pipeline system.[38]

Given the initial positive trajectory, it is striking that Turkmenistan became prone to gambling on its long-term energy security future by the end of the decade. What infuriated Western investors and statesmen, who were otherwise predisposed to pursue the TCP, was the quixotic shift in Turkmenistan's strategic frame and seemingly irrational impatience with steady improvement in the viability of alternatives to the Russian-dominated export pipeline system.[39] On closer examination, however, there is good reason to believe that Ashgabat took the gamble largely at Russia's behest to avoid future losses.

Changing Frames

By the middle of the decade, Turkmenistan's strategic predicament shifted for the worse. The country experienced mounting economic hardship and successive years of negative economic growth from 1993 to 1998. Real GDP declined by approximately 65 percent, dropping 26 percent alone in 1997. The socioeconomic impact of the downturn was enormous, as the unofficial unemployment rate reached an estimated high of 24 percent in 1998.[40] There was even hint of domestic instability, as evidenced by the regime's overt crackdown on political opposition, repression of alternative political and economic outlets, and Niyazov's own erratic behavior.[41]

At the crux of Ashgabat's predicament was the significant slide in the Turkmen gas industry. After experiencing an upturn with independence, natural gas production and export revenues declined precipitously from 1993 to 1999, while continuing to account for as much as 74 percent of national export earnings and 60 percent of the central budget through 2000. Gas exports, in particular, declined to a post-Soviet nadir of 1.8 bcm in 1998, equivalent to roughly a quarter of the previous year's earnings.[42] In response, international energy companies downgraded estimates of the magnitude and cost effectiveness of accessing Turkmenistan's probable gas reserves, as well as became more sensitive to the regime's arbitrary and opaque regulatory system.[43] They began to balk at the restrictive rules that prevented foreign access to the country's export pipelines, and the fixed prices for domestic sales that were substantially below world market levels.[44]

Although gas production and exports showed signs of modest recovery by 1999–2002, Ashgabat's strategic predicament nonetheless continued to deteriorate.[45] This was accentuated by Moscow's efforts at manipulating key regional markets to devalue the prospects for Turkmenistan's natural gas exports. While in principle there were alternative options for piping Turkmenistan's gas, Russia successfully exploited comparative advantages to lower the appeal to Ashgabat of breaking into established European and Turkish markets. By preempting demand in these regional markets, Moscow reformulated the value of the different pipeline options and strategically pit the anxieties associated with Turkmenistan's future dependence on Russian pipelines against the declining prospects of the alternatives. These actions effectively altered Ashgabat's energy security reference point and reframed the *status quo* as a losing one.

First, Moscow preempted Ashgabat's prospects in Western Europe. This was accomplished by flaunting Russia's proven track record as a reliable supplier and exploiting available "ramp-up advantages." Russia expanded deliveries to Europe by 46 percent between 1992 and 2000, effectively supplying 42 percent of the EU's demand for imports.[46] This was complemented by securing additional contracts to service 80 percent of the East European market that, depending on the breadth of the EU's future enlargement, stood to increase significantly Gazprom's footprint among the member states. With an energy accord signed between the two parties in 2000, and given the projected scope of EU enlargement and consumption, Russia also was poised to capture as much a 60 percent of the future market.[47] Accordingly, the Russian government pressed Gazprom to expand deliveries to Europe by 5 percent through 2020. In an effort to increase the economies of scale for delivering gas from its huge reserves, Russia explored several new transit options to Europe that included the 2,500-mile and 83-bcm Yamal pipeline (and its 60-bcm Poland-Slovak link that circumvented Ukraine); the North TransGas Project that offered a land-and-sea route to Germany via Finland and Sweden; and an alternative Nordic Gas Grid project that was slated to increase Russian deliveries from the Barents Sea fields to the EU, including a pipeline segment under the Baltic Sea extending to the U.K. Operating at full capacity, the various pipelines could enhance Russia's competitive advantages at meeting the projected 400-bcm annual increase in global demand by 2010 vis-à-vis traditional rivals from Scandinavia, North Africa, the Persian Gulf, the Middle East, and East Asia.[48]

This posture rendered future gas deliveries from Turkmenistan virtually uneconomical. Ashgabat's only hope for breaking into the already crowded European gas market over the ensuing 30–40 years depended on decisions by foreign consumers to opt for supply diversity over lower prices. This would entail improbable concessions from European consumers to cancel commercially favorable long-term "take-or-pay contracts" with traditional suppliers, as well as a willingness on the part of Ashgabat to sell gas at bargain basement prices despite paying higher transit costs to reach European customers.[49] Russia's aggressive market strategy also ensured that Ashgabat's ability to compete for the future residual demand, even with the liberalization of regional gas markets, would be impaired by disproportionately acute transportation costs associated with piping small amounts of gas at a greater distance from Turkmenistan. It was widely accepted among industry specialists that Turkmenistan would have to land at least 200 bcm per annum for it to be cost effective for EU member states to import gas from such great distances.[50] Given these prospects, the cost of gaining independent access to European markets became greater for Ashgabat compared to servicing consumers in the former Soviet Union. This was reinforced by expectations of Moscow's burgeoning appetite for cheap substitutes for its own subsidized deliveries to the CIS that could free up domestic supplies for guaranteed hard currency payments in global markets. Moscow's EU gas diplomacy, therefore, effectively converted gas markets of last resort (e.g., Belarus, Georgia, Moldova, Russia, and Ukraine) into Turkmenistan's best available export options.

Moscow also exploited its global market power to lower the opportunity costs of using the Russian pipeline system. The focus here was to beat Turkmenistan to the fast-growing Turkish market and transit hub. This was pursued by offering potentially high volumes of gas via an unusually low cost Russian alternative that benefited from shorter distance and more reliable supplier financing. As proclaimed by the chairman of Gazprom, "we are a competitor of Turkmenistan in the gas market and we act on the basis of this precept." The target was met by sealing a backroom agreement in 1999 to construct the 720-mile Blue Stream pipeline between Russia and Ankara—including a 210-mile segment laid on the bottom of the Black Sea—that effectively locked in 80 percent of Turkish customers by 2010.[51] This squeezed the commercial value of Ashgabat's TCP option out of the region's biggest prospective market. It also sapped the commercial appeal of running the technical risks of constructing a trans-Caspian undersea link, and accepting conventional "take or pay" contracts with an increasingly insolvent Turkish government. At the same time, it exacerbated Ashgabat's anxieties related to Baku's demands for equal access to the TCP that jeopardized the cost effectiveness of delivering Turkmen gas over 1,200 miles. Because the Turkish gas distributor had so many supply options, the price of Turkmen imports was expected to decline, further reducing the long-term cost effectiveness of relying on the TCP option for access to Turkish markets and retransit services to Europe. Combined with the uncertainty associated with the magnitude of Turkey's projected demand for gas, these actions strengthened the long-term appeal of the Russian pipeline system as Turkmenistan's only commercially viable conduit to emerging hard currency gas markets in the Balkans and Europe.[52]

Finally, Russia maneuvered to beat Turkmenistan to the punch in Iran. With the opening of spurs to Iran, Moscow lost control over Turkmenistan's access to both the Iranian and Turkish markets. By affirming its commitment to the Blue Stream project, however, Russia effectively restricted the commercial viability of a pipeline linking Turkmen gas to Turkey via Iran. Realistic forecasts of the volumes and profits associated with this Iranian link paled in comparison to the costs incurred by Ashgabat of canceling existing orders serviced via the Russian transit network. The prospective payoffs of using these shorter alternatives were negated by the greater costs of competing directly against future large-scale Iranian and Russian deliveries in both the Turkish and West European markets.[53] Similarly, Moscow smoothed the way for Gazprom's megadeal for gas exploration in Iran's South Pars field by officially endorsing the contract, wooing support from Western Europe, and defending it against U.S. sanctions. This was reinforced by Moscow's activism at exploring a "north-south" energy corridor linking Russia to India via Iran.[54] By increasing the stakes for Tehran of doing business directly with Russia in searching out new markets in Southwest Asia, Moscow effectively discouraged expansion of Turkmen gas exports to Iran.

Russia's preemptive actions significantly shifted Ashgabat's reference point for assessing each pipeline option. With its export prospects dimming, Ashgabat no longer associated the status quo with burgeoning opportunity. Faced with choices among highly risky prospects, the leadership was more concerned about locking in future profits than with diversifying potential outlets. Niyazov exhibited mounting frustration with the uncertainty of gaining access to alternative gas export pipelines and markets. By the end of 1999, he started to send mixed signals, at times dismissing different export options as "pipedreams." Niyazov became especially impatient with Western investors and statesmen who were responsible for arranging financing for the TCP project. Notwithstanding the commitment by Royal Dutch/Shell to finance 50 percent of the TCP in 1999, Niyazov had all but given up on the project, chastising his foreign partners for dragging their feet and threatening to look elsewhere for buyers. In a heated exchange with the Turkish Energy Minister, Cumhur Ersumer, he opined about the declining prospects for exporting Turkmen gas.

> I cannot understand some of your politicians. You can take the very cheap gas from your brothers. That gas will be delivered to you at a cost of $70 per 1000 cubic meters on your border. At the same time, your politicians are buying it from Russia for $114, and you are not thinking about the fate of simple Turkish people. Exporting Turkmen gas will be of advantage to you. Turkmenistan has 23 trillion cubic meters of gas. It will be enough for the Turkmen and Turkish people for 500 years! What are you waiting for, then? What do you want? What are you afraid of?

He further complained that: "For the past eight years I have been running after you, for the purpose of bringing Turkmen gas to your homes. From now on, you should be running after me."[55] This fueled investor anxiety, as well as political concerns in Washington about Niyazov's own fickleness. By Fall 2000, frustrated by Niyazov's exhortations, successive delays in approving the project and his threats of delivering 50 bcm of gas per year via the Russian pipeline system, the consortium of Western energy firms that was gearing up to construct the trans-Caspian pipeline to Turkey closed its office in Ashgabat.[56]

The sense that Turkmenistan's strategic trajectory was declining also was reflected in Niyazov's assessment of alternative export pipeline projects as a choice between risky prospects. Initially, the primary concern in Ashgabat was to diversify Turkmenistan's export routes and regain access to credit worthy markets. A premium was placed by 1997 on attracting support for the construction of the Korpedzhe-Kurdkui gas trunk line to northern Iran that was scaled for delivering 8–12 bcm per year with an initial capacity of 4 bcm per annum.[57] This was complemented by interest in developing the TCP with an initial capacity of

16 bcm that could be expanded to 30 bcm. Niyazov seemed content with the promise of these pipelines for political reasons and for diversifying exports, even though neither project was a sure thing or unambiguously cost effective.[58] By the end of the decade, however, Ashgabat no longer measured success in terms of securing international political and financial commitments for alternative pipeline routes. Instead, what mattered most were projected windfalls to Turkmenistan, calculated in terms of the payoffs received for specific quantities of gas exported via each pipeline. This was made explicit in a press release issued by the Turkmen foreign ministry in 2002.

> We can deliver gas to the north, south, and west in required amounts. There are no politics in this, only economic expediency. That is how the Turkmen government assesses all gas-export pipelines. In this respect, any gas-export route is only a commercial profitable project for Turkmenistan if it is aimed at exporting amounts of not less than 30 bcm of natural gas annually.[59]

The criticism reached an apogee when Azerbaijan began to lay claim to at least 50 percent of the prospective TCP pipeline. Without options for delivering his designated annual minimum, Niyazov became increasingly desperate and prone to gambling on the country's energy future.

Manipulating Risk

Russia's aggressive strategy for landing greater volumes in regional gas markets not only circumscribed Turkmenistan's export prospects, but effectively manipulated the riskiness (i.e., variance in outcome values) of each major pipeline option. The first choice, the short pipeline to Iran with a transit link to Turkey, had a very low variance. Although the segment to Iran was opened in 1998, Turkmenistan consistently delivered volumes well below the 8–12-bcm capacity.[60] Russia's saturation of the Turkish market effectively stemmed the commercial prospects for further development, rendering moot Ankara's initial interest in importing 28 bcm from Ashgabat. Given that Iran was a potential competitor for the Turkish market, there was little hope that Ashgabat could ramp up deliveries via this pipeline even with future recovery of the Turkish market. Another negative consequence was the potential for a rift with the United States over relations with Iran. The relatively small capacity of the pipeline and competitive limits to Iran's demand for Turkmen gas imports, however, attenuated this risk. In practice, this option offered the lowest utility but the lowest risk for Turkmenistan.

A second transit option, the pipeline to Pakistan (and possibly onto India) via Afghanistan, offered a somewhat wider variance in outcome values. The upside was much more attractive than the Iranian variant, given the projected 30 bcm throughput capacity. Yet, the project was expected to cost upward of $2 billion, with highly dubious financing prospects owing to the political uncertainty during and after the Taliban's rule in Afghanistan and withdrawal of the consortium headed by UNOCAL in 1998. Given that the source of gas for the pipeline project would be the same Dauletabad field slated to supply Turkmen gas transited via the Russian pipeline system, there also were serious doubts about the long-term commercial viability of the trans-Afghan pipeline.[61] At the same time, the more likely negative outcomes for Turkmenistan were potentially much greater, as exploration of the pipeline ran counter to the containment strategy pursued by Russia and the other CIS members prior to 2001. Although the political costs of dealing with the interim Afghan government significantly lowered with removal of the Taliban, the lingering political instability in the country and technical challenges of constructing the long-distance pipeline undermined its appeal. Therefore, this option remained highly risky, marred by the wide variation between unlikely high payoffs and the more likely negative effects.

Alternatively, the TCP option offered moderate risks and rewards. Due to Russia's proactive courting of the Turkish market, Turkmenistan could expect to confront intense competition for a declining demand. Given project costs estimated at $2.5–3 billion, the limited absorption capacity of the Turkish market, and Russia's head start at meeting the demand, Ashgabat confronted high entry costs with guaranteed sales prices only slightly above production costs. Azerbaijan's rival claims to the pipeline also threatened to reduce Turkmenistan's maximum deliveries by 50 percent. Taken together, the TCP offered Ashgabat at best an alternative transit option that could deliver slightly more than the Iranian variant but considerably less than the optimistic scenario for the Afghanistan-Pakistan link.[62] The possible negative outcomes of the project stood to be much worse than either route, given Russia's exclusion and hostility toward the project.

By comparison, reliance on the Russian controlled export pipeline system presented the option with the greatest potential variation in positive and negative outcomes. It was well established and offered prospects for exporting large volumes of gas. Should Gazprom make available considerable capacity for delivering Turkmen gas and agree to even modest increases in prices, Ashgabat stood to reap huge profits without requiring significant investment. The potential payoff would be enormous and immediate, given the large capacity of the existing infrastructure and low start-up costs. Conversely, the negative outcome was equally considerable, and based on Russia's previous practice, could entail a complete cutoff of hard currency and barter sales over the short- and long-terms. Accordingly, this gamble presented the most extreme difference between positive and negative payoffs. What was unknown, however, was just how enticing Russia would make this risk for Turkmenistan.

This was where the Russian government's domestic institutional influence proved most significant. In general, the executive leadership of Gazprom was inclined to extract highly concessionary terms for delivering Turkmen gas. Vested with near complete discretion to ramp up domestic exploration and development projects to exploit comparative advantages at gas-on-gas competition in global markets, Gazprom set its sights on seizing an equity stake within the Turkmen domestic gas industry and extracting monopoly rents at the border for exporting Ashgabat's gas. The formation of the Turkmenrosgaz joint venture gave the gas monopoly a 45 percent stake in Turkmenistan's gas projects, over and above control of the export pipeline system. This presented Gazprom with a potential veto in internal Turkmen deliberations concerning prospective exploration, production and export projects. The leverage was not lost on the chairman of Gazprom, who threatened repeatedly to exercise the firm's corporate veto to discourage Ashgabat's construction of competing export pipelines to European markets. Gazprom also was intent on acquiring a strategic presence in negotiations between Turkmenistan, Iran, Turkey, and Pakistan over viable alternative gas export routes.[63] The preference for coercion was made painfully clear in August 1997, when the chairman of Gazprom threatened explicitly to quit partnership with Turkmenistan and cut-off gas outlets unless Ashgabat came to appreciate that "he who controls the valve controls everything."[64] The threat was amplified in 1998, when he publicly declared that Ashgabat had to accept Gazprom's price of $36 per 100 cubic meter "or else let the Turkmen starve!"[65]

The domestic challenge for Moscow's statecraft was twofold. First, it had to sustain an energy strategy aimed aggressively at locking in dominant shares of lucrative foreign markets. This meant inducing Gazprom to bolster the appeal of Russian gas for European customers, while restricting competition from rival NIS suppliers. Second, once situating Ashgabat in the domain of losses, Moscow had to ensure that Turkmenistan's burgeoning inclination to run risks for energy security would favor Russia's control over the regional gas network. This required that Moscow coax Gazprom into relaxing its grip over Turkmenistan's future exports enough to improve the direct positive payoffs, while lowering the opportunity costs of compliance.

To realize this strategy, Moscow relied heavily on the government's discrete regulatory authority to adjust domestic gas prices in a manner that affected the ease of collecting payments at home, as well as the relative value of domestic deliveries and exports for Gazprom.[66] It was precisely by setting domestic prices at significantly higher levels in 1995–1998, that the Federal Energy Commission (FEC) initially facilitated the company's predation in internal and external markets. This was initially challenged by Gazprom executives who complained that these measures aggravated the firm's nonpayments problems. They petitioned Moscow to freeze, if not lower domestic prices.[67] The government, however, held firm by raising the stakes for the Russian gas company of increasing foreign market shares. At the same time, the purchasing power of Russia's export earnings on the domestic market steadily declined, prompting Gazprom to increase its sales on the domestic market. The chairman of Gazprom, for example, boasted that "the internal market is the most profitable one for us, even in today's prices."[68] With domestic gas prices escalating to 60 percent of global prices and profits increasing from domestic sales, Gazprom had strong commercial incentives to recoup mounting domestic losses by restricting Ashgabat's competition in international markets while simultaneously preventing Turkmen gas sales on the home market.

The Russian government's pricing policies noticeably shifted as the price of natural gas in Europe skyrocketed from an historical low in 1999 to contemporary highs in 2000–2001.[69] Facing new demands from Gazprom to increase domestic prices to keep pace with rising global prices, Moscow was careful to take action that made it cost effective for the company to loosen its grip but still contain Turkmenistan's gas exports. This entailed lowering domestic prices at rates that barely kept pace with inflation and that were 30–50 percent below that of coal and 75 percent lower than domestic oil prices. This policy also widened the ratio of export to domestic prices from less than two-to-one in 1996 to over nine-to-one in 2001.[70] By significantly lowering and then holding domestic gas prices down amid booming international energy prices, the Russian government improved the stakes for the company of securing access to Turkmen gas consistent with strengthening Moscow's international leverage. Gazprom had a strong commercial incentive to offset the losses of domestic nonpayments and make up the difference between low domestic and rising international prices by substituting cheap imports from Turkmenistan and exporting its own gas to lucrative foreign markets.[71]

The Russian political leadership also repeatedly pressured Gazprom to make modest concessions on the terms for delivering Turkmen gas. When Russia allowed Ashgabat to resume deliveries via the Russian pipeline system in 1999, Gazprom was encouraged to boost payments from $32 to $36 per 1,000 cubic meters for 20 bcm of gas purchased at the border. Gazprom also was pressed to increase the share of payment in hard currency to 40 percent from 1998. In return, Putin reassured Niyazov that he would broaden the bilateral dialogue in the future, and that both the government and Gazprom would "exclusively recognize and respect Turkmenistan's neutral status."[72] Similarly, the Kremlin leaned on the gas giant to break the stalemate with Ashgabat by accepting a slightly higher purchase price of $38 per 1,000 cubic meters, as well as 30 bcm in imports, 40 percent of which were paid in hard currency. This time Putin met with both Vyakhirev and Niyazov in May 2000 to broker the compromise, intimating that Russia was prepared to purchase 50 bcm of gas annually over the ensuing 30 years.[73] This dynamic was repeated in 2001, as Putin nudged Gazprom to pay $40 per 1000 cubic meter for an additional 10 bcm of gas. Although the volume was significantly lower than the 30 bcm of gas demanded by Niyazov, the terms were more favorable to Turkmenistan, as Russia agreed for the first time to pay for half of the deliveries in hard currency.[74]

Throughout 2002, Gazprom and Turkmenistan remained deadlocked over the terms of a long-term deal. Gazprom, for example, claimed that the "fair" price for Turkmen gas was no more than $25–$27 per 1,000 cubic meters and that the company should not pay more than the $16 per 1,000 cubic meters charged to Russian consumers. In contrast, Turkmenistan demanded as much as $46 per 1,000 cubic meters of gas. Putin once again intervened to broker a compromise, whereby Russia agreed to pay $44 per 1000 cubic meters for the purchase of possibly 2 tcm of Turkmen gas through 2028. Although payment up through 2007 was slated to consist of 50 percent in goods and services, Russia promised to accelerate the volumes purchased annually on renegotiated terms for the out years that could earn Turkmenistan as much as $200 billion. Putin sweetened the deal by relaxing demands for dual citizenship, signing a bilateral security accord, and confirming the Friendship Treaty between the two states. Coupled with Gazprom's obvious concessions to Turkmenistan, these gestures were widely regarded in Russia as evidence of the government's hand at securing a "political gas contract" with Turkmenistan.[75] The terms of the deal presented Ashgabat with the greatest variance between the most extreme positive and most extreme negative payoffs; a gamble that given the country's strategic predicament it could not refuse.

Gambling on Russia

Dissatisfied with the status quo and ensnared by Moscow's gestures, Ashbagat increasingly gambled on conceding to Russia. As mentioned above, Ashgabat made successive concessions over the volumes and prices charged for gas deliveries via the Gazprom-controlled pipeline system. By the beginning of 2002, Niyazov was determined to renegotiate more favorable terms of trade with Russia. He demanded that the price for Turkmen gas exports increase significantly and that Russia pay cash for all deliveries. He also advocated boosting annual exports to Russia from 30 bcm to 50 bcm, with the stipulation that contracts would be negotiated annually to avoid locking in future low price deliveries to Russia and Ukraine.[76] By the end of the year, however, Niyazov abruptly settled for a highly concessionary 25-year contract that potentially committed Ashgabat to delivering 2 trillion cubic meters to Russia.

On the surface, the deal seemed to redress Turkmenistan's basic concerns, earning the country a projected $200 billion with Russia agreeing to pay higher prices for annual deliveries rising to 80 bcm by 2009. Yet, the specific terms of the deal betrayed a series of unprecedented concessions by Turkmenistan, including acceptance of only 50 percent payment in cash through 2007, after which Russia could renegotiate the price for future deliveries. Ashgabat also acquiesced to Russia's gradual consumption of Turkmen gas, with purchases beginning as low as 5–6 bcm in 2004 and climbing only to 10 bcm by 2006 with no contractual obligations for additional deliveries thereafter.[77] By signing the long-term deal, Ashgabat strengthened Moscow's monopoly over its exports. The deal allowed Russia to earn a projected $300 by substituting cheap Turkmen imports to fill a growing percentage of domestic consumption, while freeing up a corresponding amount of Russian gas for delivery to more lucrative foreign markets. It also effectively reduced the estimated supply of Turkmen gas available for alternative pipeline routes, diminishing the commercial appeal of exploring non-Russian options.[78]

Turkmenistan's energy security concessions were multidimensional. At the regional level, Moscow acquired Turkmenistan's support "in principle" for a formal convention on the division of the Caspian Sea, and at several key junctures convinced Ashgabat to temper sporadic claims to off-shore resources, further obfuscating the legal status of the sea to Moscow's advantage. In 1996, for example, Ashgabat sacrificed potential control of the largest gas reserves on the Caspian seabed and endorsed Moscow's proposal for creating a joint company to develop energy in the coastal waters. This was followed up in the 1997 joint communique, where President Niyazov explicitly acknowledged a common understanding with Russia that "issues related to vital activities on the sea must be addressed in strict compliance with the earlier concluded agreements between the USSR and Iran on the basis of a consensus, and unilateral actions of any kind must be rejected."[79] This marked a significant concession by Turkmenistan, as Ashgabat previously favored the establishment of 60-mile exclusive economic zones in the Caspian. Explaining this concession, Turkmenistan's Foreign Minister lamented that Russia's unilateral control over the Volga-Don waterway and access to heavy drilling equipment "made it childish to think that Azerbaijan, Kazakhstan, and Turkmenistan could join hands to exclude the Russians and Iranians."[80]

Although Ashgabat became more erratic at apportioning ownership to undersea reserves, it was careful to avoid outpacing shifts in Russia's policy regarding the legal status of the Caspian Sea. Notwithstanding Turkmenistan's initial endorsement of national sectoral division of the sea that predated Moscow's own revision, Ashgabat took care to follow Russia's lead by amending its position consistent with the 1998 Russia-Kazakh agreement to delimit the Caspian seabed, and avoided supporting U.S.-preferred transit routes.[81] Similarly in 2001, Turkmenistan's agreement "in principle but not method" with Russia's modified median approach to ownership of Caspian resources, and the renewed dispute with Azerbaijan over competing claims to off-shore oilfields hued closely to Moscow's competitive line. Niyazov also was careful to consult directly with Putin before signing on to Iran's proposal for postponing the March 2001 Caspian Sea summit, and quickly backed-off of a proposal at the April 2002 summit for creating 25-nautical-mile national economic zone for each littoral state that Putin flatly rejected.[82] Niyazov's deference to Moscow at these critical junctures in the Caspian legal contest also contrasted conspicuously with his erratic behavior toward other states and investors in the region.[83]

Finally, Turkmenistan's "moderately aggressive reorientation" quickly degenerated into strategic isolation and deepening dependence on Moscow.[84] As early as 1997, Ashgabat began to backslide on its "neutral" pipeline policy, and to cast its lot increasingly with Russia. During that year's Moscow summit, Turkmenistan reluctantly committed to piping gas to Kazakhstan and relying on the Russian pipeline system as its main export route to new markets in Central and Western Europe at prices below world market levels.[85] Frustrated by the failure of Western powers "to match words with deeds" in transporting Turkmenistan's gas, Niyazov issued a series of extravagant demands to prospective financial backers and restricted the internal movement of foreign nationals temporarily working in the country.[86] In the face of mounting instability along the borders with Afghanistan, Ashgabat also embarked on a concerted program to engage the Taliban regime, wooing support for a new pipeline to transport Turkmenistan's natural gas across the country to markets in Pakistan and India. This step defied the concerns of Ashgabat's Central Asian neighbors about the regional threats posed by the Taliban regime, and together with Niyazov's fickle pipeline diplomacy, stymied talks over east-west gas transit options with foreign firms and the U.S. government by 2000.[87] After alienating potential foreign partners, Ashgabat drifted increasingly towards Russia, with Moscow becoming Turkmenistan's primary source of imports and second largest customer for Ashgabat's natural gas dominated export profile by 2001. The strategic tilt was cemented by Niyazov's reluctant endorsement of the Russian-dominated Eurasian Gas Alliance and personal invitation to Russian energy companies to take the lead developing Turkmenistan's gas industry throughout 2001–2002.[88]

Reassuring Kazakhstan

The record of Russia's gas diplomacy toward Kazakhstan was substantively different but equally impressive. Although Kazakhstan too became increasingly deferential on gas issues—as exports fell deeper under Russia's sway—it did so as the leadership gathered confidence about independence and future energy security. Unlike Ashgabat, reliance on Russia intensified without overt threats from Moscow and amid the Kazakh leadership's professed impatience with sacrificing "sovereignty for security" and eagerness to pursue meaningful strategic relations across the region and with the PRC and U.S.[89] Why Astana was seemingly content with sustaining gas dependence on Russia when presented with viable alternative energy projects and long-term prospects for diversification is difficult to explain strictly on the basis of an expected utility calculus.

Kazakhstan's behavior becomes more understandable in light of Russia's manipulative gas diplomacy. Unlike the relationship with Turkmenistan, Russia took action to accommodate Astana's gathering confidence in the status quo. Rather than relying on threats of closing access to pipelines to coerce a dramatic departure from its optimistic energy outlook, Russia used its market power and domestic regulatory authority to breakdown Astana's natural gas equation and to alter the respective pipeline and ownership decisions in its favor. By exacerbating the uncertainty of foreign demand for Kazakh gas (i.e., framing Astana's choice between uncertain and certain gains) and inducing Gazprom to expedite development of the Kazakh gas sector (i.e., reducing the risks of compliance), Moscow orchestrated a series of gradual concessions that had the combined effect of rendering Astana's default to long-term gas dependence on Russia as a wholly rational policy.

This interpretation of the success of Russia's energy statecraft stems from an appreciation of Kazakhstan's mixed strategic domain at independence. On the one hand, the status quo ante featured conspicuous vulnerabilities. Despite its sizable proven natural gas reserves of 65–70 tcf, Kazakhstan was structurally dependent on Russia not only for exporting but for processing and delivering gas to domestic consumers.[90] Astana's gas production, which was concentrated in western Kazakhstan was not connected to densely populated areas in the southeast or industrial north. Rather, there were two separate pipeline systems that distributed gas to domestic customers but that were disconnected from Astana's production, processing, and export capacities. Kazakhstan's gas production in the west was exported to Russia's Orenburg plant for processing, repurchase, and reexport; while the south was serviced largely by imports from Uzbekistan. Accordingly, Astana was in the precarious position of importing approximately 35–40 percent of its gas consumption and relying on Russia to deliver between 93 and 100 percent of the country's gas exports from 1993–1998. From 1992–1996, the country's gas production and exports declined by roughly 40 and 50 percent, respectively, due largely to the confluence of economic recession, swelling debts to foreign suppliers, and Russia's pricing policies for Kazakhstan's gas and condensate. Not surprisingly, gas shortages, especially in the north, were an annual problem and contributed directly to the 31 percent drop in real GDP and widening of the current account deficit.[91] These problems compounded the anxiety of potential foreign investors in the gas sector who were generally reticent about supporting long-term projects until commercial and energy transportation issues with Russia were guaranteed.[92]

Despite obvious structural disadvantages, Kazakhstan's strategic gas predicament was far from desperate and tempered by the secondary role that the sector played in the country's overall energy portfolio.[93] Astana relied more on imports from Uzbekistan than Russia to cover the unfulfilled domestic gas demand in the southern part of the country. Though dependent on the energy sector as the general engine for economic growth, state budgetary revenues from the gas sector paled in comparison to the 55 percent generated by the oil sector.[94] Furthermore, Kazakhstan embraced early on an ambitious program for developing three domestic pipeline projects (totaling 3,300 miles in new transmission lines) to link gas fields in the west with consuming centers in the south and east.[95] Though the government was hobbled by persistent corruption and prone to issuing sporadic threats of renationalizing energy projects, the country remained relatively successful at attracting foreign investment. Kazakhstan also outpaced the other southern NIS by luring almost 6 percent of GDP from outside sources from 1989 to 1997, the bulk of which was concentrated in the energy field.[96]

By the end of the decade, confidence in the status quo was buoyed by an upturn in the sector. With reform of the oil industry, domestic gas production nearly doubled from 1999 to 2000, reaching its highest levels since independence. This was followed by an additional 16 percent growth in 2001. With consumption basically static, Astana was able to reduce gas imports by nearly equal percentages. This mirrored macroeconomic trends, as the Kazakh economy registered three consecutive years of significant economic growth (1999–2002) at 9–10 percent.[97]

These circumstances were complemented by initial progress at independent state-building. Notwithstanding the formal separation of powers codified in successive 1993 and 1995 constitutions, the cornerstone of regime stability rested on the arbitrary personalization, centralization, and cooptation of presidential authority. Eschewing a "cult of personality," President Nursultan Nazarbaev nonetheless was positioned as *primes inter pares* within an essentially dynastic political structure under his heavy-handed leadership.[98] Nazarbaev arbitrarily exploited patronage and executive authority to apportion energy rents that balanced contending clan interests, infused the leadership with a new class of technocrats, and placated different factions among the ruling elite. Although the discretionary balancing of formal and informal political mechanisms arguably retarded the process of political pluralism, raised uncertainty over leadership succession, and undermined the long-term popular legitimacy of the regime, it effectively insulated the Kazakh leadership from both domestic and external sources of political instability during the country's first decade of statehood.[99]

These factors tempered Kazakhstan's initial anxiety over energy security and sustained early optimism about the country's trajectory in the natural gas sector. As early as 1992, the Kazakh government confidently launched a concerted campaign to solicit foreign investment to develop the large Karachaganak gas and condensate deposit that was the source of 35–45 percent of domestic production.[100] The government wooed British Gas and the Italian state oil company, Agip, into forming an international consortium with the expressed long-term objective of enhancing Kazakhstan's self-sufficiency by upgrading refining capacity and increasing production at the deposit 150 percent by 2000. Nazarbaev publicly heralded the project as the springboard for modernizing the country's natural gas infrastructure, and as a lure for Chinese and Western support for political and energy independence.[101]

As depicted in Appendix 2, the Kazakh government actively explored alternative pipeline routes to growing markets in Europe and Northeast Asia. This included proposing construction of a new pipeline to Europe via Turkey that would bypass Russian territory altogether. In a conspicuous gesture of diffidence toward Russia, Nazarbaev intimated that this pipeline was one of his highest, long-term priorities during an official visit to Moscow in 1993. This sentiment was subsequently parlayed into exploration of Kazakhstan's participation in the TCP project.[102] In addition, Kazakhstan was inserted into discussions between Turkmenistan, China, and Japan regarding construction of a possible 3,700–4,800-mile gas pipeline. Together with these partners, the Kazakh government contracted in 1995 with an international consortium—headed by Mitsubishi, Exxon, the China National Petroleum Corporation (CNPC)—to conduct a feasibility study of the project. This effort was followed in 1997 by engaging Tokyo and encouraging Japanese energy companies to explore an alternative 2,590-mile oil and gas pipeline eastward from Turkmenistan across Uzbekistan and Kazakhstan to China and Japan.[103] The enthusiasm for tapping into the

rising Asian demand for gas was affirmed in the April 1999 joint statement with Turkmenistan that advocated closer gas relations with China
and Japan. The leaders of both states unanimously championed the
effort "to provide a great incentive for the development of our infrastructure and transit links to the Asia region."[104]

Optimism on the natural gas front was symptomatic of Astana's
burgeoning confidence in regional and strategic diversification. As sum-
marized by Martha Olcott, over the course of the first decade of statehood the Kazakh leadership steadily transitioned from initial
antagonism toward exiting the Soviet Union, to a strong commitment to
preserving a single post-Soviet space, to greater self-assurance in promoting independence that balanced accommodation of Moscow with
institutionalizing all-azimuth multilateral links. Ever cautious about
preserving warm relations with Moscow, Nazarbaev nonetheless
became less optimistic about crafting common economic policies via an
inter-governmental body, such as the Euro-Asian Union that he initially
broached in 1992, and less enthusiastic about reinvigorating the Russian-dominated CIS security and economic mechanisms. This was
replaced in 1994–1995 by a stronger preference for managing these
issues on a bilateral basis. Astana relied more heavily on a series of
strategic agreements to delineate the parameters governing military
assistance, Russia's access to military bases in Kazakhstan, border
delimitation and security, and trade relations.[105] By the end of 2001,
Nazarbaev had set his sights firmly on establishing partnership with
Russia and had all but given up early hopes of creating an all-encom-
passing Eurasian institution for the foreseeable future. "Yes, we love
you and respect you," he was prone to telling his CIS peers, "but there
can be no full-fledged integration with such different economies. But
Kazakhstan and Russia can already do this."[106]

Complementing the bilateral thrust to relations with Russia was
the promotion of interlocking multilateral regional institutions. This
was epitomized by the gradual pursuit of common economic space with
the other southern NIS that began in earnest with the formation of the
interstate council of the Central Asian Union in 1994 and lead to the
creation of the Central Asian Economic Community in 1998. This complemented Astana's membership in the Customs Union with Russia and
Belarus in 1995, and served as a non-Russian model for regulating sub-
regional interests that included development of infrastructure and transportation networks, interstate financial investment, management of
natural resource exploitation, and border security. At the same time,
Astana broadened strategic ties with extraregional powers, actively

participating in NATO sponsored military exercises from 1994, and ingratiating itself with China by participating in the Shanghai Group's Asian security initiatives. Although these efforts were not intended to denigrate Russia's regional stature, they reflected Astana's self-confidence at asserting its independence as Russia's junior partner and preference for diversifying relations to insulate the regime from Moscow's domination.[107]

Lowering Expectations and Building Confidence

If Astana was gathering momentum in pursuit of alternative gas pipeline options and diversified energy security relations, why did it deepen dependence on the Russian-dominated regional infrastructure? How was Moscow able to manipulate the Kazakh leadership into becoming less anxious about reliance on the Russian gas industry for developing and exporting the country's natural gas? Answers to these questions become apparent from scrutinizing how Russia exploited comparative advantages to consolidate Astana's situation within the domain of gains. In short, by positioning itself as a potentially more viable and cost-effective supplier to key regional markets, Moscow inflated the uncertainty of future gains associated with alternative pipeline projects. This effectively reduced the opportunity costs and affirmed the desirability of sticking with the status quo for the Kazakh leadership.

The centerpiece of Moscow's diplomacy involved preempting the potential demand for gas in emerging markets and luring potential customers to Russia. In addition to saturating the Turkish demand for gas and its consequences for the expected success of the TCP project, Russia directed its "gas tentacles" at capturing a large share of the future incremental demand in Northeast Asian markets. As early as 1994, Moscow positioned itself to woo Chinese, Japanese, and South Korean groups to invest in the development and export of Russian gas reserves, as opposed to alternative projects with Kazakhstan. The objective was to solicit Asian capital investment for jointly developing and exporting 10 bcm of gas annually from Russia's large Kovykta fields that were located in Irkutsk and separated from the Russian gas system. After a series of intergovernmental meetings and the commissioning of a joint feasibility study, the prime ministers of Russia and China signed an agreement for gas exploration and construction of an export pipeline valued at $8–$10 billion in 1997. The deal, which immediately attracted significant investment from South Korea's East Asia Gas Company, envisioned building a 1,800-mile pipeline from Irkutsk to China via Mongolia that would deliver approximately 20 bcm of gas annually and facilitate additional exports to South Korea over 25 to 30 years.[108] By Fall 2000, Moscow announced the priority of expanding development and export potential of gas fields in east Siberia and the Russian Far East. This was followed in 2001 by consideration of several additional projects, including a 2,400 mile pipeline from the Kovykta fields to China and South Korea at an estimated cost of $12–$14 billion; and a 1,450-mile pipeline from Irkutsk to Beijing valued at $1.7 billion. In addition to the Kovykta projects, Russian and Chinese companies discussed construction of a $6–$10 billion, 1,700-mile pipeline to deliver gas from Far Eastern fields in Sakha to the Xinjiang province in China, a gas pipeline from Tomsk to China, and a pipeline from Yakutia following along the China Eastern Railway to Shanghai.[109]

Russian officials simultaneously campaigned to expand exports to Northeast Asia from the more proximate but slightly smaller proven gas reserves (915 bcm) located along the eastern coast of Sakhalin Island. Moscow championed construction of several pipelines, as well as development of an LNG facility on the island tailored to delivering gas to emerging markets in Japan and South Korea (possibly via North Korea) on a long-term commercially competitive basis. By the end of 2002, a feasibility study was underway to assess the prospects for building a 120-mile pipeline from the Sakhalin-1 field to Hokkaido by 2008.[110] Immediately following Kazakhstan's gas discussions with Japan in 1997, the Kremlin accelerated efforts at brokering a peace treaty with Tokyo (signed in 2000), and at linking joint exploration of natural gas reserves in Sakhalin to Japan's long-standing goal of securing the return of the territories off Hokkaido. The diplomatic offensive effectively finessed Japan's gas interests in Central Asia and refocused them on Russia. It also placed a Russian imprimatur on Tokyo's ambitious program to broaden political and economic relations with nonenergy producing Central Asian states. Coupled with the knock on effects of the 1998 Asian financial crisis, these diplomatic initiatives helped to reorient Japan's interests in Kazakhstan from gas issues to deeper engagement in the oil sector and overall development assistance.[111]

As reflected in the *Energy Strategy*, Russia looked forward to tapping into the expected quadrupling of China's gas demand and the concentrated global LNG trade in Japan and South Korea to compensate for the projected plateau in Europe's gas demand by 2005. This coincided with Moscow's expectations that one-third of the growth in

Russian production after that time would take place in the Eastern Siberian and Far Eastern fields.[112] Given the proximity of the rising supply to growing Asian markets, Russia was especially sensitive to heading off potential competition from Central Asian suppliers. The anxiety was intensified by the pressing need to develop new Russian gas fields and to modernize the export infrastructure that, together with long distance, rugged terrain, and climactic challenges threatened to undermine the feasibility of building pipelines to Asia. Lingering uncertainty over Russia's abilities to attract large-scale foreign investment and to craft proposals competitive with those offered by established Asian gas suppliers and LNG exporters—such as Indonesia, Malaysia, Qatar—compounded Moscow's concerns about keeping Kazakhstan at bay.[113]

The short-term effect of Russia's energy offensive was to displace early interest in developing tangible Kazakh-Asian gas relations. Such offers were more commercially attractive, given the economies of scale, shorter distances, and less challenging technical constraints of constructing pipelines and LNG terminals for delivering gas produced in Russia's East Siberian and Far Eastern fields to emerging Asian markets. In light of the preliminary investor interest in the different pipeline projects, and estimations that the commercial viability of the Kovykta projects would turn on China's capacity to absorb a least 20 bcm of gas per annum from Russia, Beijing too looked more favorably on its potential leverage in dealing with Russia than Kazakhstan.[114] Not surprisingly, none of the proposals for Northeast Asian gas pipeline projects initially promoted by Kazakhstan matured beyond feasibility assessments by the end of 2002. With this all-azimuth solicitation of potential Asian partners, Russia was positioned as the main incremental competitor in the future Asian gas bonanza, pushing Kazakhstan's prospects in these markets further out to an uncertain future. Consequently, even before the end of the decade, and in response to China's repeated delays at pursuing an alternative Central Asian project involving the Karachaganak field, the Kazakh leadership began to intimate that it held little hope for constructing main export pipelines to Asia for the foreseeable future.[115]

Settling on Russia

Russia's play for a greater share of the incremental demand in regional markets altered the relative riskiness of the different pipeline options for Kazakhstan. The first option, expanding into Northeast Asia, was converted into the riskiest choice, as the variance between the best and worse possible scenarios for Astana became the greatest. Successfully landing large volumes of gas in the fast growing Chinese, Japanese, and South Korean markets presented a potentially lucrative but very long-term prospect. Although proposals for Russian pipeline projects certainly complicated the commercial appeal of Kazakh gas for Asian customers, the expected long-term expansion and diversification of the Chinese economy and rising regional dependence on gas imports offered access to the fastest growing markets. In the event that the Russian gas sector became overextended in Europe and Asia, there could be even more room for Kazakhstan to compete against traditional Asian suppliers over the long haul.

Yet the probability that Kazakhstan would realize the potential payoffs of selling gas in Northeast Asia was very low for the foreseeable future. The downside for Astana was significant and likely due to Russia's comparative advantages in proximity and scale at wooing potential customers in Asia. Even modest success at attracting Asian customers could inflame relations with Russia. The probable disruption of key gas imports would intensify over time as the Russian gas industry refocused on Asian markets. Astana also would run dangers of: entering into disadvantageous gas-on-gas and LNG competition with traditional Middle Eastern suppliers; subjecting Kazakhstan's gas future to uncertain strategic relations in Northeast Asia and to the vagaries of the regulatory mechanisms and price controls imposed by the Chinese government; and conceding bargaining leverage to Asian customers that enjoyed many different supply options over the short-, medium-, and long-terms. If Astana committed wholeheartedly to pursuing this option, and it subsequently failed due to commercial and technical obstacles, the political fallout would very likely disrupt the delicate balance that it strived so hard to broker between establishing close bilateral ties with Russia and diversifying strategic energy relations. Thus, gambling on Northeast Asia's gas potential did not appear to be worth the likely sacrifices to Astana's strategic trajectory over the short and medium terms.

The second option, participation in the TCP or other European oriented projects that bypassed Russian pipelines altogether, offered more modest positive outcomes with even higher negative payoffs. As was the case with Turkmenistan, Russia's ability to capture large shares of the incremental demand in Europe and Turkey, as well as Azerbaijan's claims to 50 percent of the TCP's potential capacity, significantly reduced the commercial appeal of the project to Kazakhstan. With payments owed to Ashgabat for previous gas deliveries, Astana had the added concern of not alienating Turkmenistan's own aggressive claims to the TCP. Such a political row would likely prompt Ashgabat to cut off gas supplies to Kazakhstan, as it had done on several occasions when Astana was delinquent at meeting its obligations.[116] The negative outcome also was potentially more damaging than pursuit of the Northeast Asian options, given Russia's strong antagonism toward competition in the established European market. In short, active exploration of European options presented Astana with moderate utility but at considerable risk.

The final option—sustaining reliance on the Russian gas pipeline system for exports and critical imports in the north—offered the least variance between positive and negative payoffs. Perpetuation of the status quo had the advantages of stabilizing relations with Russia and positioning Astana to continue the late decade surge in gas production, exports and overall economic growth. Although prone to scaling back the demand for Kazakh gas and delaying deliveries to northern Kazakh customers, the Russian gas industry historically refrained from cutting off access to pipelines altogether. The Russian pipeline system also provided Astana with potential access to the European market with low start-up costs. In the best scenario, Kazakhstan could join Russia in ramping up deliveries to the European market. This offered immediate tangible benefits to the Kazakh leadership in contrast to the very uncertain and long-term prospects associated with other options. By contrast, the worst possible outcomes, which included the potential stranding of Kazakh gas exports and disruption of domestic deliveries, were not significantly different from the negative payoffs of antagonizing Russia via the pursuit of an independent niche in the European market. Given the need for investment and questions about Russia's own technical capacity to meet the growing long-term demand in Europe and Northeast Asia, it was likely that Russia would have interest in exploring gas prospects jointly with Kazakhstan. Thus, continued reliance on Russian pipelines potentially offered a more risk-averse choice than aggressive exploration of alternative options over the near- and medium-terms. Whether such a scenario could woo the otherwise cautious Kazakh leadership, however, depended on the interest that the Russian gas industry had in inflating the risks of Astana's available policy alternatives.

At first, Gazprom was noticeably ambivalent about enlarging its stake in Kazakhstan's energy development. Although Gazprom boasted a global strategy for exploring long-term commercial opportunities in

Asia, the firm fixated on enhancing market share in the European market under the leadership of Chairman Rem Vyakhirev.[117] To the extent that Kazakhstan's export ambitions threatened Russian gas interests in established markets, Gazprom was prepared to act with extreme prejudice to deny Astana access to Russian pipelines. "Surrendering one's market when there is a lack of sufficient capacity is, I believe, nothing less than a crime against Russia," intoned Vyakhirev in 1997.[118] However, Vyakhirev was less anxious about threats that Astana posed to the company's eastern ambitions. Convinced that Gazprom's long-term global presence rested with the "Asian vector," the chairman acknowledged. "the gas market in Asia was absolutely empty or devoid of competition."[119] Left to its devices, according to insiders, Gazprom was content with the status quo and less inclined to activate a predatory strategy toward Kazakhstan. The great lengths to which the Russian gas giant subsequently went to stave off potential regional competition in the Northeast Asian market and to manipulate Kazakhstan's future gas prospects were undertaken largely to placate critics in the industry and government.[120]

Much of this impetus followed indirectly from policies adopted for the Russian gas market. In brief, Moscow approved foreign contracts and set domestic gas prices that effectively reduced the transaction costs of Gazprom's predatory policies toward Kazakhstan. For example, the Kremlin took the lead by negotiating a 1992 deal for the transit of Kazakh gas across Russia that empowered Gazprom with unambiguous authority to regulate access to Astana's existing domestic and export pipelines. By setting domestic prices at very small percentages of prevailing world prices (between 3 to 26 percent from 1991 to 1994), the Russian government elevated the short-term commercial appeal of substituting cheap Kazakh gas imports to meet domestic demand. Gazprom was motivated to use Kazakh gas to cover Russia's spare processing capacity and to lock out competition with the company's other regional projects. This also boosted Gazprom's incentives to manipulate fees for refining and reexporting the gas to Kazakhstan at rates higher than those for which Astana could garner in Russian or alternative markets.[121] This rendered future development and export of the huge reserves at the Karachaganak field dependent on how high Moscow and Gazprom set barriers to entry to international markets. Gazprom exploited the authority to restrict exports of Kazakh gas to approximately 3 bcm per annum and to resell it independently. As lamented by one Kazakh energy official, this enabled Russia to buy "all the gas from us itself, and then after that it did as it pleased."[122]

The Russian government exploited domestic pricing authority to increase the incentives for Gazprom to bargain hard for equity stakes in Kazakh gas projects, as well as to increase the opportunity costs to Kazakhstan of excluding Gazprom from these ventures. In 1994–1996, the Russian government significantly raised the domestic price of natural gas to approximately 56–60 percent of world prices. This increased the profits that Gazprom could earn by selling gas directly to Russian consumers while conserving on the costs of importing Kazakh gas. This primed Gazprom to restrict Kazakh supplies to the Orenburg gas processing plant in 1994, as well as exerted pressure on Astana to concede an equity stake in the development of the Karachaganak field. In so doing, Gazprom inflated the costs of delays incurred by Astana of soliciting investors and transit routes to Europe for the field. This, in turn, raised anxiety among foreign investors, who became especially concerned that Russia's actions would render development of the project unprofitable. In response, the foreign partners prevailed over Astana to grant Gazprom a 22.5 percent equity stake in the venture. This proposal was endorsed in a February 1995 agreement between Moscow and Astana that conferred responsibility on Russia to oversee the "fulfillment of measures providing for legal, organizational, technical, and commercial conditions in exploration and development of the Karachaganak oil and gas condensate field."[123]

In 1996, Gazprom reversed course with Moscow's approval by refusing to meet its financial obligations and withdrawing from the Karachaganak consortium, thus increasing the commercial risks associated with the project.[124] Subsequent attempts to break out of the deadlock with Russia were frustrated by Gazprom's continued pressure via its authority to regulate access to Russia's gas infrastructure at costs that made independent development of the Karachaganak field by the international consortium unsustainable. The capacity to inflate the relative costs and risks of courting international investors and constructing alternative pipelines also made Astana reluctant to push for greater access to international gas markets, and open to a compromise with Russia regarding the ownership of contested gas fields in the north Caspian basin. This forced Astana to accept Moscow's hard-line policies for extracting rents from sales and buybacks of gas at the Russian border, and for relegating Kazakhstan's main gas exports to insolvent customers in the CIS.[125]

With the change of executive leadership at Gazprom in 2001, the firm became more closely wedded to Moscow's preemptive strategy in Asia. The new chairman, Alexei Miller, was a close political associate of Russian President Putin. According to one close observer, "his appointment marked the culmination of efforts behind the scenes to reassert state control over the gas monopoly as well as to forge deals with Northeast Asia that had previously been blocked by Vyakhirev."[126] It was under his stewardship that Gazprom began to turn words into action, successfully consolidating influence over Kazakhstan's gas industry for the foreseeable future.

Defaulting to Compliance

The capacity to sow uncertainty regarding Kazakhstan's prospects in emerging regional gas markets and to lower the relative riskiness of continued dependence on the Russia gas network played into Astana's general confidence in the status quo. By advancing a series of discrete initiatives to better position Russia to meet the expected rise in incremental demand, and by raising the stakes for Gazprom of participating in the development of the Kazakh gas industry, Moscow's salami tactics manipulated Astana's risk-aversion to its advantage. By the end of the period, Astana's pattern of small concessions regarding the development and export of its gas resources set the stage for lopsided dependence on Moscow that reduced the likelihood for future regional and strategic diversification of the Kazakh gas sector.

By May 2002, Astana viewed long-term dependence on the Russian gas sector as a safe bet for stabilizing future development and export of Kazkah gas. First and foremost, the Kazakh leadership signed on to creating a Eurasian gas cartel, with Russia at the helm, to coordinate the volume, transport and consumption of the region's natural gas output. Nazarbaev also encouraged the heads of other Eurasian gas producer states to do the same. Though acknowledging that the choice was not perfect, Nazarbaev deferred to Russia as the course of least resistance.

> No one in the world has a gas transport system as Russia, Kazkahstan, Uzbekistan, and Ukraine have. No one. There is no other gas distribution system like this in the world. It has a vast number of gas pumping stations and so on. Yet, it is a huge facility that has been used but not very effectively, I think.[127]

Although the Central Asian producers formally committed only to "intensifying large-scale cooperation and strategic partnership in natural gas production and transportation," the proposed alliance was

widely regarded by experts as a strategic concession.[128] Astana acquiesced to Putin's ulterior agenda of cementing Russia at the hub of the
emerging regional gas development and export infrastructure. Lured by
hints of an expected rise in Russian gas prices, higher Russian imports,
expanded quotas to the Russian pipeline system, and participation in
joint exploration projects, Nazarbaev willingly affirmed Astana's status
as a junior partner in the regional gas equation.[129] The nature of this
concession was brought into full relief by Putin's stated ambition of
using the alliance to perpetuate the status quo indefinitely; serving as an
instrument to orchestrate the sale of Central Asian gas on the Russian
domestic market while consolidating Gazprom's dominant position in
established markets. Commenting on the joint statement, the Russian
President openly put his Central Asian peers on guard:

> I think that the formation of such an alliance would make it
> possible to exercise effective control over the volumes and
> direction of Central Asian gas exports and would insure the
> formation of a unified and natural gas consumption and con-
> sumption balance as well as its export via a single export
> channel. Today this channel is provided by Gazprom's gas
> pipeline networks—there are no others. But I think if we
> organize this cooperation on a long-term basis, it would also
> introduce an element of stability for our partners abroad.[130]

Russian officials also made it clear that coordinating Eurasian deliveries
to the European market would complement Russia's efforts to dominate
that market. The alliance was presented as a recipe for avoiding political problems with the EU and maintaining favorable gas prices for
Gazprom, in spite of the introduction of price liberalization within the
European market.[131]

Pursuit of the Eurasian gas alliance also carried potentially deleterious long-term consequences for Astana's professed interest in energy
security. Although Astana refrained from signing a long-term bilateral
deal with Russia and initially heralded the Eurasian gas "OPEC" as an
instrument for harmonizing supplier interests, the alliance effectively
institutionalized Russia's monopoly. The sheer magnitude of the Russian gas industry, almost surely set the stage for Moscow to "have the
last word on transit tariffs and on pricing policy." By endorsing dependence on a single pipeline system and construction of future pipelines
across Russian territory, Nazarbaev together with his Central Asian col-

leagues virtually conceded to Russia's future control "at either end of the Central-Asian European route."[132] This came at the expense of pursuing the country's independent niche in emerging regional gas markets. Russian energy officials, for example, made it abundantly clear that the alliance would not only avoid "ramming Europe" with excessive Eurasian shipments, but would be oriented towards rationalizing Russia's regional gas exports to satisfy increasing demand in new markets in China and possibly India.[133]

Astana also agreed to form the KazRosGaz joint venture to manage growth of the national sector. The company, which was 30 percent owed by Gazprom and 20 percent owned by the Russian oil company, Rosneft, was set up explicitly to "coordinate supply, processing, and transit of Russian and Kazakh gas; develop joint projects to market and transport gas to world markets; manage joint exploration and operation of gas fields in Kazakhstan; and harmonize respective national gas regulatory systems."[134] According to the chairman of Gazprom, the main thrust of the joint venture was to secure licenses to a number of Kazakh gas fields, paving the way for the Russian gas giant to "lay claim to explore gas fields in the region of Karachagank" as well as others including those in the North Caspian.[135] In return, Gazprom promised to deliver Kazakh gas to the Russian and European markets. Indeed, Nazarbaev was quick to praise the deal for securing the potential export of 80 bcm per annum of Kazakh gas via a new Central Asia–Europe transport corridor. Yet both Putin and Gazprom officials were quick to dispel these illusions, promising to import only 7 bcm over the foreseeable future with a potential ceiling of 30–50 bcm in the distant future.[136]

Conclusion

The comparison of Russia's gas diplomacy towards Turkmenistan and Kazakhstan demonstrated that Moscow's statecraft was successful in ways not commonly accounted for in traditional explanations. In neither case did Moscow succeed by simply altering the expected utility calculation of compliance for the respective target. Although Russia enjoyed significant and credible power advantages across multiple dimensions, efforts to coerce or induce compliance fell flat and in several instances provoked the diversification of strategic gas relations by both target states. Instead, favored results derived from exploiting

Moscow's substantial market power and regulatory discretion. With Turkmenistan, these advantages were used to manipulate the opportunity costs and risks of preferred gas development and pipeline projects so that compliance emerged as an attractive gamble for the increasingly anxious leadership in Ashgabat. Alternatively, Moscow's gas diplomacy toward Kazakhstan proved most effective when it seized on these advantages to render preferred bilateral, regional, and strategic gas relations as safe bets for the risk-averse leadership in Astana. Yet, the success of Russia's manipulative gas diplomacy contrasted sharply with deficiencies in oil statecraft, as discussed in the next chapter.

5

❖

Russia's Petro-Diplomacy

Floundering in the Caspian Basin

In September 2002, ground was broken for construction of the Baku-Tbilisi-Ceyhan (BTC) pipeline, the first "main" link of oil reserves in the Caspian Basin to world markets circumventing Russian territory. The event punctuated a decade of frustration at leveraging Russia's regional preponderance to dictate the flow of Caspian oil and consolidate control over Eurasian supplier states. Azerbaijan and Kazakhstan not only persistently challenged Russia's legal and political claims to offshore resources, but ultimately solicited alternative egress routes and foreign partners at Russia's expense. Moscow's impotence to prevent these outcomes was lamented across the Russian political spectrum, and regarded widely as symptomatic of the country's waning influence in its strategic backyard.

The failure to dominate the Caspian oil equation was intriguing on several accounts. First, Moscow's diplomacy foundered despite Russia's undisputed regional hegemony and steady recovery as a global oil supplier. Second, the diffused ownership profile of the domestic oil sector and the state's retention of critical shares in holding companies situated the Kremlin to play off contending interests in the sector for strategic effect. Furthermore, Azerbaijan and Kazakhstan were initially primed for deepening strategic oil relations with Moscow, as both inherited an export infrastructure centered on Russia and stood to earn early energy rents from forging closer ties. That they rebuffed Russia's diplomatic gestures and eventually opted for alternative trajectories in this strategic sector is especially curious.

The chapter addresses these issues in three parts. The first section outlines Russia's approach to regaining influence over Caspian oil politics during 1992 to 2002. The next two parts trace the weaknesses of Moscow's manipulative statecraft. The second section demonstrates how Baku's strategic disposition and risk-taking propensity in the oil sector were unaffected by Moscow's diplomatic carrots and sticks. Although the operative domain for Azerbaijan's leadership changed from one of gains to losses in response primarily to deteriorating domestic conditions, Russia lacked proper levers in the sector to finesse the shift for strategic advantage. Accordingly, Baku was inclined first to avoid taking risks on concessions to Russia, and then prone to gambling on strategic diversification. The third section explicates Russia's failure at manipulating Kazakhstan's growing confidence in petroleum security. As distinguished from its gas diplomacy, Moscow was unable to steer the Russian oil industry toward exploiting the Kazakh leadership's general risk aversion.

Russia's Competitive Oil Posture

Although far smaller than the reserves of the Persian Gulf, the Caspian Basin was widely expected to hold nearly 203–233 billion barrels of oil in onshore and offshore fields, ranking it second worldwide in unproven oil reserves. Despite logistical hurdles and geological unknowns, proven oil reserves were estimated at the time at 17–33 billion barrels, roughly equivalent to those possessed by Qatar on the low end and the United States on the high end. The lion's share of onshore and offshore reserves fell to Azerbaijan and Kazakhstan. From 1992 to 2002, daily oil output in the region increased by 70 percent to 1.6 million barrels per day, and was predicted to rise to 3–4 million barrels per day by 2010 (and roughly 5 million barrels per day by 2025), comparable to Venezuela's production in 2002. At its peak, Caspian exports were projected at 3.4 million barrels per day, constituting 25 percent of the anticipated increase in non-OPEC deliveries by 2025.[1]

Caspian oil was potentially critical on the margins for diversifying future global supply. Small populations and relatively low consumption rates in the littoral states ensured that most of the extracted resources would be available for export. Concentrated undersea reserves offered relatively low unit costs of extraction ("finding costs") that increased prospective profits for global oil majors that could afford the initial investment and potential exploration failure costs.[2] Accordingly, control over the division, development and piping of Caspian oil formed the pivot for intense geostrategic and commercial competition among states and international oil companies, as well as provided the crutch for advancing national programs for economic development and political legitimacy across the region.[3]

By 1992, a rare consensus emerged across the Russian political spectrum that Caspian oil reserves and pipelines were vital national interests. Great power enthusiasts and nationalists heralded restoration of Russia's dominance of the Caspian oil equation as the linchpin for wrangling forward military bases in the former Soviet south, regaining control over CIS borders, protecting the Russian Diaspora, orchestrating regional political and economic reintegration, and recapturing Russia's lost geostrategic stature.[4] Pragmatic statists and market oriented groups alike advocated consolidating power in the Caspian to stem the spread of instability along the immediate periphery, preempt potential commercial competition, circumvent infrastructure bottlenecks in the Russian oil sector, and check "outside" encroachment into Russia's "sphere of influence." At minimum, they championed a spoiler role for Moscow, stoking controversy to gain preferential treatment for Russian energy interests in Caspian oil ventures. Notwithstanding the constant fluctuation of world oil prices and uncertainty over the size of energy reserves, "winning" the Caspian energy sweepstakes was widely understood in Moscow as legitimate cause for intervention to protect Russia's "unique interests" in Eurasia.[5]

Although these divergent impulses failed to coalesce into a coherent grand design, Moscow nonetheless adhered to three strategic objectives from 1995 onward. First, Russia sought to preserve a monopoly over the residual Soviet pipeline infrastructure and to secure a commanding position in an expanded transit network to contain regional diversification. Steering pipeline routes across Russian territory would generate additional revenues and allow Moscow to keep a hand on the spigot. Second, the Russian government formally subscribed to the legal definition of the Caspian Sea as a "unique inland lake" that stipulated collective rights to offshore resources. This could formally increase the amount of Caspian resources under its jurisdiction while securing a veto over independent development by other littoral states.[6] Finally, Moscow focused on sowing confusion and exacerbating the political and commercial risks of Caspian energy ventures that competed directly against Russia's interests.[7]

Under President Yeltsin, Russia's posture initially combined the rhetoric of environmentalism with the instigation of regional tensions and traditional forms of *realpolitik*. Moscow maneuvered to oppose foreign intrusion, insist on joint management of environmental and biological issues, garner international support for favored pipeline routes, and assert legal precedents that guaranteed access to the richest subsea energy fields.[8] By mid-1997, the approach conspicuously shifted to promoting the preferential participation of Russian companies in regional energy projects. The Kremlin broadened its definition of "territorial waters," and endorsed the principle of multiple oil pipelines for the region with the proviso that the first new "main" export routes traverse Russian territory. Simultaneously, Moscow continued to court a "rejectionist" front with Iran regarding the primacy of collective solutions to the formal status of the Caspian Sea and intolerance of extraregional interference in settling outstanding ownership and development problems.[9]

On his election in March 2000, President Putin infused Russia's Caspian posture with "aggressive opportunism."[10] The Kremlin reiterated support for "commercially viable," non-Russian pipeline routes, going so far as to moderate opposition to the participation of Russian firms in rival Western-oriented projects and regional oil swaps with Iran.[11] Moscow also endorsed separate bilateral agreements and "new legal instruments" to redress grievances and set precedents for eventual collective solutions.[12] Yet the Putin team continued to promote a distinctly "north-south" transit corridor, flowing from Russia to Asia via Iran to preempt Washington's preferred pipelines.[13] It affirmed Russia's right to veto independent midsea exploration, and joined Iran by declaring that "until the legal regime of the Caspian Sea is finalized, the parties do not officially acknowledge any boundaries on the sea."[14] Putin also put the other littoral states on guard by building up the Caspian Flotilla as "a force factor for promoting Russia's political and economic interests in the region."[15] Confident that impatience for collecting energy rents among regional producer states would outpace the willingness of investors to commit to specific main oil pipeline projects,

Moscow was intent on actively wooing business deals for Russian firms while fueling commercial and political uncertainty to discourage foreign rivals.[16]

In addition, the Putin administration was determined to lure interest away from Caspian oil investments by spearheading shifts in national oil policy. Capitalizing on booming domestic production and Russia's reemergence as a leading global oil exporter, the Kremlin hyped Russia as an alternative to unstable Middle East suppliers. In 2002, a new "Energy Dialogue" was initiated with senior American government and industry officials to promote investment opportunities and the expansion of Russian oil exports. Moscow also endorsed the commitment extended by the country's second leading private oil company, YuKOS, to deliver monthly shipments of Russian crude to the American market beginning in July 2002.[17]

This was complemented by the gentle guidance of market and administrative mechanisms in support of the country's global resurgence. After consolidating political power, Putin urged formation of a mega Russian energy consortium, comprised of leading Russian oil and gas firms vested with special privileges to prospect north Caspian energy deposits.[18] By the end of 2002, Moscow also tacitly supported construction of private pipelines to facilitate an open domestic market for Russian oil exports. According to Prime Minister Kasyanov, the government was intent on "directing" private interests in the oil industry to bolster Russia's future global competitiveness, and was prepared to rely on "some other way to regulate the industry's export potential" besides retaining monopoly ownership of national pipelines.[19] Championing the Russian oil industry as a "guarantor" of global market and strategic stability, Moscow responded with ambivalence to OPEC's campaign to prop up world prices in early 2002. It made a faint show of cooperation by temporarily cutting 150,000 barrels per day in foreign oil sales while capitalizing on the cartel's self-restraint to boost national exports by 9 percent during the first quarter of 2002.[20]

Yet much to the chagrin of policy makers in both the Yeltsin and Putin administrations, successive efforts at coercing or inducing Kazakhstan and Azerbaijan were ineffectual. Both countries persisted with respective campaigns to attract foreign-financed energy projects and to develop main oil export routes beyond Russia's physical or political grasp. Similarly, Russian oil firms eagerly participated in energy consortia with an array of international partners that defied Moscow's attempts at stewardship.

Azerbaijan: Defying the Odds

Following the Soviet collapse, Moscow moved swiftly to cast a shadow over Azerbaijan's strategic oil relations. Determined to preclude independent energy development and uphold joint ownership of the "special water reserve," Moscow threatened Baku with "unpleasant consequences" if it ignored Russia's claims to offshore Caspian oil deposits.[21] This was underscored by denying Baku access to the Volga-Don waterway to the Caspian Sea in 1994. Russia later shifted course by extending bilateral concessions on the delineation of common boundaries, seeking preferential participation in energy exploration projects brokered by Baku, and enticing the transit of Azerbaijan's "early" and "main" oil through Russia's Black Sea port of Novorossiisk. At the same time, Moscow continued to tout the dangers of excluding Russia from Baku's major oil export plans by championing a north-south regional energy corridor and flexing local naval superiority.[22]

Russia's Caspian policies fell on deaf ears, as Azerbaijan persistently diversified strategic petroleum relations. Ironically, the Soviet government's earlier decision to create a "steel umbilical cord" of dependence on "mother" Russia left Baku with untouched oil reserves that it seized on to break out of Moscow's energy vice.[23] Baku neither let Russia's pressure tactics nor legal ambiguities stymie campaigns to woo strategic patrons or to initiate oil development projects based on the de facto division of the Caspian Sea. Although welcoming Russia's commercial participation in multilateral energy consortia, Azerbaijan avoided the seduction of Moscow's post-1997 inducements, gambling instead on a "use it or lose it" posture in its self-proclaimed national sector. By the end of 2002, Baku had extracted bilateral concessions on ownership of contested undersea energy resources, restricted participation of Russian energy firms in joint ventures, and surmounted a series of financial and logistical delays to construct the east-west BTC main oil pipeline that circumvented Russian territory altogether.

Curiously, Baku's dogged pursuit of Western oil markets and the BTC was not an obvious utility maximizing strategy. Markets in Iran, Romania, Bulgaria, and Asia held greater commercial promise for Azerbaijani crude, in terms of relative demand and transportation costs, than either Turkish or West European markets.[24] There were alternative pipeline options that were cheaper and offered higher prospective profits under various commercial scenarios. Oil swap arrangements with Iran, for example, presented the cheapest option for delivering small volumes of available Azerbaijani crude. The decision to transport early oil through Black Sea terminals using both Russian and Georgian pipeline segments cost potentially twice as much as consolidating shipments in a single wide-diameter pipeline that crossed Russia.[25] Given the economies of scale of large pipelines, transport distances, and tanker costs, the BTC was projected to deliver oil costing at least one dollar per barrel more than the cheapest alternatives using large volume pipelines from Baku to Supsa (Georgia), and then shipping oil either by tanker through the Bosporus Straits to the Mediterranean or via a bypass pipeline across Turkish Thrace to the Aegean Sea. Even piping Azerbaijani oil via an updated Russian pipeline to Novorossiysk and then shipping it by tanker to Italy through either the Bosporous Straits or Turkish bypass pipelines was expected to be significantly cheaper than using the long-distance BTC. Furthermore, by opting for the BTC, Baku predictably priced itself out of emerging markets in the Far East, despite projections of a swelling Asian demand and relative price premium of Persian Gulf oil sold on those markets.[26] In short, there was a big difference between the "commercially viable" BTC that at best offered only marginal economic payoffs, and other "commercially attractive" pipeline proposals for diversifying Azerbaijani exports that portended relatively safe returns on international private investment.[27]

In addition, Baku's defiance was steadfast despite changing decision frames. Amid the initial turmoil that followed independence in 1991—the ouster of the first president, a military coup, loss of 16 percent of the country's territory and dislocation of 800,000 citizens in the war with the Armenian enclave of Nagorno-Karabakh—political power was eventually consolidated in 1993 under the semi-authoritarian rule of President Heidar Aliev. Maneuvering within state institutions and exploiting patronage networks, Aliev gained control over the power ministries and national oil sector, curtailed legislative oversight, and resubordinated media outlets that together enabled him to subdue a weak political opposition and civil society. The regime then launched a strategy for expanding large-scale oil production and exports to fuel socioeconomic stability and to bolster public confidence. With proven oil reserves estimated at the time at 7 billion barrels (seventeenth among world oil producers), the regime expected that rising production and declining domestic consumption rates would eventually attract foreign capital and secure a world-class export profile.[28] Despite the precipitous decline in oil production from 1992–1994, the strategy registered early success, as the economy gradually began to grow by 1996.[29]

The upturn in the domestic oil sector from 1994–1997 buoyed Baku's early optimism. The steep decline in oil production leveled off at approximately 180 bbl/day, and oil exports accelerated from several thousand barrels per day to roughly 50,000 bbl/day (due primarily to declining rates of domestic consumption). Foreign direct investment in the oil sector reached new heights in 1994 with the signing of the 30-year, $8 billion "contract of the century" with the Azerbaijani International Operation Company (AIOC) for development of the country's largest offshore oil field (estimated at 4.3–5.5 billion barrels). Emboldened by the imprimatur of this international consortium and endorsements by the United States, European Union, Japan, and Turkey, Baku aggressively concluded an additional six offshore and five onshore production-sharing agreements with foreign operating partners by 1997. Aliev also locked in roughly $40 billion from the United States and United Kingdom for future exploration and development of the nation's oil reserves by the beginning of 1998.[30] The regime inextricably linked these achievements to economic revitalization. Mounting confidence in the recovery encouraged senior officials "not (to be) especially worried about encountering short-term delays in the beginning of oil transportation," and to be patient and risk averse in deciding among the country's energy security options through 1997.[31] Several officials intimated that these trends provided the "trump card" to eventually reclaim the territory lost to Armenian influence since 1994 and to redress the attendant sociopolitical trauma.[32]

Baku's decision frame dramatically changed during 1998–2002. The promise of oil-led economic growth no longer seemed inevitable, as declining international oil prices reduced export earnings by 11 percent in 1998, and perpetuated successive deficits in the state budget. Combined with the frequency of hitting dry holes, Western operated consortia began to reassess the utility of tapping Azerbaijani reserves. By 2002, 6 of the original 12 offshore production-sharing agreements operated by foreign companies were either suspended or terminated after encountering technical difficulties and generating commercially poor results. Only 14 of the 24 international oil projects signed after 1994 remained active at the end of 2002.[33] Even the AIOC cut costs by 20 percent and abruptly scaled back its workforce in Azerbaijan by 1999. There were signs that the country could not rely on oil export earnings and had to import energy to compensate for electricity shortages prompted by rampant corruption and theft in the energy sector. This precipitated a national energy crisis that lasted through 2000. By the middle of 2002, the once bullish prospects for Caspian oil had conspicuously dissipated, as some Western oil firms downgraded estimates of total recoverable offshore reserves to less than 8 billion barrels.[34]

Trouble in the oil sector dimmed the prospects for stability in Azerbaijan. Corruption penetrated all levels of government, with oil smuggling schemes reaching approximately $1 billion annually and bribes constituting a significant "unofficial tax" on foreign oil transactions. The leadership's arbitrary use of oil revenues to fill budget gaps and buy political support ranked Azerbaijan as one of the world's most corrupt nations that, in turn, stoked investor anxiety. By 2002, no more than $8.3 billion was actually invested of the $42 billion promised in foreign direct investment, and capital flight had reached $1 billion per annum.[35] With foreign investment concentrated (75 percent) in the oil sector, the country began to succumb to the "Dutch disease," as higher-priced oil exports inflated the value of the national currency. This lowered productivity and competitiveness of the already underperforming industrial and agricultural sectors, leaving the national economy at the end of the decade both less diversified and more dependent on precarious oil revenues than it was in 1991.[36] Downward socioeconomic trends—coupled with the "frozen" Nagorno-Karabakh conflict, and Aliev's frequent hospitalization beginning in 1999 and growing preoccupation with orchestrating his son's political succession—sapped political patience with the status quo and gave impetus to gambling on the country's energy future.[37]

Romancing a Stone

Despite the commercial temptation to rely on Russian pipelines and the changing decision frame in Baku, Moscow was unable to manipulate compliance in the contest for legal ownership of offshore resources. At first, Moscow was intent on delegitimizing rival interpretations of the Caspian as a "unique inland lake." In a formal letter to the British government in 1994, the Foreign Ministry asserted the validity of the 1921 and 1940 treaties with Iran that explicitly recognized the sea as joint property. Moscow stated that steps taken by a state, littoral or otherwise, "aimed at acquiring any kind of advantages with regard to the areas and resources . . . cannot be recognized."[38] In a memorandum to the UN General Assembly, Moscow dismissed the relevance of concepts such as territorial waters, continental shelf, and exclusive economic zones to closed inland bodies of water, and claimed that any benefits derived from developing offshore resources were the shared property of the littoral states. This included a harsh reminder to "outsiders" that Russia:

reserve(d) the right to take proper measures at the proper time
in order to reinstate the legal order and eliminate the consequences of unilateral actions. All responsibility, including
potential material damage, rests with the parties who initiate
unilateral action, thus disregarding the legal nature of the
Caspian and their obligations under international law.[39]

This was followed by a note sent to Azerbaijan that threatened force to
protect Moscow's interests, and that intimated the vulnerability of
Caspian drilling projects launched before the sea's legal status was settled. The Kremlin also contemplated imposing economic sanctions and
using Russia's Caspian Sea Flotilla to interdict ships operating under the
Azerbaijani flag.[40]

In practice, the forceful imposition of a strict legal interpretation of
collective ownership to undersea resources backfired. Contradictory
international precedents concerning the status of closed water basins
muddled Moscow's claims and invited political challenge by the very
outside powers that it sought to exclude from the region.[41] Determined
to avoid isolation and to acquire offshore concessions, Tehran moder-
ated policy towards Baku, negotiating a 10 percent stake in the "contract of the century" in 1994.[42] Moscow's hard-line approach
intensified synergy between the legal positions staked out by Baku and
Astana. The harder that Moscow pressed Baku for a narrow interpretation, the more that Kazakh offshore ventures appeared as lucrative substitutes to investors, thus generating competitive commercial incentives
for Baku to stand firm.[43]

Moscow's subsequent concessions on delineating the Caspian also
did not readily ease the anticipated pain of compliance for Baku. In
November 1996, the Russian Foreign Ministry unexpectedly acknowledged national sectors out to 45 nautical miles (divided into territorial
waters and exclusive economic zone), with the remainder falling under
joint ownership and management. This introduced a "double tender"
system, under which littoral states would have first dibs on future contracts offered by any one of them. The proposal was intended to break
the stalemate without conceding to Baku's midsea claims, while simultaneously coopting Turkmenistan and Kazakhstan that asserted national
jurisdiction primarily within the 45-mile zone.[44] In response, Baku
tightened national pollution controls over oil firms operating at its offshore fields that undermined Moscow's legal rationale for claiming joint
sovereignty to the middle of the sea.[45]

Moscow's legal gambit also provoked Washington to support Baku. Concerned that Moscow's proposal would indefinitely shut in regional oil to the detriment of America's energy security, and geostrategic and commercial aims, Washington made clear in a 1996 diplomatic communiqué to Aliev that it fully "endorsed the efforts by American investment companies and upheld the idea of a national sectoral division of the Caspian Sea."[46] Accordingly, Baku stood to gain more from rejecting the Russian compromise, even at the cost of isolating itself in a legal imbroglio with other littoral states.

Tehran too categorically rejected Moscow's new proposals. Committed to an equal division of the sea and unwilling to accept step-by-step settlement of contested claims, Tehran rebuffed Moscow by canceling attendance to expert meetings scheduled for 2000–2001 in Russia, and by threatening to lay claim to half of the sea's resources if the Kremlin persisted with its "modified median line" proposal for dividing the seabed while keeping water columns, surface, and airspace under joint control. President Khatami proclaimed that "Iran would not allow insulting actions of hostile states on its borders." This was underscored by Tehran's naval diplomacy in the summers of 2000 and 2001, as Iranian warships removed signal buoys demarcating the Caspian border and confronted survey vessels operating in Azerbaijan's territorial waters, respectively. At the same time, Tehran began to explore bilateral rapprochement with Baku, acceding to joint exploration in border areas on the very day in August 2000 that Moscow's Caspian envoy arrived to press the case for the modified median formula.[47] For Baku, signing onto any deal with Moscow would not secure an advantageous legal solution that could not otherwise be redressed in separate negotiations with the littoral states.

Lacking market power, Moscow also was unable to alter the opportunity costs of exporting Azerbaijani oil across Russian territory. Urging investors to give priority to "commercial calculations over political considerations," Moscow could not decisively manipulate the terms of financing for the pipeline options considered by Azerbaijan.[48] As depicted in Appendix 3, Baku was courted aggressively from 1994 by potential transit states and foreign investors for at least two early and one main oil alternate routes (with different variants) that would bypass Russian territory altogether. The main oil options were more direct and typically offered larger economies of scale than Russia's northern route.[49] Furthermore, the projected long-term glut in the world oil market presented producers with little incentive to rush to decision.

From a strict commercial perspective, Azerbaijan and its international partners were in the driver's seat regarding the selection of main oil pipelines, and stood to incur significant opportunity costs by committing early to use Russia's existing pipeline system.

Given that the burden of financing pipelines ultimately would fall on producing firms and the Azerbaijani government, the ground rules for selecting main export routes were set largely by commercial and technical criteria beyond Russia's grasp. The BTC, for example, became economically viable when Britain's BP oil company in October 1999 committed to "ensuring that companies work with governments and multinational financial solutions to find a way to finance such a pipeline."[50] This provided the imprimatur for the November 1999 framework accords signed between the heads of state of Azerbaijan, Turkey, and Georgia that set in motion reciprocal commitments by international financial and political backers of the pipeline through Fall 2002. It was the Turkish government's September 1999 guarantee against cost overruns for the projected $2.4 billion pipeline, and the Sponsor Group's subsequent commitments in October 2000 to fund the follow on engineering and construction studies that got the ball rolling. This was complemented by gradual commitments to the project by individual member companies of the AIOC from October 2001 to August 2002, and the consortium's upgraded estimates of Azerbaijan's main oil field in fall 2002.[51] These efforts smoothed the way for the EBRD's offer at the end of 2002 to fund 10 percent of the project, and to work closely with the World Bank's International Finance Corporation in arranging future financing for 70 percent of the pipeline.[52]

The competition for export routes also generated commercial incentives among transit states, investors, and "outside" benefactors to offer competitive rates of return for constructing wide diameter pipelines. In March 2000, Turkey and Georgia settled on lower tariffs, while Baku agreed to forgo such fees altogether. This was followed in April 2002 by the Trabzon Agreement that obliged the three states to resolve all outstanding operational and safety problems associated with the BTC, and to remove respective domestic obstacles to the pipeline.[53] Simultaneously, the Clinton administration strengthened the appeal of investing in the BTC by making available financing through government agencies, such as the Overseas Private Investment Corporation (OPIC) and the U.S. Export-Import Development Bank. These sweeteners mitigated critical commercial liabilities associated with constructing the pipeline segment in Turkey that had initially spooked private investors. Once in place, they effectively inflated the opportunity costs incurred by

Baku of defaulting to greater reliance on the existing Russian pipeline infrastructure that otherwise required lower start-up investments.[54]

In addition, the integrated structure of the global oil industry, combined with Russia's relatively small footprint in regional markets, made it difficult for Moscow to saturate Europe's demand for Caspian oil. Russia's success at servicing 18 percent of the EU demand for oil by 1999, for example, ensured neither a dominant nor growing share of the expanding community market. Frustrated by Moscow's intransigence toward accepting transparent transit protocols and fixated on diversifying future supply networks, Brussels targeted technical assistance to develop new pipelines from Caspian and southern Mediterranean suppliers.[55] Similarly, the success in 2002 at wooing the United States with prospects for direct sales and investment in the Russian oil industry did not distract Washington from promoting the diversification of Caspian exports. Instead, the oil sales were seized on by the Bush administration as vindication of the mutual benefits to developing multiple Caspian oil pipelines.[56] Furthermore, the United States frustrated Moscow by training Georgian forces to protect against future assaults on the BTC.[57]

Manipulating Risk

Azerbaijan's choice for developing and exporting Caspian oil involved risks, as well as costs. Laying exclusive claim to a self-described national sector offered both higher expected value and lower risk than conceding to Moscow's preferred condominium or modified median formulas because the variance in outcomes was narrower. The latter two options would probably satisfy Moscow, but would neither resolve the sea's legal status nor likely earn Baku additional regional or strategic allies. Although the modified median proposal was more attractive than the condominium model, as it would legitimate Azerbaijan's claim to undersea resources, it nonetheless presented the most severe negative prospect—Baku would be at the mercy of Russia's local naval superiority within its own economic zone. In light of Turkmenistan's contested claims to offshore oil fields and uncertainty about Kazakhstan's resolve at promoting the median-line division, Baku could neither count on support by deferring to Moscow's inducement nor generate profits from collective legal remedies.[58]

Conversely, if successful at asserting national jurisdiction, Baku stood to reap the full benefits of exploiting Caspian reserves located from the shoreline to the middle of the sea. It also would not preclude bilateral compromises with Iran and Turkmenistan over contested reserves. Given the conflicting commercial and legal claims, the negative consequences of standing firm were neither certain nor much worse than the benefits of acceding to the modified median proposal. Baku probably would not further jeopardize access to the northern route for early oil exports due to the commercial benefits that would accrue to Russia by sustaining operations. Even by asserting national claims, Azerbaijan would unlikely provoke a strategic union between Russia and Iran, given their contested offshore claims.[59] Thus, clutching to a national sector delineation of Caspian offshore oil reserves offered the highest potential utility at the smallest risk, due to the relatively low probability of coordinated retaliation from the other littoral states.

Similarly, the choice of main oil export pipelines was marred by risk. Of the variety of fantastic export routes proposed, Azerbaijan focused consideration on three main trunk lines. The first option was the BTC that posed the greatest risk as well as exorbitant construction and transportation costs. If successful, the option would present the best outcome: Russia's pipeline monopoly would be broken; Baku would cement close political support and future security guarantees from Ankara, Washington, and Tbilisi; Baku would gain ready access via Turkish terminals and pipelines to hard currency markets in Western Europe; and Baku would lure additional oil shipments from Kazakhstan, securing the long-term commercial appeal of Caspian oil in international markets.[60] However, the negative outcomes of this option were significant and quite likely: Russia would be alienated; Iran and Turkmenistan would intensify pressure to lock up contested reserves; the pipeline would be vulnerable to sabotage by Kurdish and Armenian rebels; transit fees would arbitrarily escalate; and the fluctuation in world oil prices would discourage future investment and deliveries of landlocked Azerbaijani oil to faraway markets.[61]

The second main export option—refurbishing the Baku-Supsa pipeline and using the related Black Sea tanker or bypass pipelines—offered the narrowest gap between prospective outcomes. It presented the cheapest set of scenarios for transporting large volumes of Azerbaijani oil westward and was the early favorite of the foreign parties to the AIOC.[62] Like the BTC, it had advantages of diversifying transit and circumventing Russia. Depending on the extension segment, it offered access to emerging markets in the Black Sea region en route to Europe. Yet even the best scenario would be slightly less appealing than the BTC, as it would receive at most tepid support from both Washington and Ankara. Turkey adamantly opposed increasing tanker traffic through the Bosporous Straits, and was committed to developing trans-shipment terminals and storage facilities located in Ceyhan.[63] The worst negative outcomes, however, portended to be slightly less probable and severe compared to the BTC. Although Turkey and the United States were not likely to confront Baku over the choice, neither state would likely extend security guarantees or offer dramatic political support for a pipeline that would almost certainly antagonize Moscow. The pipeline also courted danger by passing through at least four potential conflict zones in the Caucasus.

The third main oil export option—using Russia's preferred north-ern route—offered a somewhat narrower variance in outcome values than the BTC, but dramatically less expected utility than the Baku-Supsa line. The commercial costs of refurbishing and delivering main oil via this pipeline were anticipated to be significantly lower than those associated with the BTC, but higher than the Baku-Supsa route owing to longer pipeline distances across Russia.[64] Yet the risks were quite sig-nificant. Under the best scenario, Baku not only would curry favor with Moscow, but would get an early jump on exporting large volumes of oil, using the operational Baku-Novorossiysk pipeline. All that was required was "to reverse the direction of flow along the Baku-Grozny section, impose a certain degree of order in Chechnya, repair holes, and settle transit fees."[65] Selecting this route would likely provide Moscow with stronger incentives to curb military assistance to Armenia, and to negotiate in earnest settlement of the Nagorno-Karabakh conflict.[66] The downside of choosing the Baku-Novorossiysk pipeline, however, was predictably at least as detrimental as the BTC. It would expose Baku both to the Russian Foreign Ministry's hard-line strategy, as well as to obstructionism on the part of Transneft and respective transit regions of Russia. The protracted instability in Chechnya, with the specter of esca-lation looming in the background, was especially disconcerting. Opting for the Baku-Novorossiysk pipeline would most likely antagonize Turkey that, at minimum, would raise the costs of transporting oil from the Black Sea. Such a decision would unlikely ingratiate Baku with the United States or Iran, further isolating Azerbaijan.[67]

Unable to alter viable option sets, the success of Russia's energy diplomacy hinged on manipulating the relative riskiness of preferred courses consistent with Baku's prevailing decisional frame. Given that the Aliev regime was ensconced in the domain of gains through 1997, the main challenge was to accommodate Baku's general risk aversion that otherwise favored pursuit of a national sector delineation of the sea and the Baku-Supsa main pipeline. This necessitated accentuating the risks associated with these options, or narrowing the gap between the potential positive and negative outcomes of Russia's preferred alternatives. Either course required Moscow to prevail upon the domestic oil industry to line up behind state policy; resisting behavior that reduced the outcome variance of rival Caspian development and export options while embracing policies that made even the marginal benefits of conceding to Russia a sure bet for Baku. Yet, this is precisely where Russia's early statecraft stumbled.

The combination of low prices, declining productivity of existing fields, limited pipeline capacity, and rampant transit bottlenecks diminished the value of domestic sales and spurred Russian oil firms to explore opportunities in Azerbaijan. LUKoil, in particular, was eager to contract with Azerbaijan for the joint exploration of contested oil fields, despite Moscow's injunction against unilateral resource grabs.[68] In a direct slight to the Foreign Ministry, LUKoil petitioned for a 10 percent stake in the AIOC in 1994 that effectively conceded to Baku's national sectoral claims and set a precedent for the company's participation in four other Azerbaijani oil projects during 1996 to 2000. Company officials (with the support of the Ministry of Fuel and Energy) openly acknowledged differences "with part of the government," and all but told the diplomats to keep to "their own business" and not to interfere with the firm's "main commercial concerns of earning dividends for LUKoil shareholders" in dealings with Azerbaijan.[69] Even with Moscow's official abrogation of the contract signed by LUKoil and Rosneft with Azerbaijan for development of the contested Kyapaz/Sedar field in 1997, neither company terminated energy cooperation with Baku. Instead, both chose to put the contract on temporary hold for "commercial reasons" until Azerbaijan and Turkmenistan resolved their dispute.[70] LUKoil also actively pursued an independent role managing pipelines to export Azerbaijan's main oil via Georgia and Turkey, notwithstanding Moscow's pressure to support the Russian route.[71] Although these corporate actions were endorsed by select elements within the government, the lack of coordination among respective positions effectively attenuated the risks to Baku of defying Moscow's early coercive tactics.

Similarly, Moscow confronted significant domestic impediments to lowering the risks of the Baku-Novorossiysk pipeline. Not only did the 1994–1996 Chechen war result in large-scale oil theft and destruction of the pipeline, but the administrative uncertainty that ensued from the 1996 cease-fire agreement compounded the risks of doing business with

Moscow. Emboldened by de facto autonomy to sign state-to-state agreements with Moscow and Baku in May 1997, Grozny asserted claims to 25–31 percent of the total transit fees for each metric ton shipped through the pipeline.[72] Yet, the breakaway republic was not formally accountable to Moscow for honoring its commitments to repair and secure the pipeline. Furthermore, Moscow's plans for constructing a 311-km bypass segment via Dagestan in September 1997, provoked Chechen officials to demand ten times more in transit fees for Azerbaijani oil and to threaten the safety of repair crews working on the pipeline.[73] Although the parties sustained the flow of early oil until the Chechen segment was officially closed in June 1999, senior Azerbaijani leaders viewed with trepidation the internal polemics and lingering opacity surrounding administration of the North Caucasian segments. What they experienced purportedly made them, together with the AIOC, increasingly wary about the prospects for shipping larger volumes of main oil to Novorossiysk.[74]

At the same time, Transneft arbitrarily intensified the political risks for Baku. As the AIOC prepared to deliver early oil in May 1997, the pipeline monopoly unilaterally reneged on allowing Azerbaijani oil to pass through to the Novorossiysk terminal on a legal technicality. Transneft capriciously revised the existing transit agreement with Grozny, demanding the latter to assume greater responsibility for repairing and securing the pipeline. Shortly after Azerbaijani oil was loaded into the pipeline, Transneft balked again by imposing new customs requirements that further delayed shipments. The pipeline monopoly then pressed to renegotiate transit fees with both Baku and Grozny, as well as resisted financing construction of the bypass pipeline through Dagestan.[75] These actions complicated trilateral negotiations and reduced investor confidence in the Russian government's capacity to honor transit agreements that, in turn, raised the political risks of boosting Azerbaijan's future throughput in the Baku-Novorossiysk pipeline. As summed up by the head of SOCAR, Natik Aliev, "Russia's offers were tempting but not credible," as there were no guarantees against "the lion's share of profit arbitrarily going to defray tariffs."[76]

Sensing Baku's shifting frame to the domain of losses and growing frustration with the trajectory of Caspian oil developments in 1999–2000, President Putin attempted to bolster the appeal of gambling on Russian main oil pipelines. He openly declared, "we believe that the key question to resolving this [Caspian energy] problem is defining the balance of interests of the state and companies. . . . We will not be able to achieve anything by the power of the state alone."[77] Accordingly,

Putin sought to realign domestic oil policies to inflate the potential pay-offs for Baku of relying on the Russian route for transiting main oil. Rather than penalizing Baku for not meeting its oil shipment obligations in 1999 and 2000, the Kremlin pressed Transneft to increase the capacity and lower transit fees for the northern pipeline. Putin also expedited construction of a revised 300-km ($160 million) bypass segment that looped around the territory consumed by the outbreak of the second Chechen war.[78]

Putin's efforts to woo an increasingly antsy Baku were subverted from below by Transneft's discretionary policies. Authorized as the sole state pipeline operator, Transneft was loath to relinquish de facto monopoly over the Russian oil transit system. Despite Putin's appointment of a new company chairman in Fall 1999, Transneft regarded both the BTC and Baku-Novorossiisk options as threatening and potentially "poking a hole it is pipeline fence."[79] Company officials rebuffed pressure to extend concessionary terms of access to foreign suppliers. What mattered was making operation of the pipeline "profitable or at least recoupable" for the company, not enticing Baku into long-term contracts.[80] Furthermore, throughout 2000, Transneft was determined to squeeze Baku as it faced a mounting energy crisis, and to level fines ($29 million) on SOCAR for failing to meet Azerbaijan's annual transit obligations. This completely backfired, as it lead Baku to cease all deliveries through the pipeline during the last quarter of 2000. According to SOCAR officials, Transneft's predatory behavior made it almost impossible to avoid Russian penalties, even if Baku met its delivery target, as it subjected Azerbaijan to both an energy crisis and fines.[81] This insolence utterly exasperated the Kremlin, with Putin's Caspian envoy publicly lamenting that Transneft's fines were "simply wrong and not worth discussing."[82]

Similarly, Transneft's autonomy to operate the new Chechen bypass undermined Putin's promises of uninterrupted oil flows. Although construction of the bypass was a priority, Putin had little input on arranging the necessary financing. Instead, Transneft asserted its prerogative as a spoiler. At first, the pipeline monopoly tried to foist the burden onto foreign suppliers, demanding that Azerbaijan and the AIOC purchase shares in the project. Finding no takers, Transneft officials then moved to allay investor anxiety by falsely claiming receipt of loans from the EBRD. It then demanded that Russian oil companies purchase 75 percent of the pipeline, which antagonized domestic producers. Notwithstanding the public furor, the operator finally arranged to pay for the pipeline via tariffs charged to Russian companies. Lacking confidence in the viability of this arrangement, however, Baku balked at ramping up deliveries. At the same time, Transneft alienated Baku from future dealings by openly disparaging Azerbaijani suppliers and soliciting incremental deliveries from Turkmenistan and Kazakhstan.[83]

The administrative problems persisted throughout 2001, as Transneft arbitrarily imposed higher transit fees. Determined to maximize company profits, the pipeline monopoly unilaterally abrogated Moscow's 1996 agreement to hold transit prices steady (in return for Baku's pledge for boosting annual supply), charging fees that were five times higher than those incurred by piping oil across Georgia. Although SOCAR honored the commitment to deliver 2.5 million tons for 2001, Baku flatly resisted remitting additional payments and rejected successive petitions advanced by the Kremlin to double deliveries for 2002 and beyond.[84]

From Settling to Gambling on Defiance

While Baku was operating in the domain of gains through 1997, Moscow's shortcomings in market and regulatory power stymied efforts to circumscribe the opportunity costs and risks of its favored Caspian oil policies. Instead, Baku took great pains to keep options open and to solicit a broad range of potential partners in order to secure cost effective oil ventures and outlets. The challenge for Aliev was to tact carefully between rival Western and Russian proposals without prematurely committing to unduly risky options. There was keen appreciation in Baku that Russia was "too strong to be elbowed out" of the oil game, and that advantages to working with Moscow "extended beyond money and technology."[85] Yet, there was no impetus to bargain from weakness. Enthusiastic about its ascending trajectory and inherently distrustful of Moscow's ambitions, senior officials focused Caspian diplomacy on preserving independence and diversifying relations in order to prevent Russia from dealing "a knockout blow to Azerbaijan's energy security."[86]

Baku's early opportunism was reflected by the consistent rejection of Russia's interpretations of the Caspian's legal status. Upholding the precedent set by the 1982 UN Convention on the Law of the Sea, Baku asserted sovereignty over offshore fields extending to the middle of the sea. Confident that Russia's position would continue to evolve and undeterred by the relatively small risks of persevering with its national claims, Baku staunchly maintained that "not only the floor but also the water and the surface of the sea should be divided up."[87]

Baku also cast the net widely in soliciting international support for multiple oil export routes. From 1994–1997, the leadership successfully wooed commercial partners from over 10 countries in 12 development projects. This included negotiations with Russian companies over specific oil concessions and shares in future international consortia, the most significant of which involved the transfer to LUKoil of a 10 percent stake in the AIOC in 1994.[88] This was followed by decisions in October 1995 to diversify Azerbaijan's early oil exports to European markets. Convinced that "Azerbaijan's oil resources were enough to make everyone happy," Aliev openly endorsed overhauling both the Baku-Novorossiysk (from 100,000 bbl/day to possibly 300,000 bbl/day) and Baku-Supsa (from 115,000 bbl/day to possibly 600,000 bbl/day) pipelines for the westward delivery of early Azerbaijani oil via the Bosporous Straits. Yet, the message to Moscow was clear:

> I'll say frankly that if you do not open this route up for us, we will have the opportunity to move oil along two other routes: one through Georgia to the port of Supsa, the other to the port of Ceyhan through Turkey.[89]

At the same time, Azerbaijan courted Tehran by offering a 10 percent stake in the country's first foreign oil concession and inviting Iran to participate in other consortia, including the "contract of the century." Aliev also proposed incorporating an Iranian spur into early versions of the BTC. Committed to keeping options open, Baku was unwilling to preclude cooperation with Tehran, despite U.S. political pressure and the imposition of American legal restrictions on domestic and third-party firms engaged in large-scale energy ventures with Iran.[90]

Frustrated with deteriorating domestic conditions, the West's mostly rhetorical support for its interests in the Caspian, and anxious to secure legal cover for its de facto ventures, Baku was poised to take risks on codifying claims to offshore resources after 1998.[91] At the January 2001 Baku summit with Putin, Aliev signaled that he was more amenable to exploring a bilateral agreement premised on the median division of seabed resources. At first, this appeared a success for Russian statecraft. Although careful to underscore that national zones would be geographically delineated according to a mutually accepted "equidistant median line," Aliev nonetheless agreed, in principle, to limit claims to the seabed in return for Russia's acceptance of the legality of Azerbaijan's offshore rights.[92] According to one of Aliev's close advisers, the risk was warranted because the prospective advantages to such a deal were simply too great to pass up.

> We and Russia both won because, by signing the agreement on the legal status of the Caspian, we put a legal gloss onto the contracts that we have signed with foreign oil companies, granted them international status and also, prepared the ground for the phased signing of an agreement on the surface and depths of the sea. And finally, by signing this agreement, we taught a lesson to the West, which is indifferent to our problems. I think that this is the greatest victory that Azerbaijan has gained in the international arena. Russia also won because by signing the agreement Russia secured its economic interests in the Caspian and gained a marvelous chance to strengthen its strategic influence in the region.[93]

Aliev's warming proved to be a Pyrrhic victory for Moscow. He hailed the joint declaration as vindication of Azerbaijan's sovereign claims to offshore resources. He pressed Russia to break with precedent by using terms such as "sector" and "zone" to describe delimited areas on the sea floor. Aliev also prevailed on Russia to recognize the legitimacy of de facto claims as "common contemporary practice," and to drop insistence on a joint "strategic development center" with jurisdiction to manage environmental and legal issues. He was explicit that the final agreement codified an "optimal principle for demarcation" consistent with Azerbaijan's historical claims to offshore resources, thus laying down diplomatic markers in the event of an ex post challenge by Moscow.[94] Similarly, Baku hedged the bet on accepting common navigation of the sea by securing $4.4 million from Washington to improve the combat-readiness of its navy and air force.[95]

In addition, Aliev extracted critical concessions from Moscow. He manipulated Moscow's interest in an agreement to secure favorable terms for leasing the Qabala early-warning radar station, and to strengthen his hand at negotiating settlement with both Turkmenistan and Iran on contested fields. By signing the agreement with Moscow, Aliev set a bilateral precedent for negotiating with its south Caspian rivals, as well as positioned Russia to bear the brunt of a negative reaction.[96] Baku also held off consummating the bilateral agreement until Moscow dropped explicit reference to a five-party accord as the ultimate arbiter of the Caspian's legal status. At the signing ceremony, Putin publicly endorsed the "step-by-step" pursuit of successive bilateral agreements as preferable to awaiting a final, collective resolution of outstanding problems.[97] Because these measures effectively inflated possible positive outcomes mostly at Russia's expense, even a limited agreement with Moscow on the division of the seabed, became more appealing to the Azerbaijani leadership than no agreement.

Azerbaijan snubbed Moscow by restricting participation of Russian firms in joint oil development projects. None of the international consortia formed after 1997 to explore Azerbaijan's offshore oil reserves included a Russian firm. By the end of 2002, Russian oil companies acquired stakes in only five of Azerbaijan's twenty-four onshore and offshore production sharing agreements. Moreover, Russian firms were restricted to holding minority stakes (no more than 13 percent), with the exception of two cases where LUKoil formed joint ventures with a Western partner, and were flatly denied a 50 percent stake in a third project. Yet even this standing was precarious, as the two projects in which LUKoil possessed the largest stakes were either abandoned or in jeopardy of cancellation by the end of 2002. This contrasted with U.S. oil companies that participated in 10 of the 24 projects and secured majority stakes in several productive consortia.[98]

Finally, Azerbaijan's risky defiance was epitomized by the selection of the BTC. In September 2002, after eight years of debating multiple options, political posturing, and uncertainty over technical and financial arrangements, Baku together with the AIOC and its Georgian and Turkish governmental partners launched construction of the pipeline. Although the decision marked a dramatic break from the past, it by no means assured smooth delivery of main oil to Western markets, as there remained considerable uncertainty over the volume of oil committed to the pipeline, the reliability of private financing, cost overruns, and technical delays. Almost immediately after the signing ceremony, construction was halted and the AIOC demanded additional concessions from Turkey as reassurance of its commitment to the project amid an economic downturn.

Nonetheless, the BTC decision represented a serious setback to Russia's Caspian oil diplomacy. Though Russian officials were indignant, they were resigned to rising above imposing fruitless obstacles and political friction.[99] In a poignant concession, Moscow proposed construction of a Russian connecting link to the BTC via a joint venture with Georgia. Yet even this gesture, did not give pause to Baku's decision to forge ahead with the BTC.[100] Ultimately, Russia could do little more than stand idle while Baku and its partners grappled with bringing the BTC to fruition. Frustrated with the inability to manipulate the risks of this decision, Putin's Caspian envoy wryly noted that if there was a "silver lining" to the fiasco, it was that the competition might shake up the Russian oil industry to be more responsive to broader national interests.

> The state must create conditions under which monopolies would be forced, not under direct pressure, but by themselves, to conduct flexible policy. And an oil pipeline such as the BTC must make Transneft think about what sort of tariff policy it should implement.[101]

Kazakhstan's Cautious Opportunism

Kazakhstan's Caspian oil policies marked another defeat for Russia's energy statecraft. The resort to direct coercion—imposing preferred understandings of Caspian "user" rights, asserting unilateral resource claims, and withholding oil deliveries to Kazakh refineries—as well as to offering key inducements—greater access to the Russian pipeline system and recognition of Astana's ownership of seabed resources—failed to arrest the diversification of Kazakhstan's petroleum security. By the end of 2002, Kazakhstan restricted Russia's participation in the development of the Caspian's largest-known offshore reserves, and pursued at least two major pipeline projects that sidestepped Russian operators altogether. Although Astana and its corporate partners signed bilateral transit agreements and negotiated compromises on the division of undersea resources, they did so on terms costly to Moscow.

Kazakhstan's defiance was remarkable for several reasons. First, Astana's crude production and exports were structurally dependent on Russia. Notwithstanding significant proven oil reserves estimated between 9 and 17.6 billion barrels (comparable to Algeria on the low end and Qatar on the high end), Kazakhstan inherited an infrastructure that rendered it a net importer, dependent on Russian pipelines for both foreign and domestic deliveries.[102] The country's two major refineries in the east were neither proximate nor connected to main oil deposits in the west, and were tied to feedstock from Russia's West Siberian fields. The third refinery in western Kazakhstan could not handle more than a fraction of the annual crude production in the region. In the short run, therefore, it was cost effective to pump roughly 70 percent of domestic oil production across the border to Russia's Volga region for refining, re-export, and repurchase. This subjected Astana to adverse price differentials, as well as to the knock-on effects of related bottlenecks in the Russian oil sector. The frustration was summed up by Kazakhstan's first Prime Minister, Sergei Tereshchenko: "We extract our own oil, but are forced to import petrol from Russia. This dependence must end."[103]

Second, unlike Azerbaijan, decision makers in Astana engaged Russia from the domain of gains, increasingly confident about the sector's strategic trajectory. From 1992–1995, the trends were mixed. Domestic output precipitously constricted by roughly 25 percent, resulting in a 100,000 b/d decline in production by 1995. As exports fell by nearly an equivalent amount and domestic consumption held steady, the current account deficit rose and cumulative GDP dropped by 31 percent, notwithstanding favorable world oil prices. The contraction left 37 percent of Kazakhstan's foreign oil deliveries tied to Russian end-users. Amid the downturn, the economy also became more dependent on the fuel sector, which increased the share of gross production from 15 to 25 percent over the three years.[104] Yet, the country's huge reserve base sustained optimism for an imminent recovery. This was reinforced by an influx of foreign direct investment concentrated in development of the large Tengiz onshore field (estimated at 6–9 billion barrels) and exploration of Kazakhstan's north Caspian shelf that totaled $1.8 billion by mid-1993, with future commitments that exceeded $20 billion. With growth in the nonoil sector, the percentage of total foreign direct investment in the energy sector dropped from 76 percent in 1993 to 32 percent by 1995, fueling hopes for greater diversification of the economy.[105]

By 1997, the favorable trends gathered momentum to cement the positive decision frame. The upturn accelerated during 1999–2002, as oil production expanded at an annual rate of 16 percent, doubling output (over 900,000 b/day) at independence. Export volumes soared from 1996 to 2002, nearly trebling the 1992 volume (700,000 b/day). With the rise in world oil prices, the total value of oil exports grew by 27 percent, generating $3.8 billion for the national budget during 1999–2002. This was buttressed by forecasts of an even steeper trajectory of expansion over the ensuing 20 years, with Kazakhstan producing nearly 55 million tons per day by 2010 and as much as 100–120 million tons by 2015–2020. The three largest onshore and offshore fields were conservatively projected to yield 30 billion barrels and investments totaling as much as $8 billion per annum. Kazakhstan's huge, underdeveloped Kashagan field, located off the north shore of the Caspian, alone was reliably expected to contain 7–9 billion barrels of recoverable oil, with prospects for generating $25 billion in investments and an additional 30 billion barrels in possible reserves.[106] Moreover, the government refurbished one of the country's refineries and had plans to build another, as well as began construction of an internal pipeline to integrate the national infrastructure.[107]

The sheer scale of the projected upturn generally overshadowed persistent macroeconomic imbalances associated with the oil industry's 20–25 percent contribution to government revenues and the economy's overwhelming dependence on oil exports. Concerns about the fiscal vulnerability of the nonoil sector were partially eased by the allocation of approximately $2 billion by 2002 to the independently managed national oil fund that was designed to sterilize oil windfalls and stabilize government spending amid fluctuating world prices.[108] Moreover, the palpable anxiety among investors—aggravated by rampant "insider trading," government threats to "rebalance oil contracts," and the capriciousness of the national pipeline monopoly—did not stifle foreign direct investment into the Kazakh oil sector through the end of 2002.[109]

With the sector's recovery, the Kazakh leadership took the long view on building up the national oil industry as a springboard for economic growth and strategic diversification. President Nazarbaev professed that Kazakhstan's vast "sea of oil" enabled the leadership to avert painful trade-offs of reform. He preferred to buttress prudent short-term initiatives by keeping options open for as long as possible, and avoiding risks that might prematurely sacrifice almost certain future gains in the oil sector.[110] "The investment potential of Kazakhstan is so large," Nazarbaev boasted, "that it would require resources that are not available even to highly developed countries. Hence, the requirement for an objectively diversified pool of investors that represents dozens of countries from Europe, Asia, and America, the centers of the world business." Although Kazakhstan could expect to encounter near-term difficulties along the way, he reassured skeptics that diversification was virtually inevitable, given the "intelligent and long-term balance of interests" among foreign investors and customers.[111]

A third surprising feature of Astana's diffidence toward Moscow was that there were initial advantages to relying on Russia's preferred options for servicing lucrative markets in Europe and Asia, as compared to the six new pipelines that were under consideration. Russia's existing Atryau-Samara trunk line and the CPC main oil pipeline—that ran west from Tengiz to Russia's Novorossiysk port and then onward via alternative Turkish bypass routes—involved shorter distances and larger capacity pipelines than links to the BTC or routes to China and Pakistan. The latter, in particular, required construction of very long distance pipelines that were commercially suspect, given the intense incremental competition from established suppliers in the Middle East, and demand for comprehensive deregulation and investment in domestic infrastructures that would allow Northeast and South Asian oil markets to develop.[112] Although oil swaps and pipelines involving Iran offered potentially the quickest and least expensive options for transiting Kazakh oil to Europe (comparable to Saudi transit costs), the smaller throughput capacity limited the scale advantages of ramping up deliveries compared to using large volume Russian pipelines.[113] Accordingly, Moscow seemed well positioned to strengthen the already compelling commercial appeal of shipping large volumes of Kazakh oil via the Russian pipeline system.

Finally, Astana's mounting confidence in the sector's trajectory prompted cautious but firm diplomatic maneuvering to contain Moscow's energy predation. As described by Vladimir Babak, the Kazakh leadership viewed its role as "Caspian peacemaker," brokering compromises with the other littoral states to reduce legal and political uncertainty in the region that, in turn, could unlock foreign investment, exploration, and export of the region's vast oil potential. Yet, unlike the gas sector, Astana increasingly shied away from accommodating Russia's attempts to impose collective control of offshore resources, secure access to Astana's onshore deposits, limit extraregional investment in the Kazakh oil sector, and dictate the terms and routes for the country's long-term oil exports. Instead, the Kazakh leadership embraced small, sure gains that advanced independent development and shipment of its crude oil without inflaming foreign relations.[114] Accordingly, the residual effects of Moscow's carrot and stick oil diplomacy were quite modest, and even sparked an otherwise conservative Kazakh leadership to diversify its energy relations.

Troubles with Reframing the Status Quo

With Astana situated in the domain of gains, it was important for Russia to foreclose competitive solutions to ownership, development, and transport of Caspian oil. The diplomatic tussle over the legal status of the Caspian Sea presented the first challenge. For Kazakhstan, the principal issue concerned legitimatizing de facto claims to offshore resources while advancing prospects for a "consensus" settlement. Because most of the richest reserves were expected to lie closer to shore in the northern Caspian, Astana was not as preoccupied as Azerbaijan with laying claim to midsea reserves. Kazakhstan also preferred to "internationalize" the legal dispute by inviting extraregional input and trumpeting the international "Law of the Sea Convention" as the relevant precedent for conferring sovereign rights. Because of the fragile ecological system in the north Caspian, Astana also was keenly interested in establishing joint responsibilities for fishing and navigation.[115]

That the littoral states advanced separate legal interpretations of Caspian rights effectively stymied Moscow's efforts to win over the Kazakh leadership. Azerbaijan's initiatives at prospecting offshore oil both provided early diplomatic cover and generated competitive pressures on Kazakhstan to uphold the principle of sovereign rights. Turkmenistan's persistent waffling on the legitimacy of collective versus national ownership from 1994–1996, as well as its rivalry with Baku over midsea fields, reinforced demands for delineating national sectors. As summed up by a Kazakh energy official: "We believe that we will find a solution to this problem. We are supported by Azerbaijan and Turkmenistan."[116] Ashbagat's subsequent endorsement of Astana's right to prospect for seabed minerals in 1997 and its support for a complete division of the sea into national sectors in 1999, also revealed the limits to which Russia could impose its legal interpretation.[117] Furthermore, Russia's insistence on a condominium model was thrown into relief by Washington's decisions in 1995 to endorse direct negotiations among littoral states and to challenge the credibility of Soviet legal precedents.[118]

A similar dynamic marked the play for investment in the Kazakh oil industry. Nazarbaev's oil strategy hinged on attracting direct investment from international majors. His justification was straightforward: "Major companies and money from the USA, Russia, China, Britain, and other leading states will be involved in developing the Caspian shelf and the Karachaganak field. This will increase the leading powers' interest in our independence." Moscow was not precluded from participating in these ventures but at the same time was regarded as "only one among several equals."[119] In response, Moscow first sought to divert attention from investment in the Kazakh oil sector. It urged CIS members to focus first on arresting the collapse of Russian oil production, developing Russia's West Siberian fields—coordinating "the speediest revival of oil output consistent with the interests of the entire CIS."[120] In an affront to Astana, Moscow offered stakes to Russia's oil restructuring in return for commitments to forsake independent regional oil investment and to forge a common oil strategy.[121]

Moscow simultaneously maneuvered to set the terms for Kazakh oil development. At the end of 1992, Russia pressed for a bilateral agreement acknowledging the right of Russian companies to drill on Kazakh territory and to acquire equity stakes in Kazakh oil firms.[122] This was followed in early 1994 by demands for a 10 percent stake in all offshore projects. Moscow modified this position in 1996 by agitating for the inclusion of Russian firms in joint ventures to develop Kazakhstan's onshore fields, while discouraging investment in offshore exploration orchestrated by Astana. Preying on investor dissension in developing the Tengiz oil field, Moscow offered additional financing in return for the international consortium's decision to "surrender interests" to LUKoil.[123] By 1997, Moscow sought to foment investor anxiety by unilaterally issuing tenders to oil fields that were partially situated in areas of the north Caspian claimed by Kazakhstan. Putin, in particular, prodded Russia's mega-energy company to accelerate prospecting the Caspian's northern rim as a means to "help Russia strengthen its stand in the region."[124]

Yet Russia neither was the largest investor nor enjoyed a commanding position to discourage the international majors from pursuing stakes in Kazakh oil fields.[125] This was especially poignant regarding extraction of north Caspian reserves, where not only was Western investment readily available, but international partners were prepared to consummate deals that assured Astana 80 percent of future profits.[126] This presented Astana room to pursue international consortia without actively soliciting Russian capital. In 1993, for example, Astana set up an international joint venture, *Kazakhstancaspishelf,* comprised of six international majors that were authorized to explore nearly 200 blocks in Kazakhstan's sector of the north Caspian. LUKoil failed to acquire a 20 percent stake in the prospecting consortium, and was subsequently overpowered by the other members of the Kazakhstan International Development Company (OKIOC) that held exclusive rights in 1999 to develop the massive Kashagan field in the north Caspian.[127] Although Nazarbaev claimed that he was not averse to Russian involvement, there were no commercial imperatives to enlist LUKoil's participation: "The place of Russian capital in Kazakhstan is being taken by more resourceful American, German, Turkish, Chinese, and other investors. And not because we did not invite Russian capital, but because these investors win out in open tenders."[128] Even as investor confidence in the consortium wavered throughout 2001, the internal competition for available shares precluded sales to nonmembers.[129] By 2002, Astana devised a blueprint for prospecting the north Caspian that reflected the leadership's understanding that international majors, "above all American ones," would ultimately determine the viability of offshore oil projects.[130]

Similarly, Russia lacked market standing to foreclose rival pipeline options considered by Astana. Despite early reliance on the Transneft system, Astana was aggressively solicited from 1992 by a range of prospective international investors to develop four independent export options that circumvented Russia (see Appendix 4). Foreign partners lined up to conduct feasibility studies and create joint ventures to develop alternative main oil export routes to Europe and Asia, including separate pipelines to Pakistan and China, as well as a trans-Caspian hook up to the proposed BTC. Tehran also conducted intermittent oil swaps with Kazakhstan from 1997 to 2001 that delivered Kazakh crude to Iranian refineries and sold off a similar value of Iranian crude to Kazakh clients. By 1997, China too presented a potentially important source of independent financing for development of Kazakhstan's large onshore fields, as well as for construction of both long-distance pipelines running west to China and development of southern outlets for Kazakh oil via Iran.[131] Although none of these projects were perfect substitutes for transporting early oil via Russian territory, each operated according to its own commercial logic and together presented Astana with both mid- and long-term prospects for diversifying exports.

Swelling interest among these investors sapped Moscow's ability to preserve the initial commercial appeal of its favored long-term pipeline options. Neither China nor Iran was directly beholden to Russia for setting prices, tariffs, and right of way fees collected from Kazakh oil shipped across respective territories. The commercial feasibility of both long-term projects depended on advantageous global oil prices and the priorities that Saudi Arabia and OPEC placed on exploiting swing production to capture a greater share of the growing demand in Europe versus the price premium in Asian markets; two issues over which Russia had very little impact. In spite of protracted deadlock over financing the long-distance Kazakh-Chinese pipeline, Chinese energy officials continued to stress their long-term commitment to making the project viable should sufficient reserves materialize in the future.[132]

Similarly, the United States and Turkey were strategically placed to offset Russia's initial advantages. Following Washington's official endorsement of "multiple pipelines" for Caspian oil exports in 1995, the United States together with the commercial backers of the BTC, sought to lure Kazakh participation by offering long-term access to the pipeline, arranging financing for the trans-Caspian link, and soliciting the Kazakh state pipeline agency (headed by Nazarbaev's son-in-law) to be the operator for the trans-Caspian segment to Baku.[133] By increasing both the probability and cost effectiveness of the project, this effectively neutralized Russia's cancelation strategy of pitting proposals to upgrade "ready to use" Russian pipelines against the long-term commercial prospects for a trans-Caspian link that at best existed only "on paper."[134] Much to Moscow's chagrin, the predicament worsened with the opening of the BTC in September 2001 that mitigated uncertainty about the availability of alternative non-Russian pipeline routes for the projected near doubling (60 million tons) of Kazakhstan's uncommitted export volumes by 2015–2020.

Confounding the Risks

Astana faced two critical decisions for expanding its petroleum footprint. First, the leadership had to choose among contending approaches for establishing formal rights to Caspian reserves. Second, Astana had to determine early and main oil pipeline routes. Each decision involved risks, as well as uncertainty and commercial costs. Unable to alter the Kazakh leadership's decision frame and attendant risk aversion, the success of Russian statecraft hinged on narrowing the expected variance of preferred ownership and export options so that they would appear as safe bets for realizing Astana's long-term objectives.

There were three viable options for addressing the dispute over the Caspian Sea's legal status. The riskiest course for Astana entailed getting on the bandwagon with Russia and Iran, endorsing the "lake" interpretation of joint jurisdiction over offshore activities. This offered the potential advantages of ingratiating Astana with both Moscow and Tehran, and of strengthening joint management of conservation and energy depletion rates. However, this choice would compromise Astana's exclusive claims to undersea resources that lied beyond narrowly specified territorial waters. Even if successful, there were no guarantees that either Iran or Russia would refrain from exploiting collective user rights to impede development of undersea resources. As lamented by Kazak officials, preexisting Soviet-Iranian treaties left ambiguous "regimes for exploitation of the seabed and what lies beneath it, for ecology, for use of air space over the sea, not to mention such basic questions as the territorial sea and the adjacent exclusive economic zone, continental shelf, etc."[135]

The second option—endorsing Azerbaijan's national sector delimitation of the Caspian Sea—presented nearly as wide a variation in prospective outcomes. If this option succeeded, it would offer the best possible solution: Astana's claims for exclusive ownership of undersea resources would be unambiguous and favorable; the potential political costs of exploiting offshore resources would be contained; and Moscow and Tehran would be unable to play off differences with Baku. Yet because none of the other three littoral states accepted this definition, even this choice would not likely expedite collective settlement or significantly reduce the political fall-out of asserting Astana's de facto claims to undersea resources. That Washington refrained from openly questioning the validity of the Soviet-Iranian "common use" formula, underscored the shallowness of international support for Astana's legal case.[136] Failure to realize the positive outcome would leave Astana in the precarious position of provoking a showdown with Moscow and Tehran that, at minimum, would raise the political costs to offshore exploration. That said, the downside was not potentially as grave as the first option that would place Astana at the mercy of Russian and Iranian policies without international support.

The safest bet was to protect Astana's claims to offshore resources while leaving open the sea's final status. By April 1996, a compromise materialized that involved scaling back Astana's ambitions and securing bilateral acceptance of its sovereign rights to seabed resources.[137] This option paved the way for potentially very attractive outcomes: formal acknowledgement of Astana's de facto claims to undersea resources; agreements with Russia and Azerbaijan to develop the north Caspian shelf; solicitation of international support for specific agreements without directly challenging the validity of previous conventions; and concessions to Russia to reduce the incentive for encroachment into Kazakhstan's national sector and stymie a potential Moscow-Tehran condominium. However, this option carried the potential negative consequences of muddling Astana's national sector claims. A settlement short of recognition of Astana's sovereignty over its national sector would invite future challenges to the development of offshore resources. Postponement of a collective legal remedy to the Caspian's status left open the possibility that bilateral agreements would be invalidated at a future date. Notwithstanding these dangers, wrangling a compromise that secured Kazakhstan's offshore claims at the expense of delaying a consensus settlement would expedite development of the country's new-found oil wealth.

There were slightly different risks associated with oil transit. The first decision involved selecting among available options for exporting early oil that presented relatively moderate risks. One option involved continued reliance on the functioning Atyrau-Samara pipeline via Russia. This would accommodate both the Kremlin and Transneft, on which Kazakhstan was critically dependent for delivering and refining domestic supplies. However, it would empower Transneft to determine Kazakh export volumes and tariff rates, as well as subject Kazakh consortia to direct competition with Russian oil firms for pipeline capacity. Similarly, the second option, using swap arrangements with Iran, offered the positive outcome of enlisting political support from Tehran while diversifying Kazakhstan's immediate outlets at a discount, as this was the shortest and cheapest available route. Yet selecting this option invited rebuke from Washington that adamantly opposed significant transit projects involving Iran. Coupled with Tehran's ambivalence toward pursuing this option in a depressed global market and upgrading Iranian refineries to process the high sulfur content of Kazakh crude, Astana expected to incur regular disruptions of small-scale exchanges.[138] The third option available to Astana was to use a westward trans-Caspian route, via rail and tanker, to hook up to the early oil Baku-Supsa pipeline and its Black Sea bypass routes. This presented the potential positive outcomes of establishing an alternative western outlet for Kazakh early oil, as well as of consolidating political ties with Azerbaijan, Georgia, and Western backers. It also carried the possible negative implications of provoking both Russia and Iran, and subjecting Kazakh exports to political instability in Georgia and rivalry among outside powers over Black Sea egress routes.

Alternatively, the risks associated with the proposed main oil pipeline routes varied significantly. The least risky option—endorsing the Caspian Pipeline Consortium (CPC)—offered potentially very favorable outcomes. The venture, which initially combined Kazakh and Russian land and infrastructure with foreign financing, envisioned piping at least 55 million tons of oil nearly 1,000 miles from Kazakhstan's giant Tengiz field across Russian territory to the Black Sea port of Novorossiysk.[139] In addition to presenting possibly the most cost effective main pipeline route to West European and Mediterranean markets, the CPC accommodated simultaneously the interests of international investors and the Russian government. As an independently operated project, it could discipline future predatory practices while easing competition with Russian oil firms for scarce capacity in the existing Transneft system. With significant Western firm involvement, the CPC also would likely engender support from the United States, the European Union, and depending on the selected bypass option, Turkey. Furthermore, construction of a wide-diameter pipeline that could later be linked to other Kazakh oil fields would present ramp-up advantages for delivering larger volumes projected to come on line after 2015. However, the possible negative outcomes of the CPC included costly delays over financing. Astana would likely incite a dispute with Iran, given the adverse effects on the feasibility of future southern outlets. It also would antagonize Turkey, if Kazakh oil increased tanker traffic though the Bosporous Straits at the expense of reducing future volumes available for Ankara's favored BTC. This effectively would tie Astana to Russia over both the short and long terms, should the projected growth in Kazakhstan production fail to materialize.[140] By contrast, the second option—upgrading the Atyrau-Samara pipeline and linking incremental growth in Kazakh oil shipments to other existing Russian pipeline segments to Novorossiysk—would likely amplify the risks of using the Russian early oil variant. The possible advantages of appeasing Russia were offset by the likely downside of strengthening Transneft's long-term grip over Kazakh exports. Moreover, the link to Russian pipelines traversing the North Caucasus would increase exposure to instability in Chechnya.

The alternative Western variant—constructing a pipeline link to the BTC—presented a risk that at first was slightly greater than the CPC project. If a trans-Caspian pipeline segment could be constructed, Astana would be in position to ramp up deliveries to the BTC that would enhance the cost effectiveness of delivering both Azerbaijani and Kazakh oil and cement close ties to Western supporters and the Turkish government. The success of this option also would portend lucrative commercial spin-offs, given promises by Washington to encourage further American investment in other sectors of the Kazakh economy in return Astana's participation in the BTC.[141] Construction of a trans-Caspian link would reinforce export diversification, insulating Astana from future Russian and Iranian political pressure, and provide additional scale advantages of boosting domestic production. Pursuit of this option, however, would lead to a schism with Russia and potentially Iran. It also would subject Astana to the financial and logistical problems that hounded the BTC, as well as would spark competition with Baku for access to residual pipeline capacity. Furthermore, early commitment to the BTC would be very costly in the event that the projected growth in production failed to come on line by 2015. This would saddle Kazakh oil investors with paying higher transport costs for smaller export volumes, depriving Astana indefinitely of anticipated oil revenues.

The Iranian export route presented a greater risk than either building the CPC pipeline or a future trans-Caspian link.[142] If successful, it would lower transport costs for landing oil in Mediterranean markets that were comparable to those faced by Persian Gulf suppliers. It would reduce Astana's reliance on the Russian pipeline system, and would curry favor from China, Pakistan, and India.[143] If this option failed, however, it would magnify the possible negative consequences of the swap arrangements. Given Washington's vehement opposition to enhancing Iran's regional profile and determination to redirect flows from the Persian Gulf, this was certain to provoke acrimony with the United States, making it extremely difficult to attract reliable, large-scale project financing.[144] Furthermore, this option would likely bump up against Tehran's concerns for stemming regional competition for its own oil exports and purchasing Kazakh oil at a discount to service domestic demand. This raised the possibility of a very bad outcome whereby Astana's high-quality crude would be blended to lower the price premium, and its main oil exports would be subjected to the competitive interests of unreliable partners.[145]

Transporting main oil directly to expanding markets in Asia offered the widest variation in potential outcomes. Pipelines to China, for example, would eliminate interference by transit states, and would likely induce Chinese oil companies to purchase larger stakes in Kazakhstan's oil sector. With the ouster of the Taliban regime in Afghanistan, Washington looked favorably on routes to Pakistan with possible egress routes through Afghanistan and extensions to India. The exorbitant commercial costs of constructing such long-distance pipelines, however, would likely price Kazakh oil out of Asian markets, virtually stranding the country's oil wealth. Reliance on eastern pipelines would sacrifice access to more stable and cost effective markets in Europe, if the expected long-term projections for Kazakh production proved inaccurate. A strategic turn eastward also would alienate Moscow, and would relegate Kazak deliveries to compete at a disadvantage against both Saudi Arabian and Russian oil exports.[146] Given the mutual distrust between Washington and Beijing over the other's growing strategic presence in the Caspian, a decisive move to tie future main oil deliveries to Northeast Asia would court tension with Washington, especially if it came at the expense of delivering large volumes of Kazakh oil directly to markets in the West.[147]

In light of Astana's conservative decision frame and the expected variance in outcomes of respective options, the success of Russia's Caspian oil diplomacy turned on accentuating the positive benefits while reducing the full negative effects of settling on Russia's favored outcomes. The credibility of these diplomatic overtures rested on securing the compliance of domestic actors in the implementation process. Yet the obfuscation of authority within the national regulatory system deprived the Kremlin of institutional levers to prevent bureaucratic, corporate, and local actors at home from pursuing independent initiatives that inflated the expected risks to Kazakhstan of deferring to Moscow's positions.

By 1997, Moscow's campaign to impose the "inland lake" interpretation of the Caspian was countermanded by the inability to prevent Russian oil firms from negotiating stakes in the contested Kazakh sector.[148] This was evidenced in Moscow's ignominious retreat from the competition over the Kurmangazy field in the north Caspian Sea. After claiming exclusive ownership and auctioning the field in a closed tender, Moscow found itself outmaneuvered by initiatives undertaken by both LUKoil and Astana. Much to Moscow's chagrin, LUKoil acknowledged Astana's claims of ownership of the field and promised increased investments to upgrade onshore oil fields in return for an equity stake in the project. Emboldened by the opening, Kazakhstan held firm by rejecting both LUKoil's offer and the Russian auction. Not surprisingly, Moscow was forced to retract unceremoniously its earlier claims in deference to "ecological considerations," while conceding to Kazakhstan's de facto exploitation of the fields.[149]

By the same token, support on the home front was crucial to the credibility and effectiveness of Moscow's pipeline diplomacy. Given the inability to alter available options, Moscow had to lure Astana away from exploring the trans-Caspian link to the BTC. This entailed removing complications and reinforcing confidence in the CPC project, and improving prospects for smooth deliveries via other Russian pipeline segments to meet the expected swelling of Kazakh exports by 2015. Accordingly, Moscow formally agreed in 1996 to increase the quota for Tengiz oil pumped through the existing pipeline system from one million to 4.5 million tons per year. This was repeated from 1998 to 2001, as annual quotas were officially raised from 7.5 to 15 million tons. As summed up by the Russian Energy Minister: "more transit will bring more money to the federal budget," raising confidence in Kazakh-Russian joint ventures.[150] In addition, Moscow tried to encourage long-term commitments by agreeing to establish a "quality bank" for the CPC blend to compensate Kazakh suppliers for potential market losses incurred by mixing Tengiz crude with poor quality Russian oil. This was praised by senior Kazakh and Russian officials for establishing fair market returns for each shipper, and laying the foundation for long-term transport agreements "devoid of restrictions."[151]

Yet from the beginning, Moscow's attempts at reassurance were subverted from below. Transneft was loath to sacrifice monopoly rents by reducing throughput of Russian oil via the existing pipeline system in order to make room for large volumes of Kazakh oil or to accommodate the rival CPC pipeline. In 1995, Transneft rejected appeals by both LUKoil and the Ministry of Fuel of Energy to free up pipeline capacity, as well as reneged on a promise to increase Kazakh deliveries by 50 percent. The recalcitrance discouraged Kazakh officials from exploring proposals to increase throughput via the Atyrau-Samara pipeline. Kazakh suppliers were simply befuddled by the Kremlin's deference, grumbling that "the government should not care who pays taxes, CPC or Transneft."[152]

In addition, Transneft stymied Moscow's diplomatic gestures by consistently delaying negotiations and reducing the cost effectiveness of Kazakh exports. As early as 1993, Transneft claimed that it could not expand Kazakh deliveries until additional processing facilities were constructed to lower the mercaptan level in Tengiz oil. Although initially this enabled Russian firms to wrangle larger participation in the Tengiz project, it ultimately affirmed doubts among Kazakh and foreign investors concerning the feasibility of Russian export options. According to the Kazakh field manager of the Tengiz project, Transneft's objections were disingenuous and made no technical sense, "since the Russians were accepting 330,000 tons a month of mercaptan crude in the old days, and now all of a sudden they have become environmentally conscious."[153] Transneft officials also snubbed the Ministry of Fuel and Energy's efforts to expedite the opening of the CPC pipeline by delaying commissioning of the oil quality bank in June 2001. They complained that this mechanism set a dangerous precedent that compelled the pipeline operator to compensate Russian firms for blended crude shipped through the existing Russian pipeline system.[154] Moreover, the construction of an altogether new Russian-Kazakh-Iranian pipeline motivated Astana and its foreign partners to diversify outlets. It caused them to question future transactions with Russia as "clearly it would be hard for both Iranian and Russian options to be the best."[155]

The risk of future dependence on Russia was accentuated by the lack of coordination among rival tax authorities. In 2000, for example, customs officials independently reclassified and taxed Kazakh shipments as "imported crude," on grounds that the quality of Russian oil was altered by the blending of the foreign crude, thus voiding "pass through" discounts extended in bilateral transit agreements.[156] Similarly, only two months after oil started to flow through the CPC system, pumping was stopped due to confusion between the Ministry of Foreign Affairs and the State Customs Committee. Russian customs officials withheld deliveries for "additional documentation," notwithstanding the very public endorsement by both presidents of the opening of the pipeline. Kazakh officials viewed this as a pretext for restricting annual throughput, and precursor to "future problems at the border of the Russian pipeline segment."[157] Despite assurances from the Ministry of Fuel and Energy and the prime minister's cabinet that the CPC project would be exempt from double taxation schemes, the Federal Energy Commission (FEC) unilaterally threatened in Fall 2002 to strip the pipeline of its special tax status. Fully cognizant of the anxiety that the move would excite among CPC operators, the commission nonetheless proposed to register the CPC pipeline as a "natural monopoly," subjecting it to arbitrary tariff hikes. Not surprisingly, Kazakh energy officials increasingly soured on assessments of the cost effectiveness of significantly ramping up CPC deliveries in the future.[158]

Divisions within the Russian oil industry further complicated calculations for both Moscow and Astana. On the one hand, LUKoil frustrated Moscow's diplomacy in 1998 by declaring: "one of its priorities today is transport of Kazakhstan oil through Azerbaijan." On the other hand, Russian oil executives lined up behind Transneft to oppose the extension of transit quotas for Kazak oil. Outraged that the concessions came at their expense, Russian oilers complained to the Ministry of Fuel and Energy that:

> the pipeline is not elastic and it cannot serve all interested parties. If someone gets more, others' pumping rights are diminished. Furthermore, for every million tons of (Russian) crude oil which does not go into export, the budget loses around $50 million.[159]

By 2000, industry officials claimed that the increase in Kazakh quotas had cost Russia nearly $500 million, due to the dramatic surge in world prices. This provoked a bureaucratic battle between the Ministry of Fuel and Energy that sought to uphold concessions to Astana; and the State Customs Committee and FEC that insisted on boosting state export revenues. The confusion was compounded in mid-2000 when Deputy Prime Minister Khristenko openly declared that "shipment terms, which are fairly advantageous for Kazakhstan, may be reconsidered."[160]

Competition among Russian provincial administrations directly engaged by the CPC project also reduced the credibility of Moscow's promises to levy uniform transit fees.[161] Several transit regions could not resist the temptation to exploit the "CPC trump card" in political gamesmanship with Moscow. For the President of Kalmykia, for example, the pipeline was more than a commercial deal.

> It turns out that among us the pipe for the transit of oil isn't simply a pipe. It is an independent, political factor. The regions obtained a lever for pressuring Moscow. Right now, while oil is not flowing through the pipe, they do not want much, but as we know, appetite comes with eating.[162]

Other regional leaders seized on the opportunity to advance local interests. They imposed local environmental standards and fees, as well as conditioned access to local egress routes on receiving promises from both Moscow and pipeline operators to increase regional employment and investment. Each transit region looked to the CPC to cover budgetary shortfalls, placing elastic demands on the operators to fund sundry services, such as the upgrading of local telecommunications, new schools, and local beautification projects.[163] The governor of Astrakhan, for instance, boasted that "the project brought our oblast new jobs, well paid work, and investments in the oblast's budget, and we have been supporting it from the very beginning."[164] Still another group of provincial leaders virtually encouraged "illegal cuts" into the pipeline to satisfy local rackets and meet local demand. Although none of the transit territories could dictate the fate of the CPC, their arbitrary interventions undermined confidence in the project.[165] They also raised questions about the credibility of Moscow's diplomatic inducements and future profitability of committing larger volumes of Kazakh oil to the Russian route.[166]

Hedging to the Future

Increasingly confident in the oil sector's trajectory by 2002, Astana settled on safe policies to keep options open. Although the hedging neither was as extreme as Baku's defiance, nor precluded cooperation with Russian state and private entities, it nonetheless marked a bitter defeat for Moscow's Caspian oil diplomacy. The results were especially disconcerting in light of Moscow's initial demands for significant equity stakes in all onshore oil fields and vigorous opposition to independent offshore exploration. Throughout the period, Astana succeeded at forging development and transit partnerships with state and private entities with home bases in 15 countries. Yet, Russian commercial entities participated in only four of 43 joint oil development, production, and export projects established by the end of 2001; and not one Russian company acquired shares in the consortium assigned to develop Kazakhstan's largest proven and unproven oil reserves. By contrast, foreign rivals secured controlling stakes in numerous projects, including the prized Kashagan, Tengiz, and Karachaganak oil fields.[167] Although Astana signed bilateral agreements in May 2002 to develop jointly three contested offshore fields in the north Caspian, it secured sovereignty over the largest field (Kurmangazy) without officially committing to a time table for exploration or specific terms for joint operations. Furthermore, the most vexing problems that threatened to derail the Kashagan and Tengiz projects stemmed from internal disputes among international investors, domestic operators, and the Kazakh government that had little to do with Russian intervention.[168]

Astana's strategic insolence also crystallized with the signing of the bilateral agreement in July 1998 that divided the seabed in the north Caspian between the two countries, while relegating the waters, surface, and airspace to joint control. This compromise not only was the least risky option for Astana, but also marked a significant concession from Russia's demands from 1992 to 1997 for a condominium and comanagement of offshore resources beyond strictly defined territorial waters. The failure was especially conspicuous, given that Astana had earlier attempted to distance itself from Baku's provocative "sea" interpretation, and offered in September 1995 to drop insistence on application of the "Law of the Sea." This softening was codified in that year's joint declaration, where Nazarbaev agreed that both countries "are convinced that joint participation in exploiting the Caspian natural resources suits their mutual interests."[169] Astana's subsequent backsliding and steadfast commitments from 1998 "to fiercely uphold" national oil interests and to secure favorable terms for dividing and exploiting undersea resources in a "civilized and diplomatic way," underscored the extent to which new realities had slipped from Moscow's grasp.[170]

The terms of the compromise agreement also represented a significant setback for Moscow's battle for joint control of offshore resources. Although the agreement did not make explicit reference to "national sectors," the Kremlin acquiesced to the division of the seabed and Astana's de facto resource claims. It also codified bilateral deals for resolving outstanding issues and framing the final settlement. Similarly, Astana's pledge to "coordinate construction" and solicit collective support in pursuing future undersea pipelines, by no means constituted an ironclad deference to Moscow's opposition to a trans-Caspian pipeline.[171] As summed up by one close observer, "Russia's defeat over the seabed's status is of cardinal importance, for a condominium would have meant power over site development and pipeline routing as well. Moscow's failure, therefore, severely handicapped its efforts in the other two areas."[172]

The 1998 compromise also set a precedent for follow-on agreements that raised the costs to Moscow of reneging on a seabed division. While the original agreement did not specify the exact coordinates for delineation and left open the potential for encroachment at the surface, Putin signed an additional bilateral declaration in October 2000 that affirmed Moscow's concession. This joint statement officially acknowledged the previous division, and urged the littoral states to accept the principle of a "modified median" delimitation of the seabed as the model for settling outstanding disputes. The accord also dropped reference to creating a Caspian Strategic Development Center to coordinate exclusive littoral state exploration and exploitation of the sea.[173] This was followed by a series of bilateral agreements in May 2002 that specified sovereign rights and joint development to three contested oil fields in the northern Caspian. This effectively conceded to Kazakhstan exclusive legal ownership of the Kurmangazy oil field, as well as authority to supervise the timing and operations for its joint development. Furthermore, the 1998 agreement laid the basis for the 2001 Moscow-Baku delineation accord and a Fall 2002 trilateral agreement that upheld the technical delimitation of the north Caspian as the prevailing governing norm until the five littoral states reached a final solution.[174]

Finally, the limits to Moscow's oil statecraft were manifest in the results of its pipeline diplomacy. On the surface, the balance of short-term costs and risks confronting Astana favored upgrading oil transit via Russian pipelines, irrespective of Moscow's actions. The routes were operational, incremental competition was weaker, distances to established international markets were shorter, and the variance between likely positive and negative outcomes was no greater than that associated with rival Iranian and westward options. Therefore, initial decisions by Astana and its partners in the Tengiz consortium to rely on the Transneft system for the lion's share of oil exports (15 million tons per year by 2001) made sense from both expected utility and risk perspectives, notwithstanding repeated frustrations over delays and the failure to meet export quotas. Yet Moscow's manipulative diplomacy failed to raise significantly these deliveries or to reduce the riskiness of the Ayrau-Samara pipeline. In return for using the Transneft system, Moscow accepted transit fees that were lower than the world standard.

Moreover, decisions to continue oil swaps with Iran that reached 120,000 barrels per day in 2001–2002, as well as to construct a pipeline across Georgia capable of carrying 7–8 million tons per annum of Caspian oil to the Black Sea, were consistent with Astana's overall risk aversion. Moscow's inability to lock in these residual exports did not leave Russia in a more advantageous position and reflected Astana's early commitment to keeping options open, despite the initial commercial advantages of upgrading deliveries via the operating Russian transit system.

The real impotence of Moscow's pipeline policy, however, was captured by the battle for main oil deliveries. From both utility maximizing and risk minimizing perspectives, the Russian variants offered relatively cost effective and safe medium-term bets for Astana. Faced with the need to export as much as 55 million tons of oil per year by 2015, it came as no surprise that the risk averse Kazakh leadership settled on spreading commitments among rival Russian options and operators. In 2001, for example, Astana agreed to deliver up to 28 million tons of oil per year via the least risky CPC pipeline. However, for Nazarbaev the commitment was limited and guided "first of all by economic expediency in the search for profitable directions."[175] This was followed by successive agreements in 2002 that committed Kazakhstan to distribute as much as to 27.5 million tons of oil per annum via alternative Russian pipelines. In return, Russia guaranteed transit of at least 17.5 million tons per year of Kazakh oil, as well as acknowledged that tariffs would be set between "executive authorities from each country."[176] By spreading midterm deliveries among independent Russian pipelines, Kazakh producers increased commercial discipline on rival operators.

Moreover, Kazakhstan and its foreign partners refused to commit to meeting Russia's long-term transit capacity. The Tengiz consortium left open prospects for filling the extra 40-million ton capacity of CPC pipeline that was scheduled to be available by 2010. Concerning the additional 60 million tons of oil per year that were expected to come on line by 2010–2015, Nazarbaev remained conspicuously noncommital: "Time will show whether it will be an oil pipeline via Russia to the Baltic states, or via Iran to the Persian Gulf, or via Baku-Ceyhan."[177]

To underscore Kazakhstan's commitment to keeping its options open, Nazarbaev repeatedly professed interests in a trans-Caspian link to the BTC. Just weeks after the signing of the BTC agreements in November 1999, and caught between pressure exerted by Washington and Moscow, Nazarbaev put the Kremlin on guard by openly declaring that "the project could not take place without including annual

shipments of 20 million tons of Kazakh oil," and that Astana was prepared to sign up after the extra oil materializes.[178] The hedging on
future deliveries of Kashagan oil via the CPC pipeline persisted, as Naz-
abaev reiterated support for the BTC while promising to boost deliveries to 10 million tons per year via interim rail shipments through
Georgia. This ambivalence reflected the distrust of a long-term commitment to Russia among senior Kazakh oil officials who acknowledged:
"If they can give it to us, they can also take it away."[179] Responsive to
these concerns, as well as to Moscow's inability to prevent the BTC
from becoming a reality, Astana officially requested to ship oil through
the pipeline and to become a sponsor of the project at the end of 2002.
Hoping simultaneously to allay anxieties in Moscow while preserving
flexibility, Nazarbaev defended his interest in the BTC as a long-term
option that complemented short-term transit agreements with Russia.[180]

By the same token, Astana proved difficult for Washington and
other backers of the BTC to pin down. Though they endorsed the project, Kazakh officials were in no rush to commit oil to the pipeline.[181]
The ambivalence frustrated American officials, who by the end of 2002
openly stated that the BTC would be implemented with or without
Kazakh future shipments. Astana's wavering, however, did not create an
opening for Moscow. As intimated by the Minister of Energy and Mineral Resources, the hedging served Astana's instrumental objectives.
"The Baku-Ceyhan route is useful for Kazakhstan now, as an argument
in the talks with Russia, from which Kazakhstan is demanding that the
tariffs for pumping oil be reduced."[182] Though this strung along BTC
investors, it was definitely a blow to Russia's Caspian diplomacy.
Kalyuzhniy, for example, openly conceded that although Russia "takes
a severely negative attitude toward construction of a trans-Caspian
pipeline, it will not object to this. This is business."[183]

Conclusion

The analysis of Russia's Caspian oil diplomacy supports the hypothesis
for strategic manipulation outlined in chapter 2. The inability to manipulate the prospects for compliance, and not other explanations that feature the capacity to punish or reverse target state behavior, accounts for
the limits to Moscow's energy statecraft. As in the gas sector, Russia
enjoyed relative, relevant, structural, and issue-specific power to credibly raise the costs of noncompliance, yet wielded such advantages for
only marginal diplomatic success. The lack of market power alone does

not explain why Russian oil firms were indeed able to invest and divest in Caspian oil ventures at strategic junctures, yet operated consistently beyond the grasp of the state. That said, market weakness likely mattered, not because it inhibited Moscow's capacity to impose additional costs for noncompliance, but because it prevented Russia from altering the frames and substantive appeal of preferred development and pipeline options for target states and domestic firms to preempt defection. Ultimately, failure to manipulate the opportunity costs and risks of compliance cost Russia dearly and lead to systematic gaps in Moscow's influence in the oil sector both at home and in the near abroad.

6

❖

Russia's Radioactive Diplomacy

Reconstituting Commercial Nuclear Relations with Kazakhstan and Kyrgyzstan

This chapter examines Russia's commercial nuclear diplomacy during the first decade of post-Soviet independence. Resentful of lost control over key stages of the former Soviet nuclear fuel cycle but confident in future comparative advantages over emerging Central Asian suppliers, Russian officials seized on opportunities to reconstitute commercial nuclear relations to counter extraregional encroachment and promote favorable economic and political integration along the southern periphery. Unlike the uniform successes and failures in respective gas and oil sectors, however, the achievements of Russia's nuclear energy statecraft were conspicuously mixed. Although Moscow outmaneuvered foreign competitors to woo both Kazakhstan and Kyrgyzstan into restoring nuclear fuel relations and acquiring equity stakes in extraterritorial ventures, it was unable to make use of these developments to consolidate strategic gains in the region. In both cases, Russia orchestrated joint ventures to reintegrate critical elements of the former Soviet fuel cycle that advanced mutual commercial interests; none of the deals were leveraged to secure disproportionately favorable terms or to arrest regional diversification.

Russia's nuclear energy statecraft also unfolded in a distinct manner. Contrary to the situation with natural gas, Moscow lacked significant power in front- and back-end nuclear fuel service markets to check foreign competition or to alter the opportunity costs of compliance for Central Asian targets. The Russian government also enjoyed more transparent and direct control of the national nuclear fuel cycle than was the case with regulation of the oil sector. Accordingly, Moscow had less commercial power to saturate or foreclose foreign market niches than in the gas sector, but greater institutional capacity to marshal its resources than with oil. Explication of Russia's ability to partially offset market disadvantages by manipulating the different hedging strategies of rival suppliers and the attendant risks of nonimperial ties in the nuclear sector lies at the crux of the ensuing analysis.

The chapter unfolds in three parts. The first describes the twin pillars of Moscow's strategy for reviving the nuclear energy sector and wielding it as an instrument to secure competitive advantages and mold a common Eurasian energy space. The second and third sections examine the Russian government's mixed success at putting this strategy into effect in relations with Kyrgyzstan and Kazakhstan. Each section begins by reviewing the limitations of competing explanations for Russian statecraft, and then analyzes the manner in which Moscow exploited regulatory authority in the nuclear energy sector to fashion closer cooperation with both target states. Faced with a Kyrgyz leadership more risk-prone than its Kazakh peers, Moscow compensated for its lack of market power by making commercial nuclear cooperation both a more lucrative gamble for Bishkek and a safer bet for Astana compared to other opportunities in the sector. Each section concludes by reviewing the enduring constraints on Russia's heavy-handed policies and the mutual benefits of commercial joint ventures for uranium production and nuclear fuel supply that distinguished each set of relations from Moscow's gas and oil diplomacy.

Toward Nuclear Reintegration

After years of wallowing in the legacy of the 1986 Chernobyl nuclear disaster, contending with the inherited contamination problems of the industry, suffering from government funding shortfalls, and subsisting mostly off of funds generated by cooperative nuclear weapons dismantlement programs, Russia's Ministry of Atomic Energy (Minatom) was intent on reviving the nuclear industry. Buoyed by the upturn in performance from 1998 to 1999 and personally convinced that expansion was the solution to long-term self-sufficiency, then Minister Evengiy Adamov spearheaded an initiative to elevate the stature of nuclear energy policy within the state's long-term strategic posture. At base was a program to accelerate commercial nuclear power production while downsizing and securing the nuclear weapons complex. This strategy, endorsed by the government in May 2000, called for increasing the sector's contribution to Russia's commercial electricity production from 15 percent to as much as 45 percent in parts of Russia by 2030. The strategy envisioned enhancing the load factor and extending the service life of the 29 operational nuclear reactors, as well as constructing as many as 38 additional reactors by 2020 at a projected cost of $34.5 billion.[1] Minatom projected that the strategy would produce dramatic savings in the coal- and gas-fired power generating sectors, as well as free up natural gas for additional exports worth up to $32 billion over the period from 2000 to 2010.[2] Determined not to have the sector's commercial problems "dropped at the state's doorstep," Minatom officials devised a plan for the gradual self-financing of the sector's commercial activities that rested on the twin pillars of internal reorganization and foreign activism.[3]

On the domestic plane, programmatic priorities were placed on consolidating the nuclear power generation and fuel complex. Frustrated by organizational inefficiencies precipitated by the breakup of the monolithic Soviet industrial structure and attendant proliferation of parochial administrative intermediaries that siphoned off an estimated 10–30 percent of federal outlays to the sector, Minatom officials aggressively integrated the sector's commercial activities.[4] Nuclear production and fuel services were reorganized into separate but closely coordinated state-owned corporate entities. This entailed combining all but one of the nuclear power plants into the single state-owned Rosenergoatom, as well as merging manufacturing and trading organs under the commercial management of TVEL. Minatom also solicited new contracts to capitalize on prospective comparative advantages at integrating different services from uranium production to spent fuel imports. As explained by the ministry's director of the fuel cycle department:

> The world market for nuclear services is very large. Russia is able to compete in that market. But for that purpose we need to offer not simply an end product, but integrated services across the whole technological process for its production. Today, when the enterprises are operating separately, we

cannot implement that approach. There is no vertical management structure within Minatom. . . . Resolution of this problem would fundamentally improve state management.[5]

The second phase of internal reorganization, however, was left
ambiguous, with debate stalemated over the practical boundaries of
centralization. Adamov pressed for the formation of a single state commercial nuclear corporation, Atomprom, as a "corporate" version of
the super-centralized Soviet ministerial apparatus that would function
as a natural monopoly, akin to an entirely state-owned Gazprom. The
entity was intended to root out costly internal competition, and to be
sufficiently nimble at coordinating nuclear fuel cycle services to restore
Russia's competitive advantages in the global industry.[6] However, critics
inside the ministry and opponents of nuclear expansion questioned the
technical, commercial, and administrative merits of additional consolidation. They feared resurrection of an opaque, hydra-headed state
structure that would stifle autonomy at lower echelons, and that would
provide cover for bureaucratic corruption and the eventual arbitrary
privatization of strategic assets. Officials in charge of producing and
managing fissile material for the weapons complex also were loath to
endorse a plan that threatened to subject support for defense-related
activities to the vagaries of the ministry's commercial ventures. As one
deputy minister sarcastically noted: "Not a single country in the world
has come up with that (private funding to maintain the nuclear
weapons complex) as yet, that is our unique invention."[7]

Yet Minatom remained undeterred. Amid the change of leadership
at the ministry in March 2001, Minatom steadfastly promoted "vertical
integration" as a panacea for improving self-financing and competitiveness in domestic and foreign fuel service markets. As underscored by
Adamov's successor, Alexander Rumyantsev, Russia's rightful place in
the global nuclear industry and the sector's value as an instrument of
diplomacy could be guaranteed "only by the consistent and accurate
consolidation" of nuclear fuel service enterprises under state auspices.[8]

The second pillar of the strategy involved enlarging Minatom's
international footprint. A premium was placed on expanding foreign
nuclear energy sales to grow the ministry's commercial customer base
and fund the domestic agenda without increasing financial pressure on
the national budget. This featured completing construction of the
Bushehr reactor, as well as pursuing demand for additional nuclear
energy ventures—including fresh fuel sales, feasibility studies for three
more reactors, development of a uranium mine, construction of a uranium enrichment facility and an isotope enrichment facility—all with

Iran. This was buttressed by plans for building and supplying two reactors each to China and India, as well as for pursuing reactor deals with Cuba and Indonesia and expanding electricity sales to Europe, Japan, China, South Korea, and Turkey.[9] In addition to these front-end services, the Minatom leadership set its sights on capturing 10 percent of the international spent nuclear fuel processing market. Although the initiative flouted the public's misgivings toward importing irradiated waste and was subjected to critical review by the parliament, the government nonetheless succeeded at amending national legislation in 2001, to earn a projected $21 billion eventually earmarked for environmental remediation, social, and regional projects within the sector.[10]

A crucial component of Minatom's foreign strategy was reintegration with former Soviet uranium and fuel service suppliers. By the end of the 1990s, ministry officials and technical experts openly acknowledged that Russia lacked sufficient low cost reserves to support the projected long-term growth of the sector, despite accounting for nearly 10 percent of uranium produced in the world. With fresh uranium production expected to meet less than half of the annual demand for domestic consumption and exports through 2030, Russian officials urged more efficient prospecting of existing mines and greater reliance on secondary supply sources, such as reprocessed plutonium and blended-down HEU from military stockpiles.[11] Because these resources were not expected to provide reliable relief over the long haul, Minatom fixated on reclaiming lost assets. The recovery of the once closely integrated nodes of the former Soviet nuclear fuel cycle—world-class uranium reserves in Kazakhstan and Uzbekistan, research reactors in Kazakhstan, as well as processing and fuel fabrication facilities in Kyrgyzstan, Kazakhstan, and Tajikstan—was considered crucial for augmenting supply security and solidifying favorable regional harmonization. As summed up by the then head of TVEL, Valentin Shatalov:

We need to worry about the reintegration with Russia of those types of production in the nuclear fuel cycle that ended up outside our territory in CIS countries following the break-up of the USSR. This goes first and foremost for the enterprises involved in the extraction of uranium concentrated by and large in Kazakhstan, Uzbekistan, and other republics of the former USSR. After all, the planned increase in nuclear fuel production volume associated with the construction of new AES *(nuclear power stations)* in this country and the possibility of gaining access to fuel markets in China, India, Iran, and other countries requires vigorous efforts to solve this problem.[12]

Russian officials promoted an assertive regional strategy to arrest the encroachment of Western companies into Russia's traditional market. Minatom officials warned that they "were forced to deal with stepped up integration processes between the largest Western companies. . . . that indicate a desire to redistribute existing markets and restrict TVEL's presence in the European market."[13] Adamov went further by intimating that the reclamation of residual Soviet nuclear assets would undoubtedly carry the geopolitical advantage of bolstering Russia's influence along its vulnerable southern periphery.[14] Similarly, Rumyantsev "mourned the collapse of the Soviet Union's nuclear infrastructure," and openly advocated reconstituting former nuclear fuel cycle relations in Central Asia as a springboard for creating a "confederation of the former Union republics" and preserving Russian control over vital regional transport infrastructure and power grids. "A common home must be built," he urged, with "consolidated nuclear relations" as its foundation.[15]

Minatom's Central Asian Gambit

The first practical steps toward realizing Russia's nuclear energy ambitions in Central Asia involved engaging Kyrgyzstan and Kazakhstan in joint ventures. In June 2000, Minatom presented both leaderships with a three-step plan for extracting and converting Kazakh uranium into nuclear fuel for use in Russia. The first stage involved mining fresh uranium at Kazakhstan's Zarechnoye deposit that was owned by the national nuclear company, Kazatomprom.[16] The uranium would be extracted via a chemical leaching process overseen by Kyrgyzstan's Kara-Balta Ore Mining Combine that, in return, would acquire a concession to the deposit. The uranium slurry then would be sent to the Kara-Balta facility in Kyrgyzstan for purification into concentrate. Finally, the nuclear fuel would be delivered for use in Russian nuclear power plants. Minatom projected that the arrangement would produce 500 metric tons of uranium concentrate per annum, thus compensating for expected medium-term shortfalls in domestic production and long-term depletion of the nation's low-cost deposits.

Russian officials regarded the deal as critical for securing a veto over the use of Kazakhstan's large uranium reserves and for curbing Astana's flirtation with engaging Western nuclear energy firms. The venture was expected to strengthen Moscow's influence over Kyrgyzstan's strategic energy relations and eventually over the activities of nuclear facilities in Tajikistan that could be incorporated into the arrangement. Regaining control over Central Asian fuel service providers in this manner would secure a springboard for Minatom to exert pressure on global nuclear fuel markets.[17]

In April 2001, the three states formed a joint uranium supply venture. Kazatomprom secured a 45 percent stake in the arrangement, while Minatom's Atomredmetzoloto and the Kara-Balta Combine acquired 45 percent and 10 percent shares, respectively. The deal was sealed against the backdrop of the Kremlin's escalating demand for debt-equity stakes in Central Asian industrial enterprises. Prior to signing the agreement, Putin put the Kyrgyz leadership on guard by endorsing Russian claims to equity stakes in 23 of Kyrgyzstan's most significant industrial enterprises—including Kara-Balta's uranium and gold mining operations and several of Kyrgyzstan's prized defense industrial companies—as payment for Bishkek's outstanding $150 million state debt.[18]

Minatom then moved to secure control over the supply of low enriched uranium fuel pellets. The Ulba facility in Kazakhstan produced uranium dioxide pellets from Russia's enriched uranium that, at peak production, were used to fabricate nearly all of the fuel rods and assemblies for Soviet RMBK and VVER reactors spread across the former Soviet Union, Eastern Europe, and Finland.[19] With the Soviet collapse, almost overnight Moscow lost control over these critical supplies that could not be readily compensated by other foreign or domestic fuel service fabricators. At the same time, Russia faced intense competition from Western firms in the Ukrainian fuel market. Although Russia supplied nearly all of the fuel for Ukraine's Soviet-type reactors, it had difficulty collecting cash payments and contending with Kiev's mounting complaints about the quality of Russia's fuel supply and need to diversify commercial nuclear relations.[20] By the end of the decade, Minatom was committed to resolving both problems by demanding a 51 percent stake in the Ulba plant and long-term supply contracts from Kazakh uranium producers.

Russia's strategy for reconstituting nuclear fuel relations with Kazakhstan showed early signs of success. Between 1998 and 2000, the states committed to establishing an integrated nuclear fuel cycle and acquiring shares in respective nuclear fuel cycle companies. Minatom also signed a five-year contract that guaranteed TVEL a minimum number of nuclear fuel pellet deliveries from the Ulba facility. In November 2001, Russia succeeded at brokering a second trilateral agreement, creating the Joint Ukraine-Russia-Kazakhstan Fuel Enterprise, Ltd. to produce fuel pellets for use in Ukrainian reactors. The arrangement called for Ukraine to supply uranium and zirconium, Kazakhstan to produce uranium fuel pellets, and Russia to use the pellets and zirconium to manufacture fuel assemblies for its reactors. This was complemented by TVEL's acquisition in 2001 of an equity stake in the Ulba plant. Similarly, an agreement was reached in July 2002 between TVEL and Kazatomprom to cooperate on radioactive waste treatment and development of Kazakhstan's uranium deposits.[21]

As was the case with Kyrgyzstan, Minatom attempted to exploit commercial transactions to reintegrate the nuclear fuel cycle and gain leverage over Kazakhhstan's uranium and fuel pellet production. The joint venture, in particular, was intended to keep Western firms at bay and perpetuate Ukraine's primary dependence on Russia's nuclear energy supply in a manner both that eased Moscow's vulnerability to Kiev's future default and defection, and that preempted future competition with Kazakhstan.[22] Similarly, the stake in the Ulba plant secured TVEL the right to veto Kazakh fuel fabrication activities that threatened Minatom's strategic interests.[23] "Our goal," as explained by the director of one of Russia's main fuel fabrication plants, "was to restore corporate economic relations with Kazakhstan in the nuclear sector in the shortest time possible." This entailed, "creating a transnational nuclear fuel and energy corporation in the post-Soviet economic space with Ukraine and Kazakhstan (and possibly Uzbekistan), with our [Russia's] dominant participation."[24]

These commercial transactions with Central Asian competitors seemed on the surface to be remarkable successes for Russian statecraft. In addition to narrowing the expected 5,000–6,000-metric-ton shortfall in demand for uranium, the deals set important precedents for more intimate sectoral cooperation, and generated momentum for reintegrating former nuclear supply assets that, in turn, would augur well for Russia's manipulation of global nuclear fuel markets. At the Yerevan Summit of the CIS in October 2000, uranium production and nuclear fuel supply transactions with Kazakhstan and Kyrgyzstan were singled out as bulwarks for future integration of the Caucasus and Central Asia under Russia's aegis.[25] As the deals unfolded, analysts across the region referred to their intrinsic "colonial character" as symptomatic of Russia's creeping neo-imperial domination of this geopolitically significant region.[26]

By 2000, "nuclear cooperation" emerged as a critical dimension to Russia's strategy for engaging the southern NIS. This included a concerted push to plug Kazakh and Kyrgyz uranium producers and fuel service providers back into the Russian nuclear fuel cycle in the hopes of yielding significant strategic dividends.[27] The pattern of success was mixed, however, corresponding neither with Moscow's stated ambitions—the uniform effectiveness of Moscow's statecraft in other energy sectors—nor the expected utility calculations of otherwise weak target states.

Raising the Stakes of Cooperation for Kyrgyzstan

The trilateral joint venture with Russia seemed to come at a steep price for Bishkek. Together with Putin's coercive pressure for debt-equity stakes in key defense industrial enterprises and Russian investment in Kyrgyzstan's hydroelectric projects, the nuclear deals threatened the country's "economic sovereignty," potentially subjecting lucrative export opportunities to Russian predation.[28] Because of the range of activities performed at the Kara-Balta facility, the joint venture also opened the door for Moscow to influence the country's prized gold mining operations. This was the source of considerable consternation, as production, refining, and sales of gold accounted for nearly 40 percent of Kyrgyzstan's industrial output and export earnings by 2002.[29]

Yet Kyrgyzstan did not ultimately succumb to Moscow's pressure. Notwithstanding obvious power disadvantages, Bishkek avoided conceding equity stakes to Minatom in the strategic Kara-Balta mine, curtailed Russian encroachment into other sectors, bargained for a "fair" stake in the new joint venture, and remained undeterred in pursuing international opportunities. In short, Russia's highly favorable power position and strong motivation to assert control over lost nuclear assets did not intimidate or circumscribe Bishkek's sovereign control over strategic energy policy.

By the same token, Bishkek's decision to intensify nuclear fuel relations with Russia stood out on several accounts. First, it was inconsistent with early progress at broadening strategic relations since independence. Although Kyrgyzstan had the weakest export outlook and experienced swelling trade deficits relative to the other Central Asian states from 1992 to 1996, it gradually developed the most diverse trade portfolio. Though Kyrgyz trade was initially concentrated in the CIS, with Russia as its main exporter (27–48 percent) and importer (20–48 percent), by 1997 foreign sales to Russia were significantly scaled back (13–16 percent) and exports outside of the region outstripped those to the CIS. Although Bishkek continued to import more goods from Russia than any other single country, it purchased nearly as much from non-CIS partners as from within the bloc.[30]

The nuclear sector echoed this trend. Unlike other energy relations with the NIS, Russia did not enjoy structural power over Kyrgyzstan's nuclear fuel assets, as the latter's uranium mining and processing operations were neither owned by nor physically connected to Russian suppliers or consumers. Kyrgyzstan also retained abundant low-cost hydropower, relying less on Russia than on Kazakhstan, Uzbekistan, and Turkmenistan to service its demand for hydrocarbons via barter trades of hydroelectric energy for natural gas.[31] Furthermore, both Kazakhstan and the United States expressed strong interest in expanding commercial uranium processing arrangements with Bishkek. Coupled with America's growing presence in the region attendant to the post–September 2001 "global war on terrorism," Bishkek was poised to benefit from deepening engagement with Western partners, including in the nuclear industry.[32] Thus, given the independence and range of choices, closer engagement with Russia was neither predetermined nor without risks. In effect, Bishkek confronted a difficult paradox: the long-term prospects for diversifying lucrative commercial nuclear relations rested on potentially dangerous intimate association with the Minatom complex.[33] While both short- and long-term options offered comparative costs and benefits, they differed decisively in respective risks.

Second, the direction of Bishkek's nuclear energy risk-taking appeared out of sync with its previous hedging strategy. Although it was careful to cultivate close strategic ties with both Russia and China, Bishkek nonetheless was poised almost from independence to gamble on diversifying strategic relations with prospective Western partners. President Akaev embraced the "Diplomacy of the Silk Route," which committed the country to playing a "connecting bridge between countries and civilizations of the West and East."[34] Although friendship with Russia was "ever-lasting," he claimed that establishing close political and economic relations with the West was imperative for advancing post-Soviet modernization. He called for taking risks on international openness, arguing that the "key to the country's democratic and economic Renaissance," specifically the successes of currency reform and export-oriented growth, rested on emboldening the "Western vector" of foreign policy.[35] This went beyond expanding trade relations, as was underscored by the decision in 1993 to exit the CIS ruble zone that otherwise complicated exchange with Russia, Kazakhstan, and Uzbekistan, as well as by the concerted push for greater Western support and membership in the WTO that occurred in

1998.[36] It also was manifest on the security plane, as Bishkek asserted independence by joining the North Atlantic Cooperation Council in 1992 and NATO's "Partnership for Peace" program in 1994.[37] Catering to Moscow's requests to restore intimate nuclear relations, therefore, seemed out of step with this established trajectory of risk-taking. However, when pursued in parallel with other initiatives—that included characterizing ties with Russia as "bestowed upon us by God and history," amending the constitution to make Russian an official language of the country, inviting Russian troops to set up the first new military base on foreign soil since the Soviet collapse, and welcoming deployment on Kyrgyz territory of a CIS antiterrorist rapid reaction force commanded by a Russian general—reintegration of commercial nuclear ties formed part of a discernable gamble on intensifying strategic engagement with Russia from 2000 to 2002.[38]

Contending with Rivals

Almost from independence, the Kyrgyz leadership found itself firmly ensconced in the domain of losses. A landlocked country with mountain borders nearly impossible to defend or control, Kyrgyzstan was pinched between Russia and China and remote from Western markets. The absence of nuclear weapons, vast oil and gas reserves, or geographic proximity to Europe exacerbated this difficult strategic predicament, depriving Bishkek of diplomatic assets enjoyed by other NIS for wooing extraregional patrons. The country also inherited the weakest armed forces in Central Asia, as well as contested borders with China, Tajikistan, and Uzbekistan. While demarcation of the Chinese border was part of a broader multilateral initiative set in motion before the Soviet collapse, Bishkek contended with nearly 200 disputed border enclaves with its Central Asian neighbors. The latter, in particular, precipitated frequent clashes with both Tajikistan and Uzbekistan throughout the first decade of independence, as well as fostered terrorist infiltration and significant organized criminal and drug activity on Kyrgyz soil.[39] The sense of impending danger was made clear by Akaev's declaration in April 2002 that the southern parts of the country had been "invaded by thousands of terrorists and their gangs." Successive battles from 1999 to 2002, he lamented, had cost the lives of numerous Kyrgyz soldiers and stretched the government's capacity to manage on its own.[40]

The anxiety over the status quo was reinforced by the turbulent political transition. The political scene was especially chaotic from

1991 to 1996, as Akaev, a political outsider and ardent proponent of political liberalization and market reform, was elected amid extreme ethnic and socioeconomic tension stoked by the outgoing conservative Communist leader prior to the Soviet collapse. As president of the newly independent state, he confronted myriad political challenges over defusing interethnic hostility, unraveling and reconstituting regional and clan-based power and patronage networks, and consolidating leadership authority; all against the backdrop of weak central administration, a collapsing economy, and a national identity crisis.[41] The initial elite struggle reached a fever pitch in 1994–1995, when in the face of intractable parliamentary opposition to his campaign for building an "island of democracy," Akaev struck preemptively to rule by plebiscite. Through a series of referenda, he engineered a "quiet revolution" that degraded the constitutional powers of the legislature, promoted loyal regional allies, expanded presidential powers, and secured his reelection.[42]

After a brief hiatus during which he imposed a "republican monarchy," Akaev came under intense public pressure to crack down on rampant corruption, rebalance regional power relations, and deliver on his democratic rhetoric by seizing control over local and economic and political reforms that together threatened his political legitimacy. Yet, in the lead up to the 2000 elections, Akaev conspicuously reversed course, coercing judicial approval of his run for a third term, placing insurmountable barriers to his political rivals, and curbing media and political freedoms. Akaev created an increasingly precarious political predicament by the end of 2002 that rested on a spiral of arbitrary rule that then created the basis for more fundamental political and social instability. As summed up by Regine Specter, "the cycle of appeasing and reshuffling members of his political and personal network continue[d] to invite resort to coercive methods, which in turn increase[d] instability and weaken[ed] state capacity to effect change."[43]

At the same time, Kyrgyzstan confronted a difficult and deteriorating post-Soviet economic situation that accentuated the gap between real and expected achievements of economic reform. Irrespective of the leadership's early and bold market initiatives to dramatically transform industrial property relations, decollectivize farms, and liberalize prices, Kyrgyzstan nonetheless remained the second poorest NIS well after the Soviet collapse. Although the national economy showed signs of rebounding in 1996 (after suffering a dramatic drop of 45 percent in GDP over the preceding four years with inflation reaching 1,300 percent) and was projected to expand at nearly 5 percent through 2006, there were structural problems that dimmed near- and mid-term prospects for growth. Mushrooming unemployment rates among young males, extremely low per capita income rates, and "distressing" levels of external debt sapped optimism for economic stabilization. The challenge was compounded, as primary products (including gold, coal, and precious metals) and the agricultural sector continued to account for the lion's share of GDP and employment. With gold production and exports projected to decline significantly by 2010, it became increasingly clear that Kyrgyzstan could not rely on a narrow trade profile to fill economic shortfalls. Neither were the prospects for trade diversification promising, as nongold exports steadily lost shares in primary CIS and non-CIS markets, and the country's hydroelectric barter with Uzbekistan was subject to large swings. Despite becoming more competitive in CIS markets, Kyrgyz exports also confronted significant trade restrictions leveled by primary CIS customers from 1998 to 2002, including Kazakhstan, which controlled the transit of Kyrgyzstan's entire land-based trade with Russia and Western Europe.[44]

The challenges were not lost on the political leadership. Akaev, who personally reveled in spearheading radical currency reform and mass privatization in 1992–1994, acknowledged the tenuous nature of the early success. By 2000, he openly referred to the second stage of reforms as "the moment of truth" for the country's economic future.[45] Although dismissive of revolutionary change, Akaev nonetheless argued that failure to make urgent progress toward realizing the 2000 reform strategy—with dramatic improvements in productivity, fiscal and administrative reform, foreign investment, and export earnings—imperiled the economy and risked instilling "fear" across Kyrgyz society. "A society filled with fear," he warned, "is incapable of implementing reforms and building its own future."[46]

The difficult strategic, political, and economic backdrop accentuated the bleak situation in the nuclear industry. Unlike Armenia, Kazakhstan, Lithuania and Ukraine, the country did not inherit nuclear reactors on its soil or reliance on nuclear power generation in the Soviet divorce. Instead, it acquired 7 nuclear mining and milling facilities and a centrifuge testing site, as well as 63 nuclear dump sites that contained approximately 56 cubic meters of radioactive waste.[47] Though the country had a natural abundance of uranium, most of the mines were closed by the late 1960s. The five mines that were conserved during the latter Soviet period and activated with independence had ceased production by the mid-1990s, owing to the depletion of low- to medium-cost uranium deposits. The Kara-Balta Combine, the country's largest uranium-processing facility commissioned in 1954, had long ceased exploiting fresh uranium and fell idle until 1994. Russia had reportedly removed all uranium stockpiles from the mine shortly after independence, claiming that the combine lacked appropriate controls to operate the mill or to supervise exports.[48]

There also were real impediments to the sector's future recovery. As the mill featured only chemical (in situ) leaching operations, it was ill equipped to process uranium from conventional underground and open mines that, in turn, restricted the potential supplier base. It was not until 1994 that Bishkek restored foreign commercial relations. Reports of corruption within the national uranium industry also sidetracked recovery and provoked public outrage.[49] The prospects for wooing additional lucrative deals to process uranium from global suppliers were hampered by an early legal battle over uranium exports with the U.S. government. Prodded by an ad hoc group of American uranium miners, the U.S. Department of Commerce established a price-based quota on Kyrgyz imports and launched investigations into the dumping practices of all post-Soviet suppliers.[50]

Bishkek simultaneously was burdened with redressing a mounting radioactive waste problem. Among the polluted locations were 44 deteriorating tailing sites and slag heaps buried under thin layers of soil that presented immediate risks of environmental contamination, public health, domestic stability, and regional security beyond the government's capacity to manage alone. A landslide in May 2002 nearly destroyed one of the country's largest tailings impoundments, washing radioactive waste into the adjacent river. The near disaster provoked alarm at home and abroad about an imminent threat of radioactive contamination to the densely populated Ferghana Valley and neighboring areas in Uzbekistan.[51]

Yet Russia was handicapped at exploiting Kyrgyzstan's losing decision frame. As discussed in chapter 3, Moscow was poorly positioned to manipulate the opportunity costs of compliance, due to limited market power. As neither the uranium supplier nor consumer of last resort for Bishkek's processed uranium, Minatom was unable to fend off commercial rivals or saturate potential markets. In practice, Minatom lacked the ability to alter the costs or benefits of cooperation so that reintegrating with Russia was an obvious choice for Bishkek.

In the first place, Minatom was unable to preempt rival uranium suppliers from engaging Kyrgyzstan. Lacking sufficient supplies of fresh or converted uranium to meet its own needs, Minatom was unable to engage the full capacity of the Kara-Balta facility. Similarly, Minatom's own uranium conversion facilities were ill equipped to service the global demand. By contrast, there were commercial competitors willing and able to engage Kyrgzstan in joint efforts to revive the Kara-Balta facility. The EU, for example, signed a "partnership and cooperation" agreement with Bishkek in 1994 to develop a distribution system for Kyrgyzstan's natural and low-enriched uranium exports to member states.[52]

Similarly, Kazakhstan, which inherited the largest low-cost deposits of uranium, beat Russia to the punch by presenting Bishkek with viable offers to reconstitute commercial relations. An agreement was reached in 1994 to process 1,000 metric tons of Kazakh uranium at the Kara-Balta facility, and to sell the concentrate in Kazakhstan. The profits were divided 71:29 percent between Kazatomprom and Kara-Balta, respectively. This was followed by formation of a separate bilateral joint venture in 1997 that stipulated the eventual processing of 2,000 metric tons of Kazakhstan uranium for foreign sales. In the agreement, Kazatomprom held a 65 percent stake and Kara-Balta owned a 35 percent share in the venture. Although Russia was identified as a principal market, the venture received early interest from potential consumers in the United States, Germany, Japan, and France.[53] Astana's subsequent investment of $7 million in 2000 reinforced the promise of the venture. Coupled with the license to refine gold from the London Precious Metal Exchange, there were high hopes that the combine would ultimately turn a commercial profit without necessarily resuming intimate association with Minatom.[54]

Minatom also faced stiff competition from the United States in purchasing processed uranium from the Kara-Balta facility. Because Kyrgyzstan was no longer actively mining fresh uranium, the U.S. Department of Commerce suspended the 1992 dumping investigation and cultivated new supply relations for processed uranium. In particular, Washington considered Kyrgyz offers to revise the price-tied quota contract for delimiting exports of "yellow cake" to the United States. Although Bishkek consistently balked at counterproposals tendered by the Commerce Department, it left the door open for future negotiations.[55]

Washington outmaneuvered Russia with offers to upgrade Kyrgyzstan's border security equipment traditionally supplied by enterprises under Minatom's aegis. U.S. Secretary of State, Madeline Albright, for example, personally lobbied Kyrgyzstan to purchase American radar equipment. Russian officials took umbrage from what were perceived to be Washington's competitive tactics. Specifically, Moscow complained about Washington's underhanded practice of dispatching a special representative to Kyrgyzstan to negotiate a "mutual understanding" on the eve of the arrival of a Minatom delegation that was poised to consummate a competing contract.[56]

By the end of the decade, the future of the country's uranium processing business was ambiguous. Although it offered one of the few potential growth areas for the Kyrgyz economy and for diversifying foreign relations in general, the dormancy of the sector and legacy burdens were discouraging. Independent of Russia and courted by potential foreign customers, the domestic industry nonetheless lacked the commercial standing to set the terms for foreign engagement. Given shortfalls in the global fresh uranium supply, the Kyrgyz leadership had little sway over the pace or direction of the industry's revival, operating in the domain of losses as it considered rekindling foreign commercial nuclear energy relations.

In addition, the main options for breaking out of the status quo were problematic. The decision to reengage Kazakhstan presented an immediate opportunity to revive moribund operations at Kara-Balta, but did not offer a long-term solution for the combine's commercial viability. The bilateral agreement to process and market 1,000 metric tons of concentrate from 1994 to 2000, for example, occupied less than 30 percent of the plant's capacity.[57]

Similarly, there were tangible costs of doing business with the United States. Washington insisted on restricting imports and imposing price quotes for Kyrgyz-processed uranium significantly below Bishkek's expectations. There also were no guarantees that Kyrgyzstan would receive compensation for any losses incurred in contested transactions. This posed a tangible threat, as Kara-Balta already had to file suit in 1995 for reimbursement of $2.8 million for purchases canceled by American customers as a consequence of the initial dumping investigation.[58]

Outflanking the Competition

Kyrgyzstan's viable options presented nearly indistinguishable moderate risks. At worst, reliance on the export of concentrate derived from Kazakh uranium might complicate sales on the U.S. market absent an established price-quota mechanism. Because of Astana's own economic problems and the questionable solvency of future Kazakh consumers, Bishkek also would likely encounter nonpayments problems for its conversion services. At best, the flourishing of bilateral contracts would provide an immediate infusion of cash for the underutilized Kara-Balta facility, and pave the way for future contracts with foreign customers, including Russia. Thus, tying Kara-Balta's fate to the revival of the Kazakh nuclear industry offered safe, marginal benefits. Working with Astana to break into foreign markets, however, compounded the risks associated with other available options.

The second option—soliciting Western uranium suppliers and customers for processed uranium—offered a slightly wider variation in potential outcomes. A positive result would entail breaking into new international markets, subject to mutually acceptable price-quotas. The potential downside was that Bishkek would be saddled with unfavorable prices artificially imposed by Washington. The redirection of former Soviet energy assets also was likely to antagonize Russia. Given Akaev's commitment to advancing his "Silk Route" diplomacy, this was a very important consideration. Hence, active pursuit of extraregional markets was riskier than confining cooperation to Kazakhstan, as both positive and negative values were slightly greater.

Restoring nuclear fuel relations with Russia presented a roughly equivalent risk. The danger for Bishkek lied with being exposed to political pressure and financial problems posed by Russian partners. If successful, however, Bishkek could secure limited uranium supplies and access to well-established markets for its yellow cake. Given Minatom's growth strategy, the latter stood to expand over the coming decades, outstripping potential gains from relying exclusively on sales to Kazakhstan. Because of Russia's limited market power and the relatively small contribution of the uranium processing industry to the existing Kyrgyz economy, the initial costs of Minatom's predation also were not likely to be exacting. From a strict commercial perspective, therefore, reconstituting uranium relations with Russia was risky, but no more so than gambling on other foreign competitors.

Viewed in this light, Kyrgyzstan did not confront an obvious choice. Though Bishkek was prone to taking risks, the predictions of prospect theory are indeterminate, given that the positive and negative values of each available option were roughly equivalent. Yet this calculus was dynamic and susceptible to Moscow's manipulation. The challenge for Russia was to distinguish cooperation as the riskier option. Unable to shape the prevailing decision frame in Bishkek or to alter significantly the opportunity costs of cooperation, Minatom nonetheless improved its competitive edge at landing commercial contracts by exploiting domestic authority to hype the relative riskiness of cooperation. This was accomplished by accentuating prospective positive and negative outcomes for Bishkek via a combination of diplomatic carrots and sticks.

On the one hand, Minatom tapped its central authority over subordinate enterprises to extend material sweetners that outflanked Washington and improved the likely positive outcomes from further cooperation for Kygyzstan. This was epitomized by offers in mid-2000 to supply front-line radar equipment, border protection devices, and alarms—such as the prized Vitim-, Fara-, and Gardina-class radar systems—at a subsidized cost of 20 million rubles to cover 850 miles of the most vulnerable borders with China, Tajikistan, and Uzbekistan.[59] Minister Adamov personally instructed the leaders of key Russian nuclear enterprises to extend concessionary terms allowing components of these systems to be produced at local Kyrgyz enterprises. This was promoted to "help revive production and create new jobs," and to provide reliable access to "dual use" spare parts that also could facilitate the modernization of Kygyz aircraft. Minatom wooed the Kygyz leadership by suggesting that these ventures would facilitate future financial and industrial cooperation beyond the nuclear sector.[60] Kygyz defense officials, in particular, relished the prospects that "income from these activities would replenish the state's budget so that the government would be able to draw funds to tackle security problems." Elated by the opportunity, Defense Minister Esen Topoyev pronounced: "We have great affection for Russia's Ministry of Atomic Energy."[61]

Minatom also attempted to allay Kyrgyz concerns by proposing that the trilateral uranium processing venture actively solicit foreign investment from a wide range of sources.[62] The ministry also promised to earmark an additional two million rubles to support Russian language instruction in Kyrgyzstan, and to improve training of Russian teachers at the Kyrgyz-Russian Slavonic University.[63] This was coupled with subsequent proposals to fund reclamation projects at radioactive dump sites contaminated by Soviet uranium mining practices from 1946 to 1990. In particular, Minatom underwrote a bilateral feasibility study on the rehabilitation of three of Kyrgyzstan's largest and most precarious tailing impoundments.[64]

Simultaneously, Moscow accentuated the possible negative outcomes for Kyrgyzstan. Adamov carefully pitched proposals for cooperation as closely linked to the counterterrorism theme of successive bilateral, CIS, and Shanghai-Five summits in 1999–2000. During his visit to Bishkek in December 2000, for example, he emphasized the unique vulnerabilities of the Ferghana Valley and Kyrgyzstan's southern borders to regional Islamic terrorist groups, including the Taliban, as well as intimated Russia's extraterritorial designs. "Kyrgyzstan is becoming an outpost of our security," he proclaimed, and one that we

are "certainly interested in reinforcing." By the same token, he intimated that although subordinated Russian enterprises were willing to
deliver sophisticated radar systems, Minatom would not hesitate to
withhold assistance to an embattled Bishkek in the event that offers for
collaboration were not reciprocated. The latter purportedly would
include intensifying pressure on Kazakhstan to scale back bilateral
cooperation with Bishkek.[65] Adamov acknowledged publicly that interest in bolstering Kygyzstan's security was not purely altruistic or commercial, but that such assistance would of course garner Russia
geopolitical influence along its periphery.[66]

Minatom also issued veiled threats to complicate Kyrgyzstan's
already difficult situation in the nuclear sector. Receiving "no answer"
to the initial proposal for creating a trilateral venture, Minatom joined
several other government agencies by renewing claims to equity stakes
in the Kara-Balta Combine as payment for Kyrgyzstan's outstanding
debt to Russia.[67] This was followed by threats to withhold environmental remediation assistance. Not only was Minatom reluctant to fund the
restoration of contaminated areas, but intimated that it would not
approve the transfer of critical documents that mapped out Kyrgyz
deposits or fully disclosed the mining activities conducted by its Soviet
predecessor.[68] In short, Minatom underscored that the payoffs for
Bishkek lied more in restoring relations than in exploring new ones.

Gambling on Business with Minatom

Recognizing that the variance between positive and negative outcomes
was greater if it sided with Russia, the risk-prone leadership in Bishkek
ultimately renewed relations with Minatom. Under no pressure to make
generous concessions, the Kyrgyz leadership held out for an equitable
partnership arrangement. The final deal signed in August 2001 stipulated formation of a new venture to mine Kazakh uranium and convert
it into yellow cake at the Zarechnoye field, process it into uranium
oxide at Kara-Balta, and deliver the concentrate to Russia. Shares in the
new company were distributed in a manner that prevented a single
entity from gaining a controlling stake: Kazatomprom held 45 percent
of the shares, Minatom enterprises received 45 percent (TVEL 20 percent, Tekhsnabexport 15 percent, and Atomredmetzoloto 10 percent),
and Kara-Balta owned a 10 percent stake.[69] It was expected that the
joint venture would begin operations in 2001–2002, after financing was
secured, but no official start date was established.

The deal marked an important step toward reviving Kyrgyzstan's commercial nuclear activities. The Kara-Balta Combine acquired a direct concession to Kazakhstan's prized Zarachnoye deposit that was projected to hold nearly 20,000 metric tons of fresh uranium.[70] The production of uranium concentrate was expected reach 500–700 metric tons during the first several years, nearly doubling output at Kara-Balta in 2000, after which it was projected that the venture would support the production of 1,600–1,700 metric tons of concentrate per year.[71] Furthermore, the tripartite agreement stipulated that the venture would not settle for Russia as the primary customer. Although Russian nuclear plants would receive the lion's share of deliveries during the initial years of operation, the venture would subsequently diversify clients, specifically targeting foreign, cash-paying clients to support additional production.[72]

The deal was structured to accent mutual commercial benefits and to contain Russia's future predation. Rather than granting Minatom enterprises debt-equity stakes in Kara-Balta, the final arrangement stipulated a mutual debt exchange among enterprises within the three states. The terms called for Kazakh energy consumers to pay Ulba, for Ulba to continue supplying nuclear fuel rods to TVEL, for TVEL and Minatom to cover the costs of the border security equipment, and for Kyrgyzstan to continue to supply electricity to Kazakhstan.[73] Furthermore, Minatom was on the hook to devise a mutually acceptable business plan to support the first phase of operations. During the first two and a half years, Minatom enterprises could sell the uranium on foreign markets to recoup an initial $10 million investment in the project. Yet Minatom also was obliged to extend direct loans from its budget, as well as to arrange additional funds to cover the projected $14.4 to 14.6 million costs of constructing a new mine at the Zarachnoye deposit.[74] By all accounts, this was an onerous task for the cash-strapped Russian ministry. As lamented by one Minatom official, it was one thing for Russia to secure use of lost Kyrgyz uranium processing assets, but "the real problem is drawing up profitable business plans."[75]

Eager to demonstrate progress and dismissive of the potential negative outcomes, Bishkek heralded the venture as a commercial success. With the contours of the final agreement in hand, Kyrgyz officials boasted that the three sides had managed to parlay national security goals and strategic assets into profitable business. Akaev lauded the debt exchange mechanism for providing "not just jobs, but jobs for skilled engineers involved in the mining and defense sectors that used to form the backbone of the Kyrgyz economy."[76] The real challenges ahead were acknowledged to be commercial not political in nature. In a candid assessment, the managing director of the Kara-Balta facility stated: "The Project, which is to ensure deliveries of raw materials for the combine's uranium production as well as sale of output, will make it possible for the enterprise to operate with optimum efficiency, that is using the combine's capacities not 100 percent but in accordance with the state of the market."[77]

Playing It Safe With Kazakhstan

Reintegration of commercial nuclear relations with Kazakhstan was equally puzzling. In many respects, it seemed inevitable. Astana could ill afford to turn its back on Minatom, given that Kazakhstan's uranium dioxide fuel pellets derived from uranium enriched in Russia and were sold primarily to Russia's Elektrostal (Moscow Oblast) and Novosibirsk fuel fabrication facilities.[78] With control over the supply to the BN-350 neutron liquid metal cooled reactor, the only nuclear unit operating in Kazakhstan at independence, and with offers to construct a new light water reactor, Minatom also was well positioned to dominate nuclear power generation in the country for the foreseeable future.

Similarly, reintegration of commercial relations seemed to follow logically from the removal of nuclear weapons from Kazakhstan. With prodding and assurances from Washington, Moscow and Astana reached successive agreements to manage the repatriation of nuclear weapons and fissile material and to safeguard nuclear facilities and industrial activity in Kazakhstan. The two states also agreed to continue nuclear fuel exchanges. As part of these efforts, Astana signed on to exporting uranium to American, Australian, and European markets via Russian territory, providing Moscow with an opportunity to control the global footprint of a resurgent Kazakh uranium industry.[79] This potential was not lost on Minatom officials, who consistently identified "lost assets" in Kazakhstan as lucrative targets for reclamation. According to a leading Kazakh specialist on nuclear affairs, Murat Laumulin, Minatom's interest in regaining control of the Kazakh sector served "dual purposes":

> On the one hand, it (Minatom) seeks to preserve Kazakhstan, together with its industrial infrastructure, as a major source of uranium ore and enriched fuel after the year 2010, when primary reserves of uranium on Russian territory will be

depleted. On the other hand, Minatom would like to retain Kazakhstan as a strategic market for Russian-built nuclear power plants, as it hopes to do in China, India, and Iran.[80]

Adamov specifically linked restoration of a dominant role in Kazakhstan's commercial nuclear development to Russia's prospects for orchestrating integration of the fuel and energy complex. "The vector of transnationalization and orientation on international competitiveness," he stressed, "must encourage the Russian atomic corporation to deploy a network of joint ventures and subsidiary companies in countries of the near abroad and, in the future, the far abroad."[81]

Cooperation, let alone intimate association with Minatom, was neither readily embraced nor preordained. As discussed below, Russia had to settle for much less than it initially demanded with an ownership stake in the Ulba combine, and was unable to prevent strategic diversification of the Kazakh uranium industry. Unlike the gas industry, the structural hangover of the Soviet nuclear fuel cycle was not an automatic source of vulnerability for Astana. In contrast to Kyrgyzstan, Kazakhstan inherited significant indigenous uranium mining, milling, and conversion capabilities, as well as fuel fabrication facilities that attracted investors and customers worldwide. At independence, Kazakhstan's three active mines ranked it as one of the top ten producers of uranium. The Ulba plant, in particular, manufactured 100 and 80 percent of the nuclear fuel pellets used in Soviet made RBMK and VVER reactors, respectively, and produced most of the former Soviet Union's beryllium and tantalium. In addition to covering the bulk of the former Soviet demand for fuel fabrication, the plant maintained the versatility to process "any type of uranium-containing material," as well as capacity to manufacture different types of pellets to meet customer specifications.[82] Accordingly, it was Moscow that was initially dependent on Astana for the supply of resources "without which the uranium industry in Russia could not function."[83] Cognizant of the potential commercial opportunities afforded by this endowment and wary of creating unnecessary vulnerability, Astana unlike Bishkek was keen to adopt a "go slow" hedging strategy for rekindling ties in the nuclear industry.[84]

Astana jealously guarded independence in the nuclear energy sector and resisted ceding decisive ownership or control to Russia. Despite the constant pressure from Russia's electricity monopoly, Kazakhstan retained governing authority over the nuclear sector. According to Kazakh Deputy Prime Minister and Minister of Energy, Trade, and Industry, Vladimir Shkolnik, this was something that "the Russian side did not agree with and believes that the major Russian energy company is being discriminated against as a foreign investor."[85] That said, concerns about the possible deleterious consequences of revisiting the Soviet experience did not repel Astana from Moscow. Astana proved willing to commit to long-term commercial cooperation with Russia, and to swap equity stakes with Minatom enterprises. This was distinguished from developments in the oil sector, where the same Kazakh leadership seized on multiple opportunities to pursue an independent and diversified strategic course that was conspicuously beyond the grasp of Russia's energy diplomacy.

In contrast to Bishkek, Astana also approached commercial nuclear cooperation with Russia with growing confidence. As was the case in both the gas and oil sectors, Astana initially confronted a mixed situation. From 1992 to 1998, Kazakhstan suffered a precipitous contraction of mining activity, with annual uranium production reaching a nadir in 1997, at 23 percent of the average level during 1987 to 1990, before independence. Employment across the sector dropped by 60 percent from 1992 to 1998, creating social tensions that reached an apogee in 1996, when miners at several facilities staged successive strikes to protest a five-month delay in receiving salaries.[86]

The early travails were tempered by both the relative insignificance of the sector to the national economy, and the burgeoning opportunities for resurgence. Despite the dramatic downturn, nuclear power generation comprised only .1–.6 percent of total domestic electricity consumption from 1992 to 1997. The period was marked by emergence of a privatized and competitive domestic electricity market, as conventional power plants contended for different customers and against imports from Russia, Uzbekistan, and Kyrgyzstan.[87] The country was internationally recognized as a rising global supplier projected at holding between 15 and 20 percent of the world's known uranium deposits, positioned between Australia (28 percent) and Canada (14 percent), and considerably higher than Russia (4 percent). Half of the deposits were estimated as recoverable under low-cost scenarios, tantamount to nearly five times the amount available in Russia and 20 percent of the global low-cost reserves. By 1998, uranium production showed signs of recovery, as it accounted for approximately 3 percent of global production and became the object of growing international attention.[88]

The positive trends continued during 1998 to 2002, firmly situating Kazakh nuclear energy decision making within the domain of gains. The incorporation of new technologies and capacities at Kazakh mines boosted annual rates of uranium production 15–34 percent, allowing

Kazakh supply to constitute 5.5 percent of the global market by the end of the period. The combined yield of accessible domestic and joint ventures was projected to double by the end of the decade, positioning Kazakhstan to be an important low- to medium-cost supplier for the indefinite future. The situation was poised to improve with the shutdown of the country's sole nuclear power station in 1999, and Kazatomprom's subsequent pledge to sell all of its uranium production on foreign markets.[89]

The upturn was mirrored at other stages of the fuel cycle. Both the production and the size of the fuel fabrication workforce steadily expanded at Ulba so that by the beginning of 2001 there were no wage arrears and the plant generated considerable tax receipts for the government.[90] Kazakh uranium producers and fuel service providers were solicited by French, Canadian, Chinese, Japanese, Russian, and American firms for long-term mining and sales contracts. Despite Russia's best efforts to expand indigenous fuel pellet production, it remained dependent on supplies from Ulba. In addition, the Kazakh government invested $10–$15 million annually to upgrade uranium production, as well as extended loans approximating $62 million to modernize mining and fuel fabrication that were projected to generate roughly $180 million by 2003.[91] Similarly, the Japanese government pledged to underwrite feasibility studies for environmental remediation at the nuclear weapons test facility at Semipalatinsk. Finally, in mid-2002, the Kazakh government legally approved the import and burial of low-radioactive waste that was estimated to earn the country between $30 and 40 billion over the ensuing 25 years. Although Astana was subject to the same constraints on breaking into this global market as Russia, there were high expectations that any revenue raised would significantly augment the paltry $1 million annually earmarked in the state budget for nuclear clean-up operations otherwise projected to cost $1.1 billion.[92]

Not surprisingly, the Kazakh leadership was especially bullish on the prospects for becoming a global leader in the nuclear industry. In 1997, the government issued a decree that obliged Kazatomprom to solicit foreign investors to develop uranium deposits and to explore fuel service arrangements. These directives were reformulated in the "Strategy for the Kazakh Uranium Industry for 2004–2015" that outlined programmatic targets for positioning the domestic industry as the world's third leading supplier. The strategy also called for strengthening Kazakhstan's standing in nuclear fuel markets in Russia and the CIS, as well as in expanding American and European markets.[93] The prevailing confidence was attributed to "structural" features in the industry that discouraged the leadership from undertaking undue political or commercial risks. Neither in a rush to force the issue nor deterred by the "groundless phobia" of nuclear safety, Minister Shkolnik counseled a prudent posture for seizing opportunities to expand the product mix of Kazakh commercial nuclear exports. The steep trajectory of the industry's resurgence, he noted, was propelled by "several objective preconditions for the creation of nuclear power engineering in Kazakhstan, in particular significant reserves of uranium, development of the uranium mining and processing industry, presence of the Ulba plant which produces uranium dioxide and fuel pellets for power reactors, and the availability of skilled personnel."[94] The bumps presumed to lie ahead for Kazakhstan were expected to be commercial in nature, related to fluctuations in market prices beyond Moscow's grip.[95] In contrast to Bishkek, Kazakhstan was distinctly risk averse toward resuscitating a commercial nuclear rapport with Russia.

A Plethora of Opportunity

As was the case with Kyrgyzstan, Russia was poorly positioned in global nuclear fuel markets to alter the opportunity costs of preferred transactions. In addition to establishing joint ventures to process uranium with Kyrgyzstan and Russia, Astana attracted nearly $100 million in 1995–1996 to develop joint ventures with French and Canadian nuclear fuel cycle companies to extract Kazakh uranium. A trilateral deal with two Canadian firms netted Kazatomprom a $53 million investment in the Inkai Joint Venture to produce an estimated 2.6 million pounds of U_3O_8 by 2003. This was followed by the creation of the Katco Joint Venture with the French nuclear firm, Cogema, and several other European investors in 1996 that acquired ownership of a projected 700 metric tons of uranium produced and exported from Kazakhstan's Moinkoum deposit. Each project had a strong upside potential for annual low-cost production rates, valued at $500 million and $100 million, respectively, in 2000.[96]

Kazakhstan also was solicited by foreign investors for the rights to purchase uranium mines. In 1999, the Israeli firm, Sabton, outbid commercial rivals, including Kazatomprom, to acquire the Tselinnyy mining and milling combine. By 2000, the company committed $100 million over five years to develop new deposits and modernize Kazakhstan's chemical conversion capabilities.[97] As evidenced by these select deals,

Kazakhstan capitalized on foreign interests, broadening its customer base for uranium, and transferring user and ownership rights to uranium production and fuel processing facilities to numerous commercial entities other than Russia.[98]

An especially vexing challenge for Minatom was the lack of commercial power to preclude U.S. firms from engaging Astana. As early as 1992, American companies began exploring investment opportunities across the Kazakh uranium industry. After initial dumping probes, the two countries signed a suspension agreement that established price-quotas for Kazakh uranium sales on the U.S. market. From 1994 to 1998, Kazakh exports fluctuated with shifts in the U.S. market, as Astana secured only ad hoc waivers to compensate for low prices. Presented with a protracted price slump and a ban imposed by Washington on reexports, Astana in 1999 broke the agreement claiming that commercial conditions made transactions "unachievable."[99] Yet before Russia could capitalize on the low ebb in bilateral uranium trading, an American court ruled against the ongoing antidumping investigation, prompting the resumption of Kazakh uranium deliveries onto the U.S. market. This was followed by the U.S. International Trade Commission's conclusion that the volume of imported Kazakh uranium could not approach a "significant or injurious level" on the American market. This finding carried potentially adverse consequences for Moscow, as it exempted Kazakh deliveries of uranium dioxide converted from non-Kazakh uranium, while subjecting Russian origin uranium to a separate bilateral suspension agreement. These decisions effectively removed official barriers to Kazakh sales on the U.S. uranium market, fostering expectations in Astana that transactions with American and West European firms would begin to flourish in a "civilized and free market manner."[100]

By the end of the decade, Western firms forged new ties with the Kazakh uranium industry in fuel service and reactor markets. From 2000 to 2002, U.S. nuclear fuel companies signed several contracts with Ulba that provided direct access to international customers. One was a 10-year deal with the largest U.S. beryllium marketing company for global sales. This was complemented by Ulba's agreement to process and extract uranium dioxide powder from scrap materials provided by the American firm, Global Nuclear Fuel (GNF) that overnight created 300 new positions at the plant. The certification of the extracted powder used for producing fuel pellets for the United States marked the first step toward broadening Ulba's commercial footprint in European and South Asian nuclear fuel markets.[101] By early 2002, this was reinforced by the decision of the U.S. National Nuclear Security Administration to grant $1.2 million over three years to improve the technology available to Ulba for recovering uranium dioxide from scrap. In addition, the government lobbied American and European firms to pledge financing to streamline the production of nuclear fuel in Kazakhstan.[102] Furthermore, despite Minatom's headway at negotiating the construction of a new generation nuclear reactor, Astana continued to intimate interest in purchasing the next generation American reactor as a bargaining ploy.[103] Thus, Minatom faced stiff commercial competition on all fronts for Kazakh partners that it was unable to preempt.

Minimizing Risks

Astana's choice for expanding the strategic footprint of the uranium industry was relatively straightforward: it could focus on breaking into new global markets or on reclaiming a niche within the Russian nuclear fuel cycle. With the relaxation of restrictions on Kazakh exports of uranium and U_3O_8 to American and European markets and the growing interest shown by foreign investors, Kazatomprom stood to generate considerable commercial profits by tailoring developments toward extraregional markets. The benefits of tying the national industry's resurgence to Russia paled in comparison. Although there were familiar and available partners for Kazakhstan's uranium industry in Russia, they were neither as liquid nor as potentially large as extraregional competitors. Nor was Kazatomprom structurally dependent on Russia for cultivating new relations with potential global partners.

The preoccupation with diversifying commercial nuclear relations posed only a modest risk to Kazatomprom. If Astana cemented commercial ties with American companies, it could gain an entree to a large uranium market, as well as acquire established partners and certification for landing additional contracts in growing European and Asian markets. It also could garner the financing and technical support needed to take the lead in the novel business of extracting uranium powder from scrap that, in turn, could broaden partnerships and distinguish Kazakh fuel fabricators from global competitors. A fixation on extraregional opportunities, however, would subject Kazakh nuclear fuel sales to downturns in the market without viable recourse. Therefore, a global orientation would enable Astana to explore potentially lucrative commercial opportunities while exposing it directly to the vagaries of the marketplace.

On the other hand, reintegrating with Minatom was riskier than looking westward because the gap between expected positive and negative outcomes was wider. If successful, Astana would lock in traditional Russian and CIS buyers for fuel pellets manufactured at the Ulba plant, as well as curry favor with Minatom. Cooperation with Russia also would provide short-term relief for struggling Kazakh mines and fabrication facilities while Kazatomprom solicited global investment and partners. Yet, a preoccupation with Russia could render the Kazakh uranium industry vulnerable to Minatom's future predation. Conceding to Minatom's initial demand for a 51 percent stake in the Ulba plant, for example, would strip Astana of control over the commercial activities at this strategic facility. According to the head of Kazatomprom, "this was not a very good idea because TVEL would have gained absolute control; and Russian control would not necessary guarantee Ulba a new lease on life."[104] A deliberate turn to Russia also could dissuade rival foreign investors, and discourage the U.S. government from promoting the Kazakh uranium industry in home and foreign markets. The latter posed special concerns, given that American companies expressed interest in acquiring stakes in Ulba as an alternative to purchasing uranium directly from Russia at fixed prices. That Minatom touted renewed relations with Kazatomprom as a "nuclear strike on American interests in Kazakhstan" only raised the probability of this negative outcome.[105] Thus by the end of 1999, nuclear cooperation with Russia neither offered the greatest expected value nor safest bet for the relatively risk averse Kazakh leadership.

Unable to rely on market, structural, or relative power advantages to alter the opportunity costs of cooperation, Minatom officials acknowledged that it was simply "naïve" to assume that the Russian industry could outperform Western firms in wooing Astana via offers of large-scale investment and head-on competition in third-party markets.[106] Accordingly, Minatom tried to leverage its streamlined administrative authority to manipulate the balance of risk calculus for Astana. This entailed making commercial nuclear engagement less risky for the Kazakh uranium industry. The Russian government effectively exploited the consolidated authority within the Minatom complex to fashion proposals aimed specifically at minimizing the downside of commercial nuclear cooperation, while offering modest sure gains that were consistent with Astana's general risk-averse hedging strategy for capitalizing on the domestic industry's favorable trajectory.

First, Minatom played to its domestic strength by reducing the likelihood and severity of a negative outcome. Unlike with electricity exports, where Minatom's direct access to foreign customers was constrained by institutional and technical barriers imposed by the national grid operator, contracting for nuclear fuel services fell exclusively under the ministry's purview. In particular, Minatom exploited federal authority to relax previous demands for TVEL to obtain a controlling stake in Kazakhstan's Ulba plant. In 1998, TVEL proposed the acquisition of a 51 percent stake as a means for securing control over strategic decision making at the plant. As it became clear that this was a "nonstarter" for Kazatmoprom, however, Minatom sharply retracted demands, petitioning for a minority stake in the Ulba plant. In order to reassure Kazakh officials, Minatom sought only a limited veto over activities undertaken by the Ulba plant that directly impinged on TVEL's immediate interests. Russian officials made clear that their primary fear was that "something might happen which finally could cross the Ulba metallurgical plant off the list of enterprises that produce nuclear fuel for nuclear reactors of the Russian model." This seemed reasonable to Kazakh officials. As stated by the president of Kazatomprom:

> It is understandable if that (the loss of Ulba contracts) happened, then Russia would have to think very deeply about which plants to place its orders with, because then it would experience a shortage of nuclear fuel. It is clear that this is a vital issue for them and this is the issue of the future in which they should be certain.[107]

In an effort to dispel Kazakh anxiety about Russian meddling in the domestic uranium industry, Minatom offered complementary access to Russian front-end producers. It proposed that in return for TVEL's acquisition of equity stakes in the Ulba plant, that Kazatomprom obtain stakes in the Russian nuclear fuel industry. Because of the complexity of equating par values of the enterprises in the two different economies, Minatom granted Kazatomprom the right to acquire equivalent stakes in several enterprises subordinated to TVEL.[108] This gesture provided Astana an opportunity to diversify holdings in the nuclear fuel service industry. According to the president of the Kazatomprom, the swap provision with Minatom was especially attractive "because what we want most of all is a normal balanced market, not money."[109]

Simultaneously, Minatom exploited its domestic control to enhance the tangible benefits of reintegration. This included approving the use of reactors from decommissioned Russian nuclear submarines to generate power for small towns across southern Kazakhstan. Minatom approved production of uranium capsules at the Ulba plant that could extend the service life of these reactors beyond the initial seven to eight years of use.[110] In addition, Minatom signed "protocols of intent" to develop nuclear ties with Kazakhstan extending beyond specific fuel service agreements. This covered an agreement for TVEL to explore remedies for jointly managing Kazakh nuclear waste, including spent nuclear fuel. Similarly, Minatom authorized TVEL to share technical assessments of Kazakh deposits that could serve as the basis for identifying new deposits for development via new joint ventures.[111] Finally, the Russian government offered to construct a new nuclear power station in the Lake Balkhash region. Putin held up this agreement as a testament to deepening bilateral economic cooperation, and as a symbol of Russia's commitment "above all to consistently implementing joint projects in the nuclear sector."[112]

In short, Russia sought to compensate for the inability to foreclose commercial uranium opportunities by making cooperation seem less risky for Astana. Although it could not compete with the investment and multiple supply opportunities afforded by international competitors, Minatom tried to win over Kazatamprom by providing reassurance that reintegration would neither lead to costly dependence nor preclude future development of the Kazakh uranium industry. Minatom did not try to hide that it too would seek future remedies for its own dependence on the supply of Ulba's fuel pellets by expanding fuel fabrication operations at the Novosibirsk facility. As summed up by an official at TVEL, the focus of Russia's nuclear diplomacy was on reducing the potential risks of reintegration so that "the terms of cooperation are at least no worse than what our competitors have to offer."[113]

Settling on Commercial Partnership

After protracted deliberation, Astana formally committed to cementing closer cooperation with Minatom by early 2000. First, Kazatomprom accepted the proposal for an equity swap involving the Ulba plant. The agreement signed in January 2000 stipulated TVEL's acquisition of a 34 percent stake in the Ulba combine in exchange for Kazatomprom's acquisition of roughly 7–10 percent stakes in three Russian enterprises subordinated to TVEL. Astana also amended domestic legislation to permit TVEL to hold a "golden share" that would permit Minatom to veto management decisions over key policy changes as "insurance" against the short-term diversion of plant activities. In addition, Ulba guaranteed minimum sales of nuclear fuel pellets on the Russian market for the ensuing five years.[114]

This was followed in November 2001 by Astana's formal agreement to participate in the formation of the Joint Ukraine-Russia-Kazakhstan Fuel Enterprise, Ltd. to orchestrate a trilateral service exchange. The latter called for Ukraine to supply uranium and zirconium, Kazakhstan to produce uranium fuel pellets, and Russia to use the fuel pellets and zirconium to manufacture fuel assemblies for Ukrainian power plants. The three parties agreed to own equivalent stakes in the trilateral venture, contribute equally to the charter $450,000 statutory fund, and share revenues generated from Ukrainian nuclear power sales. The deal benefited each party: it preserved Russia's export market niche in Ukraine; offered a viable plan to pay for Ukrainian uranium and to support the acquisition of nuclear fuel for its power plants; and presented an opportunity for Kazakhstan to boost orders for its fuel pellets.[115]

On the surface, both deals represented unambiguous victories for Russia's nuclear diplomacy. Russia not only locked in future fuel pellet supplies from Ulba, but managed to acquire a veto over critical administrative issues such as: liquidation of plant assets, introduction of new products and redirection of fuel pellet sales, allocation of funds exceeding 25 percent of the statutory capital, and confirmation of the management strategy for the combine.[116] While Kazakhstan obtained minority stakes spread over three Russian companies that afforded Astana little input on respective plant decision making, Minatom succeeded at concentrating its holdings into a potentially formidable blocking stake in the strategic Ulba plant. Furthermore, Russian front-end nuclear service providers gained additional access to global markets via sales of Kazakh fuel services and Ukrainian nuclear power that complemented their expanded export portfolio in each state.[117] For many Russian observers, these provisions constituted clear evidence that the die was cast for reintegration, with Russia poised to parlay these initial advantages into decisive control over a reconstituted nuclear fuel cycle.[118] In the spirit of zero-sum strategic competition, Minatom officials boasted that the acquisition of equity stakes in the Ulba plant marked a decisive "victory over the American nuclear power industry."[119]

However, Minatom did not dominate the ensuing relationship. In practice, Astana too came out ahead in the commercial deals and did not sacrifice future opportunities for strategic diversification. First, there were specific limits placed on TVEL's golden share in the Ulba plant. Both parties agreed that the veto right would be temporary, and would not have par value or earn dividends. The golden share would be withdrawn after a year once the process of integration was complete, after which TVEL would hold a 34 percent stake in the plant.[120] Astana also succeeded at diversifying its footprint in the Russian nuclear fuel industry, and at hedging against future competition. In particular, Kazatomprom received a minority share in the Novosibirsk Chemical Concentrate Plant that was targeted by Minatom for servicing a larger share of the future domestic fuel fabrication market. Furthermore, Minatom's blocking right pertained only to "certain" issues, and did not compromise Astana's control over reorganization and diversification of the plant's activities. According to the head of Kazatomprom, the share swap did not impede the proposed privatization of the plant or other elements of the domestic uranium industry.[121] Rather, partnership with Russia was heralded as a double victory for the Kazakh uranium industry: the commitment to additional pellet sales to Minatom preserved a "long-term presence" on the Russian market and was contingent on Russian financing; and the equity exchange did not affect exports of fuel rods and beryllium used in the West. Despite the intimate association with Minatom, Kazakh authorities remained free to pursue international opportunities that would allow Ulba "to avoid being held hostage to the monopolism of a single consumer."[122]

Similarly, the trilateral venture offered little more than a commercial opportunity for Russia. To the extent that one party gained disproportionate influence over the fate of the deal it was Ukraine. Financing for the mutual exchange depended ultimately on the solvency of Ukrainian customers and the government's capacity to redress the large-scale nonpayments problems in the domestic electricity sector. The terms of the venture rested on the Ukrainian government's allocations to a special fund to promote the national fuel cycle industry. But because of persistent shortfalls in monetary payments, contributions to the fund were well below the $800 million originally considered a prerequisite for sustaining the trilateral venture.[123] At the same time, competition between Minatom and the Russian electricity monopoly for external markets played to Kiev's advantage. By pressing the case for front-end nuclear service reintegration, Minatom was presented with the difficult position once again of incurring nonpayments problems with Ukraine that even the Russian electricity monopoly was loath to assume.[124] Consequently, shortly after the venture was established Minatom officials downplayed its near-term potential, and stepped-back from the grandiose visions of a strategic alliance in the sector. Frustrated by the obstacles imposed by its foreign partners, the president of TVEL quipped that "we would like the company not only to supply products of others, but also to encourage larger uranium and zirconium output, draw investments and promote cooperation."[125]

Conclusion

The combination of its marginal standing in global front-end nuclear fuel cycle markets and Minatom's clearly delineated regulatory authority provided Moscow with mixed leverage to reinstate commercial nuclear relations with Kyrgyzstan and Kazakhstan. On the one hand, the lack of robust market power prevented Moscow from altering the opportunity costs of reintegration. Both states faced potentially more attractive and lucrative opportunities for expanding respective residual uranium industries. On the other hand, concentrated and clearly demarcated authority to shape commercial opportunities for domestic uranium producers and fuel service providers strengthened Minatom's hand at influencing the risks of cooperation. This enabled Minatom to tailor negative and positive inducements to each target's risk-taking propensity and hedging strategy. The result was that the Russian government made mutually beneficial commercial cooperation more appealing as a lucrative gamble for Bishkek and a safe bet for the more risk averse Astana. As predicted, these outcomes were in effect pragmatic commercial ventures between old partners that contrast starkly with the effects of Moscow's gas and oil diplomacy.

Part III

Conclusion

7

❖

Conclusion

Theoretical and Policy Implications

This book began by identifying the empirical puzzle of Russia's mixed success with energy diplomacy as a challenge to the conventional study of statecraft and globalization. Relative, interdependent, and structural power arguments by themselves are unable to explain Russia's variable effectiveness at exploiting advantages across sectors and states. These theories, rooted in expected utility calculations, cannot account for the costly and risky decisions made by small supplier states of Eurasia either to embrace or resist Moscow's preferred arrangements for energy security. The varied compliance in bilateral, regional, and strategic energy relations, and divergent international behavior exhibited by Russian energy firms, also do not conform to projections based on the static weak state-strong state balance of institutional strength between Russia and the southern NIS, respectively. Furthermore, different responses to Moscow's diplomacy across sectors, and shifts in energy security cooperation among target states pose problems for contemporary arguments that attribute causal weight to constructed national beliefs and identities among the NIS.

The variable effectiveness of Russia's energy statecraft also challenges claims that globalization has fundamentally reduced the scope, if not primacy, of interstate security relations. As evidenced by the case of Russia's gas diplomacy, and to a lesser extent commercial nuclear state-craft, globalization is not a "one-way street" leading to the erosion of sovereign stewardship of strategic interests. Rather, increasingly interdependent and networked relations create new vulnerabilities and uncertainty for extra- and nonstate actors as well. States, even domestically weak ones such as Russia, can prey on these vulnerabilities to influence trans- and subnational actors that are the primary suppliers, customers, or financial backers of strategic goods to do their foreign bidding. The battle for control over highly interdependent nonstate actors expands the scope of competition among states, with implications for the efficacy of sanctions and other nonmilitary instruments of international security policy.

Drawing on insights from prospect theory, neo-institutionalism and soft power, I advanced a theory of "strategic manipulation" to explain the effectiveness of Russia's energy statecraft while offering a preliminary cut at redressing these broader analytical gaps. This perspective stretched the domain of statecraft beyond discrete coercive dyads to include alternative methods of exploiting strategic choices and interdependent relations. Whereas traditional models of coercion and persuasion entail altering payoffs or probabilities by threatening/offering punishment/rewards or forcibly reversing target behavior, strategic manipulation involves (re)-structuring and exploiting a target's decision-making situation so that it values alignment choices that it would not make otherwise. The focus is on affecting how a target frames a choice problem and calculates gains and losses, altering the opportunity costs and relative riskiness of policy options to maximize the appeal of a favorable outcome or minimize the appeal of an unfavorable one. Understanding Moscow's manipulative diplomacy required research into Russia's market power and the allocation of regulatory authority in different energy sectors, as well as into different national assessments of strategic energy trajectories and the riskiness of choices confronted by respective post-Soviet target states.

Chapters 4 through 6 illustrated my argument in 6 critical cases of Russia's mixed success at reintegrating rival southern NIS suppliers across the natural gas, oil, and commercial nuclear energy sectors from 1992 to 2002. Although the cases alone do not conclusively validate the theory of strategic manipulation developed here, the overarching trends are compelling. Chapter 4, for example, demonstrated that Russia's gas diplomacy, which was buoyed by significant global market stature and a transparent domestic regulatory structure, succeeded remarkably by molding both the opportunity costs and risks of compliance for Turkmenistan and Kazakhstan that otherwise pursued very different strategic agendas. Conversely, chapter 5 revealed that the combination of marginal market power and opaque domestic institutions deprived Moscow from exploiting its vast oil resources to coerce or induce strategic compliance from Azerbaijan and Kazakhstan. Finally, chapter 6 identified how Russia used clearly delineated state authority in the nuclear sector to compensate for shortcomings in front-end fuel service markets with mixed results, striking mutually beneficial commercial contracts with both Kyrgyzstan and Kazakhstan that were conspicuously bereft of neo-imperial implications.

This chapter addresses three outstanding objectives. The first is to highlight the main theoretical findings of the study. The second is to review the consequences of strategic manipulation for extending our thinking about the domain of coercive diplomacy and the interface between globalization and statecraft, as well as for integrating the element of risk into alternative political theories of foreign policy and international security. The final challenge is to trace the policy implications of my argument for engaging and contending with Russia in future energy security arrangements in Eurasia.

Findings

My central finding is that under the right conditions states can secure strategic deference from interdependent rivals by manipulating risks rather than by directly altering the costs and benefits of specific policy options. The analysis of energy diplomacy generally supports the descriptive insight from prospect theory that decision makers respond differently to expected losses and gains in their state's relative power and security. But it goes beyond these claims to demonstrate that decision frames and risk-taking propensity can be effective objects of statecraft. Specifically, I contend that altering the range of options and gap between expected positive and negative outcomes of respective options can be more salient to the decision-making calculus of states than either improving the credibility of threats or accentuating potential costs or benefits of resistance. This finding represents a challenge to prevailing models of coercion and persuasion that hinge success or failure of statecraft on exploiting information asymmetries and a target's expected utility calculations.

I also find that the domestic institutions related to a leadership's administrative control over subnational behavior are critical to effective statecraft. The key to manipulating risks turns on the transparency and specificity of domestic regulatory authority that clarify interests and accountability within a governing structure, allowing statesmen to send precise signals that make mobilizing national resources and compliance among implementing agents and nonstate actors more rewarding. This capacity affords a state the ability to narrow or widen the variance between positive and negative values of available policy options consistent with its preferred outcomes and the target's risk-taking propensity. Accordingly, a leadership can bolster the value of favored alignment options without credibly threatening or bludgeoning compliance at home or abroad. Conversely, in the absence of clearly delineated regulatory authority, manipulative bargaining strategies are generally subverted from below by the independent actions of subnational actors that can shirk national security directives and pursue parochial interests without incurring hefty commercial or political costs. This finding challenges other domestic institutionalist arguments that implicate factors related either to state autonomy vis-à-vis society or to the concentration of national authority for the success of nonmilitary forms of statecraft. Both of these traditional explanations ignore the fact that incentives for compliance within a regulatory system can be intensified via the exercise of complementary authority as opposed to costly enforcement mechanisms.

A related finding is that a target's risk-taking orientation is distinct from its propensity to comply with another's statecraft. A target can gamble on compliance or defiance, as well as settle on respective outcomes depending on the range of risky options and the ability of other states to alter these values. Accordingly, one must first identify the variance between positive and negative outcomes associated with specific policy options; then embed assessments of risk-taking into specific theories of statecraft, identifying conditions under which states can more or less alter the expected variance of positive and negative outcomes.

These findings were demonstrated in Russia's energy statecraft toward southern NIS suppliers. Although Russia enjoyed significant, credible, and multidimensional power advantages, efforts to compel or persuade target compliance fell flat and in several instances were counterproductive to Moscow's strategic ambitions. In two unambiguous cases of success, it was only when the Russian government tapped its market and discrete regulatory advantages in the gas sector to alter the risks of viable options that compliance readily followed from targets that were otherwise confronted with more cost-effective options for diversification. Moscow exploited Turkmenistan's deteriorating confidence in the status quo, preserving the likelihood that Ashgabat would incur significant losses by pursuing extra-Russian pipeline options while coaxing the powerful domestic gas lobby to improve the terms of deepening reliance on the Russian pipeline system. As a result, compliance became a lucrative gamble. By the same token, Moscow manipulated strategic deference as a safe bet for the increasingly risk averse Kazakh leadership. Moscow exploited discrete domestic pricing authority in a manner that encouraged Gazprom to offer Astana successive short-term tangible benefits for using Russian pipelines and jointly exploring Northeast Asian markets. This enticed Astana into a series of incremental concessions regarding the development and export of gas that had the cumulative effect of locking the cautious Kazakh leadership into long-term dependence. The results were impressive in both cases, as the two states incurred great sacrifices by becoming junior partners in the Russian dominated Eurasian gas cartel that effectively tied their fate increasingly to Moscow's neo-imperial solutions in the sector, irrespective of the differences in their hedging strategies.

In the nuclear energy sector, Moscow compensated for limited market power and wooed mutually beneficial commercial contracts by manipulating the riskiness of cooperation. In the Kyrgyz case, Russia relied on clearly delineated state authority to proffer joint venture proposals that widened the gap between potential positive and negative outcomes for Kyrgyzstan of reconstituting nuclear fuel relations with Minatom. By approving the delivery of radars and other material inducements in conjunction with renewing equity claims to Kyrgyz uranium processing facilities, Minatom induced subordinate enterprises to take action that distinguished restoration of nuclear fuel relations as a slightly more attractive gamble for the risk-prone leadership in Bishkek. Moscow proved equally adept at exploiting Minatom's streamlined authority to fashion deals aimed at minimizing the downside of commercial cooperation while authorizing equity swaps in Russian nuclear facilities that offered modest but sure gains to Astana from reintegrating nuclear relations. By reducing the risks of collaboration for the conservative leadership in Astana, Moscow displaced potentially more lucrative proposals proffered by foreign rivals and secured commercial access to Kazakh uranium and fuel fabrication facilities.

Conversely, the inability to manipulate the relative riskiness of compliance cost Russia dearly in its oil diplomacy. The analysis of Moscow's Caspian oil politics in chapter 5 illustrates both the futility of Russia's efforts not only to coerce or induce direct compliance, but to alter the variance in expected positive and negative values confronting both Azerbaijan and Kazakhstan. In the former case, the opaque regulatory system militated against effective monitoring and enforcement of the national oil industry that significantly impeded official efforts to raise the stakes of favored joint ventures and pipeline projects for the increasingly risk-prone leadership in Baku. Presented with potentially more lucrative options, Azerbaijan was inclined to press Moscow to make painful concessions and to gamble on strategic export diversification. Admittedly, Moscow's institutional weakness did not preclude cooperation with Kazakhstan in the oil sector. Yet because it lacked the capacity to lower the relative riskiness of bilateral deals for the cautious Kazakh leadership, Moscow was poorly positioned to parlay Astana's openness to reengagement, the resurgence of domestic suppliers, and legacy of structural dependency into effective strategic levers.

Another set of important findings relates to the identification and change of reference points or expectations of targets. The empirical chapters show that perceptions of the status quo and anticipation of the state's strategic energy trajectory are neither divorced from material conditions nor obvious strictly in an idiosyncratic or ex post manner, as suggested by critics of prospect theory.[1] Within each sector, respective target leaderships assessed the status quo and expected deviations consistent with objective trends in production and technical criteria commonly associated with the political economy of energy development and export, as well as in the political, economic, and strategic landscape. By deriving these factors a priori and testing them against the record of stated expectations by respective leaderships, the findings support claims that prevailing reference points can be systematically examined.[2]

Similarly, the findings reveal that decision domains are malleable. In several episodes, the benchmarks and expectations of target leaderships changed consistently with the evolving technical and market developments in respective energy sectors. Furthermore, market opportunities were subject to manipulation by foreign states. In the case of Turkmenistan, Russia prompted a frame change that effectively lowered the opportunity costs of compliance for Ashgabat. By saturating regional markets and foreclosing potential options for Ashgabat, Russia expedited the shift from the domain of gains to losses. By contrast, Russia's lack of decisive market power in the oil sector not only inhibited Moscow's efforts to constrain Azerbaijan's evolving options, but prevented it from seizing upon mounting anxiety in Baku. There was little that Russia could do either at home or abroad to stem new oil market opportunities presented to the increasingly risk-prone Azerbaijani leadership.

The study also confirms that Russia remained a potent strategic player in the Eurasian energy equation throughout the tumultuous first decade following the Soviet collapse. Despite the multifaceted shock of transition, loss of superpower standing, weakened state capacity, and relegation to near supplicant status within the global economic order, Russia preserved its stature as regional hegemon. This was reflected in all traditional indices of absolute, relative, structural, and institutional power vis-à-vis the southern NIS. However, these advantages did not uniformly translate into influence. Not only did Russia's strategic energy leverage vary considerably across sectors, but Moscow was not always cognizant of how it affected the opportunity costs and risks of target states. Over the period of study, Moscow's energy diplomacy often was incoherent and featured simultaneously various elements of persuasion and manipulation. Yet Moscow was more or less successful in achieving its preferred bilateral, regional, and strategic energy security objectives when it either inadvertently or shrewdly manipulated the choices of southern NIS targets. Irrespective of the uniform constriction of its global reach, Russia was able on key occasions to exert extraterritorial influence over domestic firms and the hedging strategies of regional targets along its southern periphery.

Extending Theoretical Debates

What are the implications of this study for theories of statecraft and coercive diplomacy? This book explored cases involving a domestically weak but globally challenged state and domestically strong target states because those cases reduced the significance of relative power and regime types, and put more emphasis on the political dynamics of exercising leverage across different energy markets and regulatory mechanisms. Presumably for states that are stronger at home and better situated to shape the contours of globalization, strategic manipulation should be both a more accessible and effective form of statecraft. Similarly, target states with less hierarchical governance structures should be more vulnerable to strategic manipulation, with less capacity to avoid or pass down the attendant political costs of strategic compliance or defiance. Because expectations for the effectiveness of strategic manipulation should not change by applying the argument to a broader range of cases, the findings presented here offer implications that extend beyond Russia's energy shadow in the southern NIS and into the realm of theory building.

First, scholars should continue to expand the domain of statecraft to keep pace with the phenomenon of globalization. The core finding of my research indicates that national leaderships can instrumentally affect each other in ways that exceed the parameters of conventional coercive and persuasive diplomacy. As national interests become increasingly interdependent and national authority becomes increasingly circumscribed, states will be vulnerable across a wider and more intimate array of strategic issues. This opens up new dimensions for statecraft that complement, and may even supplant, traditional thinking about the mechanisms of coercion, deterrence, persuasion, and inducement. The focus on strategic manipulation extends this exploration to indirect and noncrisis scenarios that figure prominently among highly interdependent actors. It draws special attention to how states can get others to take desired action by targeting policies within a network of relationships, at home and abroad, rather than by concentrating policies exclusively on altering specific dyads of interaction. Accordingly, the book complements research on soft power and soft security by featuring elements of "soft diplomacy," illuminating how a state can actively and deliberately influence another as a byproduct of interdependent actions taken at home and towards a third party. The insight into how market power and clearly delineated domestic regulatory authority play into efforts to mold a target's risk-taking calculus, however, is by no means exhaustive of the possible dimensions and mechanisms for accessing soft power for purposes of statecraft. Given the explosion of attention to soft power and globalization, I hope that my results provide an analytical springboard for systematically exploring the influence effects of the many subtle forms of strategic interaction among highly interdependent states.

Simply identifying an interface between soft power, security, and diplomacy, however, is not enough. The challenge ahead is for scholars to specify the conditions under which soft diplomacy is more likely to prevail, and the precise behavioral consequences it should engender. I pursued this line of inquiry by highlighting both elements of manipulative energy diplomacy and the continuum of compliance pertaining to bilateral, regional, and strategic energy security. Future work could build on this analysis by identifying other material and even ideational dimensions to interdependence that can be actively molded via indirect and preemptive action to alter the value of compliance for a target. This could include analysis of both "easier" cases—involving the use of alternative economic tools by states that wield greater power in global markets and networks, as well as "harder" cases—involving military or normative instruments used by institutionally weaker states than Russia. By addressing a broader range of cases we could begin to develop an empirical boundary for understanding the strategic effects of soft diplomacy as practiced on low- and high-risk issues and by states with different market and institutional profiles.

A promising first step along these lines may come from expanding analysis of supplier state interaction to statecraft among asymmetrically interdependent states. Russia's posture toward highly dependent western NIS consumers is potentially illuminating in this regard, given the variation in modalities and intensity and effectiveness of Moscow's energy statecraft. As demonstrated by others, the divergent energy security choices made by Belarus, Ukraine, Moldova, Georgia, and the Baltic states following the Soviet collapse can in part be explained by different strategic predicaments and levels of vulnerability, alternative national interpretations of autonomy, and distinct domestic strategies for managing energy dependency on Russia.[3] These perspectives alone cannot fully account for Russia's leverage, however, as most observers agree that Moscow's success at strictly altering the expected utility of strategic diversification, reintegration, and bilateral energy relations for these states was mixed, and that such compliance shifted despite the structural dependency and persistent institutional and normative conditions confronting each consumer target. Our understanding of statecraft in this broader set of interdependent relationships, as well as in these other critical episodes of Russia's energy diplomacy, therefore, may be more complete if we take an additional factor—strategic manipulation—into consideration.

The case of Ukraine is potentially illustrative of Russia's variable effectiveness at manipulating energy dependency across sectors. Despite inheriting highly vulnerable gas and oil sectors (in terms of shares and source of energy supply and rents to key political actors), and becoming a primary target of Russia's coercive statecraft, Kiev resisted Moscow's early threats to cut off energy supply and the pressure of reduced deliveries. Notwithstanding Russia's bold attempts to link energy supply to strategic concessions, Kiev refused to capitulate fully to Moscow's terms for controlling the Black Sea Fleet and basing rights in Sevastopol and steadfastly adhered to initial postures of economic independence and strategic diversification.[4] As the national economic, political, and energy situation precipitously deteriorated by early 1993, however, Moscow's gas diplomacy conspicuously manipulated the risks of compliance for the Ukrainian leadership. Together the construction of the Yamal pipeline through Belarus, Poland, and Slovakia—that effectively broke Ukraine's stranglehold over Russian gas exports to Europe—and successive manipulation of the access, volume, and price of gas imported from Turkmenistan, empowered Moscow to accentuate Kiev's descent into the domain of losses and to lower significantly the opportunity costs of deference to favored gas policies. At the same time, Moscow coordinated diplomatic carrots and sticks with Gazprom to present the increasingly risk-prone Ukrainian leadership with options marked by the widest variance in prospective outcomes. By subsidizing gas supply (albeit at fluctuating levels) and accepting bartered transit services irrespective of Ukraine's persistent payments problems, as well as by offering terms that did not jeopardize the opaque distribution of energy rents on which President Leonid Kuchma based his domestic support, Russia offered Kiev potentially the greatest political and commercial upside for managing its gas dependence. Yet the likely negative outcomes were great as well, as Russia consistently showed determination to exact higher prices and unfavorable terms for debt valuation and rescheduling, and to accept lower percentages of bartered goods for gas deliveries than offered to Belarus, Moldova, and Lithuania.[5] Not surprisingly, from 1994 to 2003 Russia effectively stymied diversification and domestic reform in the gas sector, as well as secured critical authority to control deliveries directly to Ukrainian consumers. Gazprom also gained a leg up on reintegrating the regional infrastructure by gradually wrangling debt equity stakes in the latter's strategic gas transit and storage infrastructure.[6]

By contrast, the same set of manipulative diplomacy correlated with only marginal benefits in the oil sector, as Ukraine consistently staved off Russian control of domestic refineries and solicited alternative deliveries via Black Sea and Caspian transit routes that circumvented Russia. From 1999 to 2003, Ukraine actively sought to diversify strategic petroleum relations by playing off Russian and Caspian supplier interests in the Odessa-Brody pipeline to Europe. Although the project floundered by 2004 due mostly to problems with Ukrainian financing, successive government and legislative rulings made it clear that Kiev was not beholden to the interests of its Russian partners.[7] Unlike with gas diplomacy, Moscow's grip was conspicuously subverted by the independent initiatives of Russian oil companies that controlled nearly 50 percent of Ukraine's oil complex. On several occasions, Russian firms overtly confounded Moscow's muscle flexing by conceding directly to Ukrainian demands for higher transit fees, despite the Ministry of Energy's request "that they not do so."[8]

Russia's energy diplomacy toward the even more dependent Belarus broke down along similar sectoral lines, as well as plausibly highlights the residual significance of manipulating alternative decision frames. Though the balance between coercion and inducements in Russia's gas diplomacy was qualitatively different than with Ukraine, it too was successful at securing critical bilateral and strategic concessions. In this case, Russian policies effectively appealed to President Alyaksandr Lukashenko's strategic opportunism that followed from his increasing secure hold on power after his election in mid-1994. By setting energy prices consistently below market rates (at times approximating Russian domestic levels), constructing a new pipeline through Belarus, establishing favorable barter arrangements for Belarusian industrial goods, providing easy credit, and consistently presenting generous terms for debt repayment, Moscow offered modest but predicable benefits to Minsk of sustaining dependence on Russia. Moreover, the likely downside did not loom as large as turning to Western or Caspian suppliers, as Moscow and Gazprom threatened at most to raise prices to the subsidized levels paid by Ukraine and Moldova and to offset reduced deliveries with temporary sales from third-party Russian suppliers. Accordingly, greater reliance on Russia's gas supply became an increasingly safe bet for sustaining the perverse and noncompetitive Belarusian economy, adding meaning to the treaties establishing a "Community of Russia and Belarus," and validating Lukashenko's vision of "virtual integration" to consolidate his authoritarian regime.[9] Consequently, by early 2002 Minsk was prone to relenting to Moscow's "gas blackmail," as Lukashenko publicly conceded to Gazprom a 50 percent stake in the national gas transporter and acquiesced to toeing the Kremlin's line on the substantive terms for economic reintegration. Despite recurrent rows over Gazprom's price hikes, gas transit fees, and the valuation of Belarusian gas assets, Lukashenko made clear that he was ultimately resigned to accommodate Russia. "If Putin wants us to pay this money," he lamented, "let us try to collect it."[10]

Another important set of implications relates to extending the application of prospect theory. For the purposes of systematically constructing and testing the argument for strategic manipulation, I confined the inquiry to the evaluation of specific market conditions, institutional features, decision frames, and risk-taking propensities as they pertained to three energy sectors. This is consistent with the study of issue specific features of statecraft. However, much of the literature focuses on the widespread use of a broad range of economic and political tools of statecraft among interdependent states and firms, and there is no obvious reason to expect that manipulative diplomacy is unique to these three sectors or energy issues more generally. Future research, therefore, might compare decision frames and risk-taking across other sectors of strategic trade or security interdependence. The cumulative effect will be to provide a baseline for assessing the utility of strategic manipulation as a method for advancing grand strategy and expanding the domain of influence over target states.

Similarly, this study advanced the application of prospect theory by probing the implications for strategic interaction among states. Unlike most other applications that confine analysis to the foreign policy consequences of specific decision frames and risk-taking propensities that are otherwise exogenously determined, this inquiry explicitly addressed how one state can take action to alter another's relevant decision domain and risk-taking calculus. A direction for future research might involve extending the analytical insight on frame changes to specific bargaining scenarios, integrating domain and risk into formal models of strategic interaction. Recent advances in game theory, for example, have relaxed the conventional expected utility assumption, and incorporated subjective factors, as well as risk, into formal interactive decision-making models. Adding insights related to loss aversion, perceptions of the status quo, and expected trajectories into these models should illuminate new bargaining strategies that are especially apropos for influencing the contingent decisions by highly interdependent state actors. This would be especially useful for extending analysis of energy diplomacy beyond the traditional focus on supplier-consumer relations to explore the dynamics of strategic bargaining among rival supplier states.

Insight into the capacity of states to leverage authority at home for strategic effect also contributes to the vast literature on the domestic sources of statecraft. Much like other scholars that draw attention to the Janus-faced predicament of statesmen and to the strategic consequences associated with different regime types, I maintain that a necessary factor for shaping the effectiveness of statecraft is to be found in the domestic institutional context of policy making. The character of domestic institutions matters decisively for determining a leadership's capacity to alter the calculus for domestic agents and firms at lining up behind national sanctions and inducements, thereby affecting the value of options presented to a target state. Much like state strength arguments, I consider the outcomes of statecraft to be heavily affected by a bottom-up institutional process that interacts directly with features associated with a state's market or structural power.

Yet this analysis differs with traditional depictions of the mechanisms for strategic influence gleaned from theories of institutional strength and democratic peace. While I consider institutional features related to governing capacity and democracies to have a peculiar impact on the efficacy of statecraft, I draw attention to the effects on the formulation and implementation of a sender's policy without placing emphasis squarely on conventional measures of societal representation, concentration of power, or political competition. Promoting a different take on institutional strength related to the delineation of regulatory authority, the analysis broadens focus from the strategic effects of state autonomy and information asymmetries to explain how one state's domestic institutional composition can alter the relative riskiness of options confronted by another. Here the analysis goes beyond extant institutional arguments by offering explanation for the variable strategic potency of otherwise weak states.

My research also builds on study of the strategic impact of domestic political transparency. In addition to affecting the credibility of threats and the ability to bluff, as demonstrated by others, my analysis shows that transparent and well-specified decisional authority is critical for coherently mobilizing national resources to alter the expected values of positive and negative outcomes for a target.[11] This insight posits an avenue for extending debate over the strategic consequences of democratic diplomacy—involving study of the levels of belligerence, war outcomes, and crisis diplomacy—to examination of indirect, noncrisis and preemptive forms of soft diplomacy practiced among highly interdependent states.

Furthermore, this study illuminates the need for developing theories of statecraft that treat sanctions and inducements as complementary forms of statecraft. The extant literature has been preoccupied with demonstrating the relative efficacy of sanctions versus inducements, and with specifying the conditions under which one should work best. Yet most empirical studies reveal that states extend carrots and sticks simultaneously to advance strategic interests. My research suggests that the most important question at this juncture may be not whether or when one or the other works best, but how they can be instrumentally used together to influence the different strategic orientations of a target. Both forms of statecraft can work under the same conditions if tailored to manipulating a target's decision frames and the relative riskiness of its policy choices. In Russia's manipulative gas and commercial nuclear diplomacy, for example, sanctions and inducements constituted two halves of the same nut that were used in tandem to maximize or minimize the variance in potential outcomes rather than to punish, foreclose, or reward a target's policy choice. Regardless of the different and shifting strategic dispositions among the Turkmen and Kazakh leaderships, the availability of alternative policy choices, or the high concentration of private interests in the domestic gas sector, Moscow wielded both diplomatic carrots and sticks to gain support for favored Eurasian gas policies. The challenge for future research is to examine additional circumstances under which other forms of economic and political sanctions and inducements can be used as complements for guiding a target into making desired strategic decisions. Doing so not only will refine our understanding of sanctions and inducements, but should help to bridge gaps between theoretical insights and the actual practice of statecraft.

Moving Beyond the Great Game

Finally, my research speaks directly to the contemporary policy debate over contending with the "Russian factor" in the competition for Caspian energy. On the one hand, there are "great gamers" who discount the strategic salience of globalization and view the energy contest through the lens of classic geopolitical competition. Because of Russia's regional preponderance, continued dominance of bilateral trade flows across the NIS, and near-monopoly over the existing Eurasian pipeline infrastructure, Moscow is presumed to be intent on and well positioned to impose neocolonial solutions to regional energy security. Strategically situated to reclaim proprietorship over the region's asset specific energy infrastructure and to forge a "rejectionist front" with Iran, Russia purportedly stands poised to impose decisions for Eurasian energy and pipeline security as critical steps towards boosting Russia's economic dynamism and global competitiveness. Great power enthusiasts and "statists" of different stripes contend that "relevant" power advantages at the regional level stoke an impulse to consolidate control over the residual Soviet energy sector and to balance outside interference in Eurasia at the same time that Russia's weakness vis-à-vis the West compels it to acquiesce at the global level. Accordingly, outside powers, such as the United States, are expected to treat Moscow as a rival in commercial ventures, contain its overbearing influence, and encourage energy diversification among regional supplier states.[12]

Critics counter by emphasizing the embedded obstacles to national assertiveness that derive from increasing levels of globalization and privatization in the world economy. The play for Eurasian resources essentially boils down to commercial competition among energy companies and financial interests, as opposed to geopolitical rivalry among states. Moscow's competence to orchestrate reintegration of regional energy relations supposedly has been compromised by weak and fragmented policy-making. Some attribute this mainly to the capture of Russian politics by a parochial, market-oriented and internationally engaged energy lobby.[13] Others ascribe Moscow's waning influence over regional energy security primarily to bureaucratic infighting, Russia's mounting infrastructural problems and limited power in global energy markets, and the rising confidence of independent-minded regional leaders.[14] These policy analysts discount the relevance and effectiveness of Moscow's energy statecraft, and remain ambivalent about it participating constructively in the all-azimuth energy security arrangements gathering momentum in the former Soviet south.

My argument for the manipulative dimensions to statecraft, however, suggests that both policy strains in this debate are off the mark. The fixation on traditional gamesmanship between great powers and small states captures neither the variation in Russia's control of regional energy and pipeline politics nor the consequences of different target strategic orientations and hedging strategies. Physical control by itself guarantees little, as the value of control is determined by the interaction of multiple state and nonstate actors in global markets. The current game is no longer "great," as the contest now turns on the behavior of "weak" states and assertive subnational actors. Though it is clear that globalization has elevated the political salience of private energy interests and market considerations, it also has created new sources of vulnerability and uncertainty for firms and states alike that can be manipulated by Moscow for strategic effect. Hence, alternative presumptions that Russia has either too little or too much leverage in the region to be a constructive partner are both over- and understated, respectfully. Such views not only fail to appreciate the softer dimensions to Moscow's strategic behavior, but miss the subtle opportunities for engaging and contending with Russia's leverage over Eurasian energy politics.

On the one hand, the relative weakness of Russia's manipulative diplomacy and viable opportunities for diversification in the oil sector facing both Azerbaijan and Kazakhstan point both to the limits of Moscow's control and to the strong Russian domestic interests that constrain effective energy coercion. This development suggests that in the short run, there is no compelling strategic or economic rationale directly related to the balance of power in the Caspian Basin for extending concessionary terms to woo Russian participation in international oil consortia in the region. The costs of granting preferential treatment to Russian oil interests are not warranted, given Moscow's inability to dictate or obstruct decisively Caspian oil politics. In the long run, however, there are potential strategic windfalls to including Russian oil firms in promising commercial ventures. By engaging these interests in joint energy exploration and export projects, the international community can raise the opportunity costs of Russian retrenchment and further isolate the domestic advocates of malevolent policies toward Eurasia. Over time, these practices also should raise the costs of stalled reform initiatives in the domestic energy sector. This ultimately could bolster the standing of those in Russian politics advocating the passage of production sharing agreements, legislation to increase foreign pipeline access, and domestic energy price reforms designed to open up the national oil sector to foreign investment on commercially competitive terms. This, in turn, could make an important contribution to energy security by increasing competition among world oil producers and lowering the percentage of global oil production subject to disruption.

Conversely, the relative ease of exercising imperial control over regional gas issues vividly demonstrates that talk of Russia's strategic irrelevance is premature. There are real limits to how far vulnerable states like Turkmenistan can stray from a strategic path approved by Russia. Moreover, Russia's regional dominance undermines the commercial feasibility of diversifying the sources of Caspian gas in already crowded world markets, such as those in Europe. Given the "tyranny of distance" and specter of incremental competition in the gas sector, Europe's increasing reliance on gas piped from Russia risks undermining the practical meaning of the EU's "energy corridor" concept and effectively stranding deliveries from faraway places, such as Turkmenistan and Kazakhstan. Therefore, the challenge for Eurasian supplier and transit states, as well as extraregional investors and consumers, is to craft alternate gas exploration and export deals that do not provoke Russia but that induce the participation of Russian gas interests. This can be accomplished, for example, by easing the political and economic costs of jointly exploring emerging markets in Asia and possibly Central Europe. A reorientation of Eurasian suppliers to markets with the potential for unlimited and geographically closer demand for gas should not impinge on Gazprom's position in established markets and should reduce the opportunity costs to the firm of acquiescing to regional diversity and cooperation. Thus, a key to unlocking Eurasian energy is for international actors to pay close attention to the mechanics of Russia's strategic manipulation on oil and gas issues, and to pursue regional policies that ensure that Russia's stakes in mutually beneficial projects exceed the opportunity costs of cooperation.

Appendix

Turkmenistan Gas Routes

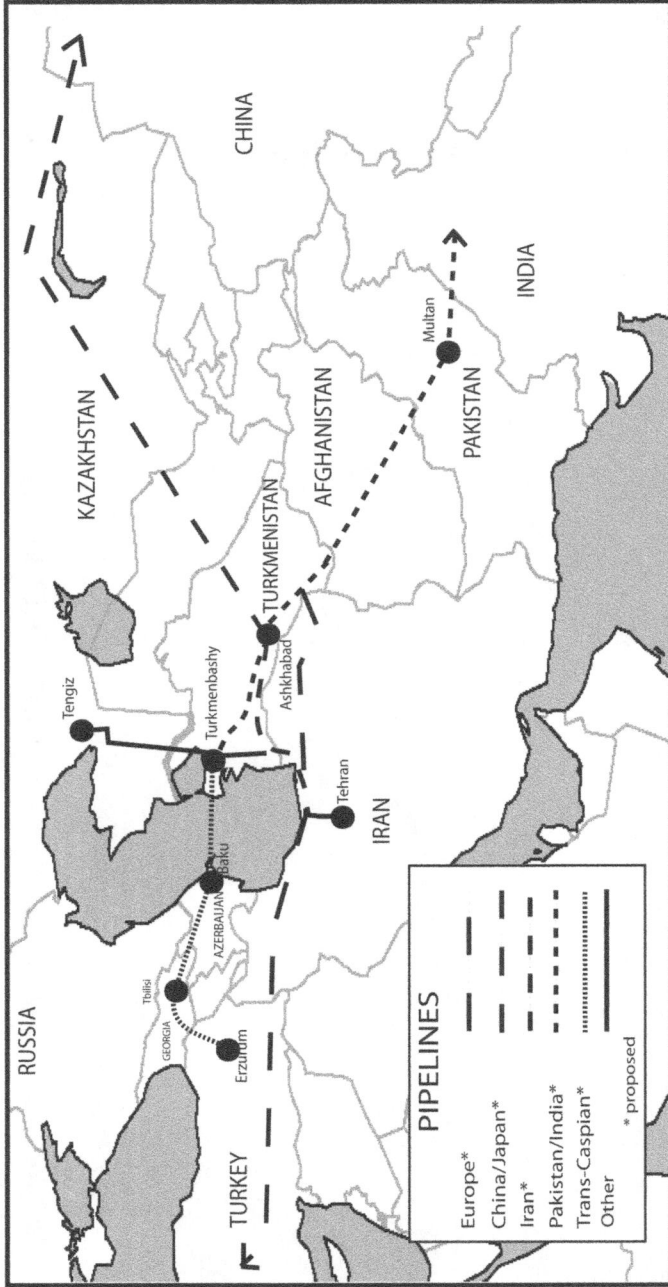

Map Outline Source: 2002 Houghton Mifflin Company

Kazakhstan Gas Routes

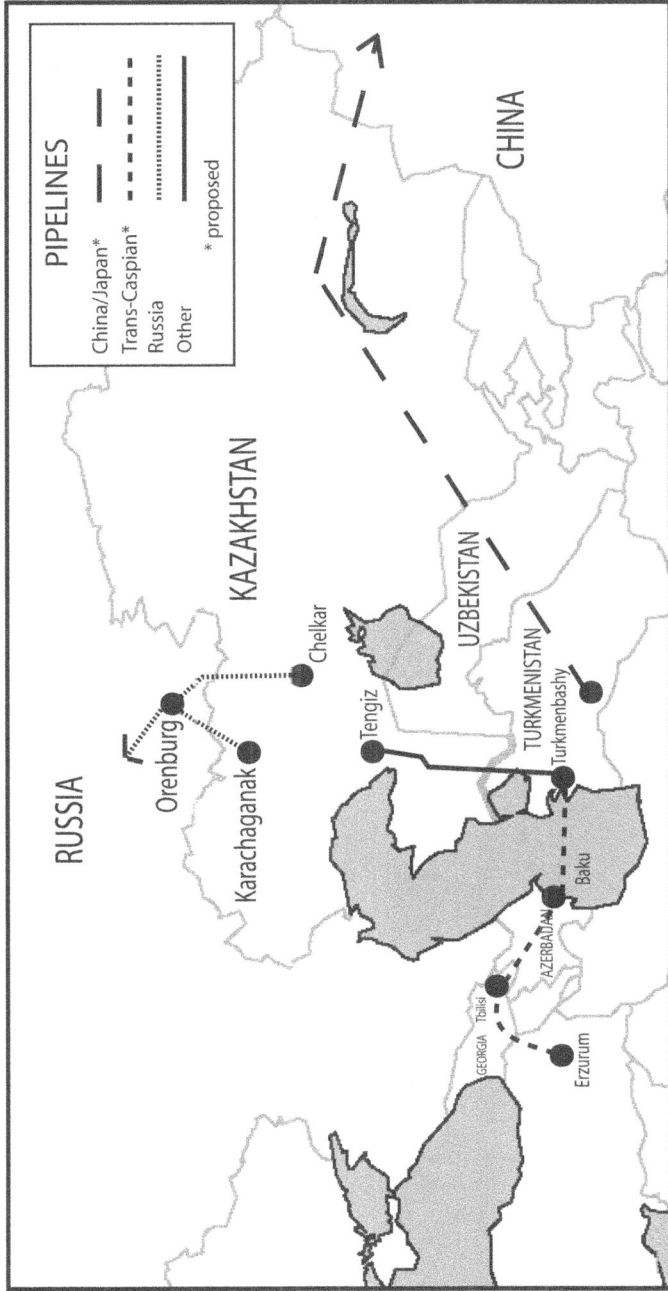

Map Outline Source: 2002 Houghton Mifflin Company

Azerbaijan Oil Routes

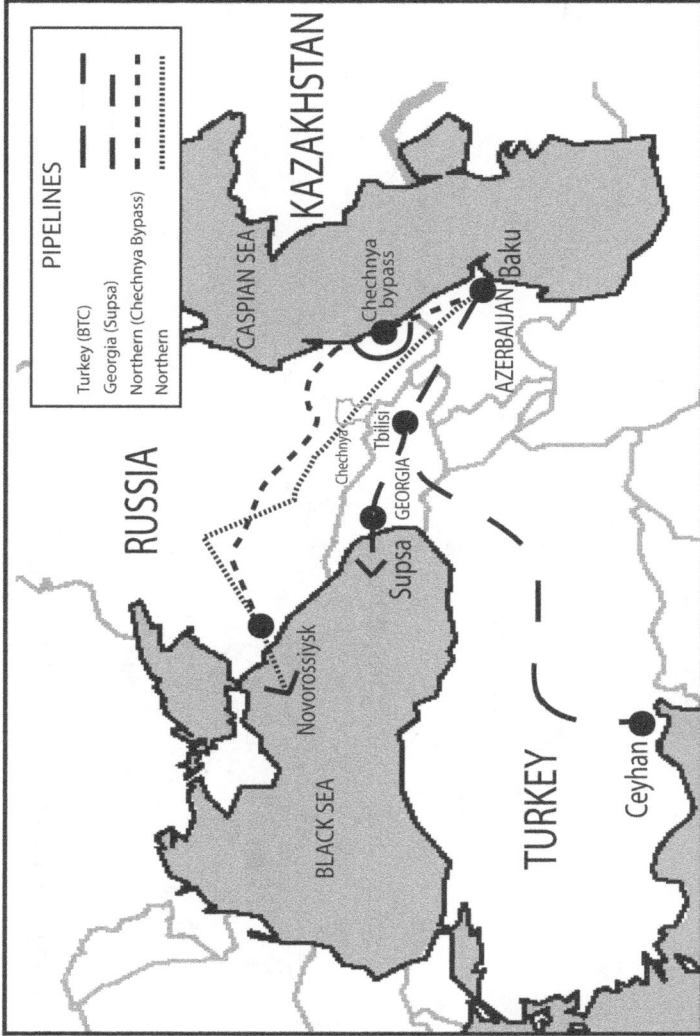

Map Outline Source: 2002 Houghton Mifflin Company

Kazakhstan Oil Routes

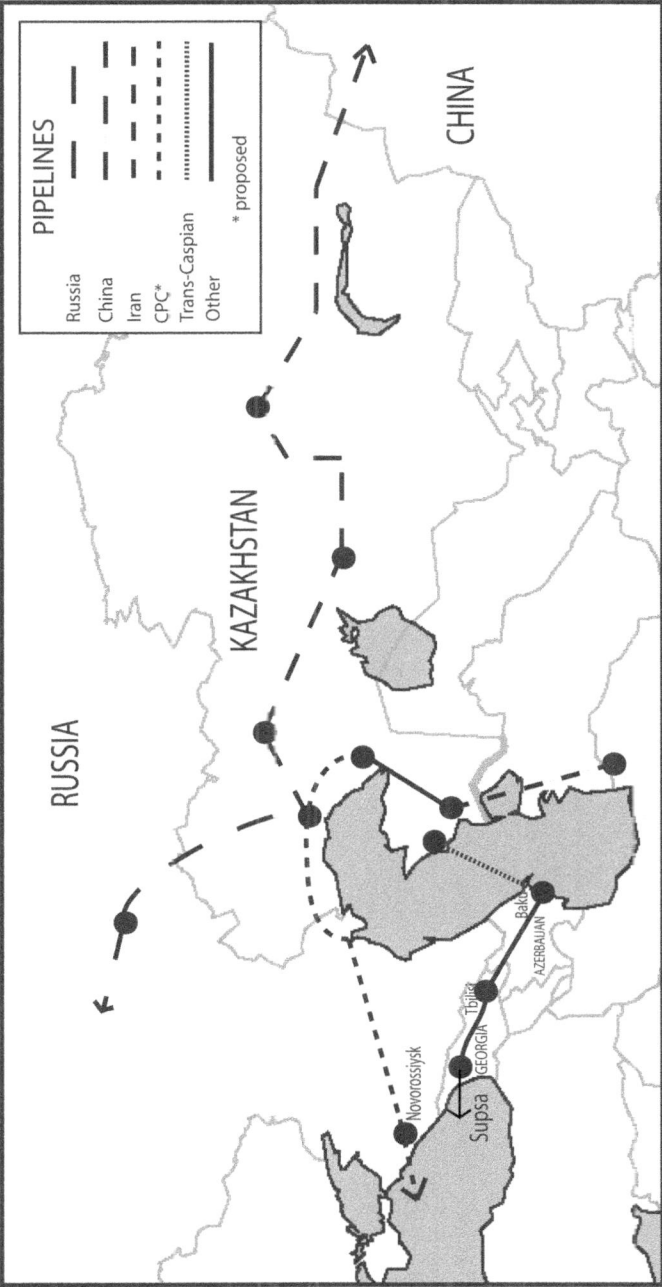

Map Outline Source: 2002 Houghton Mifflin Company

Notes

Introduction

1. Meghan L. O'Sullivan, *Shrewd Sanctions: Statecraft and State Sponsors of Terrorism* (Washington, DC: The Brookings Institution Press, 2003); George E. Shambaugh, "Globalization, Sovereign Authority and Sovereign Control over Economic Activity," *International Politics* 37 (December 2000), pp. 403–431.
2. Douglas R. Bohi and Michael A. Toman, *The Economics of Energy Security* (Boston: Kluwer Academic Publishers, 1996), p. 1; Dag Harald Claes, *The Politics of Oil-Producer Cooperation* (Boulder: Westview Press, 2001). For an exception, see Steve A. Yetiv, *Crude Awakenings: Global Oil Security and American Foreign Policy* (Ithaca: Cornell University Press, 2004).
3. Susan Strange, *The Retreat of the State: The Diffusion of Power in the World Economy* (Cambridge: Cambridge University Press, 1996); Lars S. Skalnes, *Politics, Markets, and Grand Strategy* (Ann Arbor: University of Michigan Press, 2000); Jonathan Kirshner, *Currency and Coercion: The Political Economy of International Monetary Power* (Princeton: Princeton University Press, 1995); Daniel W. Drezner, *The Sanctions Paradox: Economic Statecraft and International Relations* (Cambridge: Cambridge University Press, 1999); George S. Shambaugh, *States, Firms, and Power: Successful Sanctions in United States Foreign Policy* (Albany: State University of New York Press, 1999); Dale C. Copeland, "Economic Interdependence and War: A Theory of Trade Expectations," *International Security* 20:4 (1996), pp. 5–41; Peter Liberman, "The Spoils of Conquest," *International Security* 18:2 (1993), pp. 125–153; and Paul Papayoanou, "Economic Interdependence and the Balance of Power," *International Studies Quarterly* 41 (1997), pp. 113–140.
4. Stephen D. Krasner, *Sovereignty: Organized Hypocrisy* (Princeton: Princeton University Press, 1999); Yale H. Ferguson, and R.J. Barry Jones, eds., *Political Space: Frontiers of Change and Governance in a Globalizing World* (Albany: State University of New York Press, 2002); Robert O. Keohane, and Joseph S. Nye, "Governance in a Globalizing World," in Robert O. Keohane, ed., *Power and Governance in a Partially Globalized World* (New York: Routledge, 2002), pp. 193–218.

235

5. Robert Gilpin, *Global Political Economy: Understanding the International Economic Order* (Princeton: Princeton University Press, 2000); Shambaugh (2000); David M. Rowe, *Manipulating the Market: Understanding Economic Sanctions, Institutional Change, and the Political Unity of White Rhodesia* (Ann Arbor: University of Michigan Press, 2001); Joseph S. Nye, *The Paradox of American Power* (New York: Oxford University Press, 2002); Katherine Barbieri, "Economic Interdependence: A Path to Peace or a Source of Interstate Conflict?" *Journal of Peace Research* 33:1 (1996), pp. 29–49; and Edward D. Mansfield and Jon C. Pevehouse, "Trade Blocs, Trade Flows, and International Conflict," *International Organization* 54:4 (Autumn 2000), pp. 775–808.

6. Marian Kent, *Oil and Empire: British Policy and Mesopotamian Oil, 1900–1920* (London: MacMillan, 1976).

7. Charles K. Ebinger, *The Critical Link: Energy and National Security* (Cambridge: Ballinger Press, 1982), p. xx.

8. Dag Harald Claes, *The Politics of Oil-Producer Cooperation* (Boulder: Westview Press, 2001), pp. 96–97.

9. For contending views, see especially Michael T. Klare, *Resource Wars* (New York: Metropolitan Books, 2001), pp. 1–50; Yetiv (2004), pp. 15–16; and Hendrik Spruyt and Laurent Ruseckas, "Economics and Energy in the South: Liberal Expectations versus Likely Realities," in Rajan Menon, Yuri E. Fedorov, and Ghia Nodia, eds., *Russia, the Caucasus and Central Asia* (Armonk, NY: M.E. Sharpe, 1999), pp. 87–117.

10. On U.S. post–Cold War statecraft, see O'Sullivan (2003), pp. 11–44.

11. Shambaugh (1999).

12. Erik Gartzke, Quan Li, and Charles Boehmer, "Investing in the Peace: Economic Interdependence and International Conflict," *International Organization* 55:2 (2001), pp. 391–438.

13. Kenneth A. Rodman, *Sanctions Beyond Borders: Multinational Corporations and U.S. Economic Statecraft* (Lanham: Rowman & Littlefield Publishers, 2001).

14. This occurred despite standard projections of initial price spikes.

15. "Relevant" power comes from Brenda Shaffer, "U.S. Policy toward the Caspian Region: Recommendations for the Bush Administration" (Caspian Studies Program, Harvard University), available at www.harvard.edu/ksg/bcsia.

16. Neil MacFarlane, "Realism and Russian Strategy after the Collapse of the USSR," in Ethan B. Kapstein and Michael Mastanduno, eds., *Unipolar Politics: Realism and State Strategies after the Cold War* (New York: Columbia University Press, 1999); and Allen C. Lynch. "The Realism of Russia's Foreign Policy," *Europe-Asia Studies* 53:1 (2001), pp. 7–31.

Chapter 1

1. William M. Reisinger, *Energy and the Soviet Bloc: Alliance Politics after Stalin* (Ithaca: Cornell University Press, 1992); and Valerie J. Bunce, "The

Political Economy of the Brezhnev Era," *British Journal of Political Science* 13 (January 1983), pp. 129–158.

2. *Izvestiya,* 26 November 2003. See discussion in Andrei Konoplyanik, *Kaspiiskaya neft': na evraziiskom perekrestke* (Moscow: IGiRGI, 1998), pp. 68–71; and Timothy L. Thomas, "Russian National Interests in the Caspian Sea," *Perceptions* 4:4 (December 1999–February 2000), pp. 75–96.

3. Though realists draw attention to differences between Russia's regional versus global influence, they typically overlook Moscow's mixed success in the "near abroad." See Allen C. Lynch, "The Realism of Russia's Foreign Policy," *Europe-Asia Studies* 53:1 (2001), pp. 7–31; and Neil MacFarlane, "Realism and Russian Strategy After the Collapse of the USSR," in Ethan B. Kapstein and Michael Mastanduno, eds., *Unipolar Politics: Realism and State Strategies after the Cold War* (New York: Columbia University Press, 1999).

4. Abraham S. Becker, "Russia and Caspian Oil: Moscow Loses Control," *Post-Soviet Affairs* 16:2 (2000), pp. 91–132.

5. See especially Margarita M. Balmaceda, *Explaining the Management of Energy Dependency in Ukraine: Possibilities and Limits of a Domestic-Centered Perspective* (Mannheim, Germany: Working Paper 79, Mannheimer Zentrum fur Europaische Sozialforschung, 2004), available at www.mzes. uni-mannheim.de.

6. Andrei P. Tsygankov, *Pathways after Empire: National Identity and Foreign Economic Policy in the Post-Soviet World* (Lanham: Rowman & Littlefield, 2001); and Rawi Abdelal, *National Purpose in the World Economy: Post-Soviet States in Comparative Perspective* (Ithaca: Cornell University Press, 2001).

7. David A. Baldwin, *Economic Statecraft* (Princeton: Princeton University Press, 1985); and Lawrence Freedman, "Strategic Coercion," in Lawrence Freedman, ed. *Strategic Choice* (Oxford: Oxford University Press, 1998), pp. 20–23. This is distinguished from multilateral efforts, where a group of states strives to coordinate respective forms of statecraft toward a common objective.

8. While former serves as an instrument of statecraft, the latter represents a structural condition that theoretically governs behavior and does not privilege the role of deliberate policy choice on the part of initiators (states that attempt to influence another) or targets (states that are the object of influence attempts).

9. Baldwin (1985), pp. 8–50. Although the focus here is not on explaining why states choose one policy instrument over another, there is no reason to presume that the logic that informs the effectiveness of one technique could not apply to another.

10. Ibid., p. 32.

11. Ibid., chaps. 2–3, 7–8. For discussion of the fundamental disagreements over the meaning of hegemony, see especially Edward D. Mansfield, "The Concentration of Capabilities and International Trade," *International Organization* 46:3 (Summer 1992), pp. 735–737. Successful leverage

narrowly refers to the achievement of explicitly articulated primary policy objectives, as opposed to fulfilling a hegemon's secondary goals, such as signaling displeasure or communicating action for domestic consumption that are included in Baldwin's broader definition. In recognition that a hegemon can exercise influence "by not taking action," a distinction is made between overt and indirect forms of leverage.

12. Kenneth N. Waltz, *Theory of International Politics* (Reading, MA: Addison-Wesley, 1979); and Jacob Viner, "Power versus Plenty as Objectives of Foreign Policy in the Seventeenth and Eighteenth Centuries," *World Politics* 1:1(October 1948), pp. 1–29. On the impact of conflict expectations on strategic leverage, see especially Daniel W. Drezner, *The Sanctions Paradox: Economic Statecraft and International Relations* (New York: Cambridge University Press, 1999).

13. Robert O. Keohane, "The Theory of Hegemonic Stability and Changes in International Economic Regimes, 1967–1977," in Ole R. Holsti, Randolph M. Siverson, and Alexander George, eds., *Change in the International System* (Boulder: Westview Press, 1980). On the limits to statecraft, see Lisa Martin, *Coercive Cooperation: Explaining Multilateral Economic Sanctions* (Princeton: Princeton University Press, 1992); George E. Shambaugh, *States, Firms and Power* (Albany: SUNY Press, 1999); and Kenneth A. Rodman, *Sanctions Beyond Borders: Multinational Corporations and U.S. Economic Statecraft* (New York: Rowman & Littlefield Publishers, 2001).

14. David A. Baldwin, *Paradoxes of Power* (New York: Basil Blackwell, 1989), p. 207; and Harold and Margaret Sprout, *Men-Milieu Relationship Hypotheses in the Context of International Politics* (Center for International Studies, Princeton University Research Monograph, 1956).

15. Jeffry Frieden, "International Investment and Colonial Control: A New Interpretation," *International Organization* 48:4 (Autumn 1994), p. 570.

16. Comparisons across all major indices of power reveal that the southern NIS experienced far steeper rates of decline than Russia, leaving Moscow with "almost 26 times the GNP, four times as many soldiers, and a defense budget almost 40 times larger than the other eight states *combined*." Rajan Menon, "After Empire: Russia and the Southern 'Near Abroad,'" in Michael Mandelbaum, ed., *The New Russian Foreign Policy* (New York: Council on Foreign Relations, 1998), pp. 104–107.

17. For application to Russia's strategic posture towards the southern NIS, see Henrdrik Spruyt and Laurent Ruseckas, "Economics and Energy in the South," in Rajan Menon, Yuri Federov, and Ghia Nodia, eds., *Russia, the Caucasus and Central Asia: The Twenty-first Century Environment* (Armonk, NY: M.E. Sharpe, 1999), pp. 92–93.

18. On the variety of contractual exchange relations and concessions that Moscow made to settle rival claims to residual energy assets with the NIS, see Alexander Cooley, "Imperial Wreckage: Property Rights, Sovereignty, and Security in the Post-Soviet Space," *International Security* 25:3 (Winter 2000/01), pp. 100–127.

19. Robert O. Keohane and Joseph S. Nye, *Power and Interdependence*, 2nd ed. (New York: HarperCollins, 1989), p. 12; Joseph S. Nye, Jr., "Energy and Security in the 1980s," *World Politics* 35:1 (October 1982), pp. 128–129; and Walter Carlsnaes, *Energy Vulnerability and National Security: The Energy Crises, Domestic Policy Responses and the Logic of Swedish Neutrality* (New York: Pinter Publishers), pp. 10–11.

20. Shambaugh (1999), p. 15. See also P.T. Hopmann, "Asymmetrical Bargaining in the Conference on Security and Cooperation in Europe," *International Organization* 32:1 (1978), pp. 141–178.

21. International Monetrary Fund (IMF), *Direction of Trade Statistics: 1998 Year Book* (Washington, DC: IMF, 1998); and PlanEcon, *Review and Outlook for the Former Soviet Republics* (Washington, DC: PlanEcon, September 1999). A "principal trade partner" is the state with the highest sums of imports and exports. Notwithstanding the decline in bilateral trade flows, Russia's status as principal trade partner increased from four out of eight South Caucasus and Central Asia countries in 1994 to six out of the eight in 1997. Russia's share of total trade for Kazakhstan steadily grew from 31 percent in 1994 to 49 percent in 1996. See Neil MacFarlane, *Western Engagement in the Caucasus and Central Asia* (London: The Royal Institute of International Affairs, 1999); and Richard Pomfret, *Central Asia Turns South?* (London: Royal Institute of International Affairs, 1999).

22. IMF, *Republic of Kazakhstan: Selected Issues and Statistics* (Washington, DC: IMF, March 2000), p. 126. Kazakhstan's annual gas exports to Russia fluctuated between 93–100 percent of its total natural gas exports from 1995–1998. In addition to importing roughly the same amount of oil from Astana as it delivered to oil refineries in Kazakhstan, Russia charged additional rents for its oil exports to Kazakhstan. See Natalia V. Zubarevich and Yuri E. Fedorov, "Russian-Southern Economic Interaction," in Menon, Federov, and Nodia (1999), pp. 120–132.

23. IMF, *Direction of Trade Statistics: 1998 Year Book*; and PlanEcon, *Review and Outlook for the Former Soviet Republics*. Russia's share of Turkmenistan's total trade was less than 10 percent from 1994 to 1997. Russia's share of Azerbaijan's total exports fluctuated between 18 and 22 percent, while its share of Baku's total imports ranged from 13 to 19 percent. By contrast, the combined trade with Turkmenistan and Azerbaijan constituted less than 3 percent of Russia's total trade for the period.

24. MacFarlane (1999), pp. 35–40.

25. For statist criticisms of globalism, see Robert Gilpin, *Global Political Economy: Understanding the International Economic Order* (Princeton: Princeton University Press, 2001); Linda Weiss, *The Myth of the Powerless State: Governing the Economy in the Global Era* (Cambridge: Polity Press, 1998); and Geoffrey Garrett, "Global Markets and National Politics: Collision Course or Virtuous Circle," *International Organization* 52:4 (Autumn 1998), pp. 787–824.

26. Structural power here resembles "soft power," as discussed in Joseph S. Nye, Jr., "Soft Power," *Foreign Policy* 80 (Fall 1990), pp. 153–171. See also James A. Caporaso, "Dependence, Dependency, and Power in the Global System: A Structural and Behavioral Analysis," *International Organization* 32:1 (Winter 1978), p. 23; and Stephen Krasner, *Structural Conflict* (Berkeley: University of California Press, 1985).

27. For a dispositional approach to structural power, see especially, Susan Strange, *The Retreat of the State: The Diffusion of Power in the World Economy* (New York: Cambridge University Press, 1996), pp. 25–27; G. John Ikenberry and Charles Kupchan, "Socialization and Hegemonic Power," *International Organization* 44:3 (1990), pp. 283–315; Peter Bacharach and Morton Baratz, *Power and Poverty: Theory and Practice* (New York: Oxford University Press, 1970); Scott C. James and David A. Lake, "The Second Face of Hegemony: Britain's Repeal of the Corn Laws and America's Walker Tariff of 1846," *International Organization* 43:1 (1989), pp. 1–29; and Rodman (2001).

28. Stephen Gill and David Law, "The Global Hegemony and the Structural Power of Capital," *International Studies Quarterly* 35:3 (September 1991), pp. 313–336; David I. Kertzer, *Ritual, Politics, and Power* (New Haven: Yale University Press, 1988), pp. 179–180; and Alexander Wendt and Daniel Freidheim, "Hierarchy Under Anarchy: Informal Empire and the East German State," *International Organization* 49:4 (Autumn 1995), pp. 689–721. On the impact of national identities on the variation in NIS foreign economic policies, see Tsygankov (2001).

29. Robert Ebel, *Energy Choices in the Near Abroad* (Washington, DC: Center for Strategic and International Studies, April 1997), pp. 21; and Paul J. D'Anieri, *Economic Interdependence in Ukraine-Russia Relations* (Albany: State University of New York Press, 1999), pp. 69–96. This purportedly was accentuated after the 1999–2000 upturn in Russian energy production, as suggested by Fiona Hill, *Energy Empire: Oil, Gas, and Russia's Revival* (London: The Foreign Policy Centre, September 2004). The southern NIS' structural dependency on Russia was not unique to the energy sector. The institutional legacy of the Soviet military fostered "dual loyalties" among senior officer corps across the region, and the default to Russia of responsibilities for patrolling former Soviet borders, coordinating regional air defense, commanding CIS peacekeeping operations, and maintaining forward outposts. See Andrei Zagorski, "CIS Regional Security Policy Structures," in Roy Allison and Christoph Bluth, eds., *Security Dilemmas in Russia and Eurasia* (London: The Royal Institute of International Affairs, 1998), pp. 281–300; Menon (1998), pp. 106–107; and Martha Brill Olcott, Anders Aslund, and Sherman Garnett, *Getting It Wrong: Regional Cooperation and the Commonwealth of Independent States* (Washington, DC: Carnegie Endowment for International Peace, 1999), pp. 77–107. The region's lingering structural dependency on the Russian economy was reflected by the knock on effects of the 1998 ruble devaluation, the prefer-

ential coordinating mechanisms provided by the CIS, and the advantages that accrued to Russia as a "trade bridge" between the NIS, Europe and Asia. See especially Chris Kushlis and Ben Slay, "An Economic Overview of the Caspian Basin." *BCSIA Newsletter* (2000), p. 2; Richard Pomfret, *The Economies of Central Asia* (Princeton: Princeton University Press, 1995), pp. 41–60; John P. Willerton and Geoffrey Cockerham, "Russia, the CIS and Eurasian Connections," in James Sperling, Sean Kay, and S. Victor Papacosma, eds., *Limiting Institutions? The Challenge of Eurasian Security Governance* (New York: Manchester University Press, 2003), pp. 185–207; Abraham S. Becker, "Russia and Economic Integration," *Survival* 38:4 (Winter 1996–1997), pp. 117–136; and Boris Rumer and Stanislav Zhukov, "Economic Integration in Central Asia: Problems and Prospects," in Rumer and Zhukov, eds., *Central Asia: The Challenges of Independence* (New York: M.E. Sharpe, 1998), pp. 103–152. On Russia's "soft power" advantages in the former Soviet south, see especially K.S. Gadziyev, *Vvedeniye v geopolitiku* (Moskva: Logos, 2000); and A.P. Tsygankov, "Mastering Space in Eurasia: Russia's Geopolitical Thinking after the Soviet Break-up," *Communist and Post-Communist Studies* 36 (2003), pp. 117–120.

30. The volume and profitability of Kazakhstan's oil exports fluctuated at the discretion of Russia's "transfer pricing" system for regulating pipeline access. Pomfret (1999), pp. 9–12; and Olcott, Aslund, and Garnett (1999), p. 53.

31. For a discussion of both the high stakes and risks of depending on expected Caspian energy windfalls, see Terry Karl, "Crude Calculations: OPEC Lessons for the Caspian Region," in Robert Ebel and Rajan Menon, eds., *Energy and Conflict in Central Asia and the Caucasus* (New York: Rowman & Littlefield Publishers, 2000), pp. 29–53. On the geoeconomic opportunities that this afforded to Russia, see V. Kolossov and N.S. Mironenko, *Geopolitika i politicheskaya geografiya* (Moskva: Aspekt Press, 2001); and Tsygankov (2003), pp. 111–113.

32. R. Harrison Wagner, "Economic Interdependence, Bargaining Power, and Political Influence," *International Organization* 42 (1988), pp. 461–483.

33. On the distinct, non-Russian national identities in the former Soviet south, see especially Paul Kubicek, "Regionalism, Nationalism, and Realpolitik," *Europe-Asia Studies* 49:4 (1997), pp. 643–644; Martha Brill Olcott, *Central Asia's New States* (Washington, DC: United States Institute of Peace, 1996); and Ronald Grigor Suny, "Provisional Stabilities: The Politics of Identities in Post-Soviet Eurasia," *International Security* 24:3 (Winter 1999/2000), pp. 139–178.

34. Tsygankov (2001), pp. 165–170.

35. Extraregional industrialized states gradually displaced Russia as the primary trading partner for six out of the eight southern NIS. MacFarlane (1999), pp. 35–44.

36. Peter J. Katzenstein, "International Relations and Domestic Structures: Foreign Economic Policies of Advanced Industrial States," *International*

Organization 30:1 (Winter 1976), pp. 1–45; and Stephen D. Krasner, "United States Commercial and Monetary Policy: Unraveling the Paradox of External Strength and Internal Weakness," in Peter J. Katzenstein, ed., *Between Power and Plenty* (Madison: University of Wisconsin Press, 1978), pp. 51–87.

37. A structurally autonomous executive in a target state is expected to be less vulnerable to the demands of a sanctioning state. Jean-Marc F. Blanchard and Norrin M. Ripsman, "Asking the Right Question: When Do Economic Sanctions Work Best?" *Security Studies* 1/2 (Autumn 1999–Winter 2000), pp. 218–253.

38. Douglas W. Blum, "Domestic Politics and Russia's Caspian Policy," *Post-Soviet Affairs* 14:2 (1998), pp. 137–164; Igor Khripunov and Mary Matthews, "Russia's Oil and Gas Interest Group and Its Foreign Policy Agenda," *Problems of Post-Communism* (May/June 1996), pp. 38–48; Isabel Gorst and Nina Pousssenkova, *Petroleum Ambassadors of Russia: State Versus Corporate Policy in the Caspian Region* (Houston, TX; Center for International Political Economy and the James A. Baker III Institute for Public Policy, Rice University, 1998); and Henry Hale, "The Rise of Russian Anti-Americanism," *Orbis* 43:1 (Winter 1999), pp. 11–125.

39. On problems of "moral hazard" that inform state capacity arguments, see D. Roderick Kiewiet and Mathew D. McCubbins, *The Logic of Delegation: Congressional Parties and the Appropriations Process* (Chicago: University of Chicago Press, 1991); and Terry Moe, "The New Economics of Organization," *American Journal of Political Science* 28:4 (November 1984), pp. 754–756. On the strategic implications of "involuntary defection," see Robert D. Putnam, "Diplomacy and Domestic Politics: The Logic of Two Level Games," *International Organization* 42:3 (Summer 1988), pp. 427–460.

40. Stephen Blank, "Towards the Failing State: The Structure of Russian Security Policy," *Conflict Studies Research Center* (November 1996); and Becker (2000), pp. 100–104.

41. Blanchard and Ripsman (1999–2000), pp. 224–231.

42. William H. Kaempfer and Anton D. Lowenberg, "A Public Choice Analysis of the Political Economy of International Sanctions," in Steve Chan and A. Cooper Drury, eds., *Sanctions as Economic Statecraft: Theory and Practice* (London: St. Martin's Press, 2000), pp. 158–186; and Margarita Mercedes Balmaceda, "Gas, Oil, and the Linkages Between Domestic and Foreign Policies: The Case of Ukraine," *Europe-Asia Studies*, 50:2 (1998), pp. 257–286.

43. David M. Rowe, *Manipulating the Market: Understanding Economic Sanctions, Institutional Change, and the Political Unity of White Rhodesia* (Ann Arbor: The University of Michigan Press, 2001).

44. On basic institutional commonalities and differences of Central Asian regimes, see Sally N. Cummings, ed., *Power and Change in Central Asia* (London: Routledge, 2002); and Pauline Jones Luong, ed., *The Transforma-*

tion of Central Asia: States and Societies from Soviet Rule to Independence (Ithaca: Cornell University Press, 2004).

45. On the competing interests among Russia's multiple private and semiprivate oil firms, see Yakov Pappe, "Neftyanaya i gazovaya diplomatiya Rossiya," *Pro et Contra* (Summer 1997), pp. 61–69. On the significant differences in investment portfolios, ownership structures, and strategic orientations between the Russian oil and gas sectors, see David A. Lane, "The Political Economy of Russian Oil," in Peter Rutland, ed. *Economic Change in Post-Soviet Russia* (Westview Press, 1999); David Lane and Iskander Seifulmulukov, "Company Profiles: LUKoil, YuKOS, Surgutneftgaz, Sidanko," in David Lane, ed. *The Political Economy of Russian Oil* (Oxford: Rowman & Littlefield Publishers, 1999), pp. 111–126; and Tina Obut, Avik Srkar, and Sankar Sunder, "Comparing Russian, and Western Major Oil Firms Underscores Problems Unique to Russia," *Oil and Gas Journal* 97:5 (February 1, 1999), pp. 20–25.

46. Kenneth A. Schultz, *Democracy and Coercive Diplomacy* (Cambridge: Cambridge University Press, 2001). See also Kurt Taylor Gaubatz, "Democracy and Commitment," *International Organization* 50:1 (Winter 1996), pp. 109–140; Peter F. Cowhey, "Domestic Institutions and the Credibility of International Commitments," *International Organization* 47:2 (1993), pp. 299–326; James Fearon, "Domestic Political Audiences and the Escalation of International Disputes," *American Political Science Review* 88:3 (September 1994), pp. 577–592; Peter Gourevich, "The Governance Problem in International Relations," in David A. Lake and Robert Powell, eds., *Strategic Choice and International Relations* (Princeton: Princeton University Press, 1999), pp. 137–164; and Lisa Martin, *Democratic Commitments* (Princeton, Princeton University Press, 2000).

47. Peter Rutland, "Oil, Politics, and Foreign Policy," in Lane (1999), pp. 179–186. Some scholars argue that Russian energy firms act as agents of the Russian government. See especially, Jean-Christophe Peuch, "Russian Interference in the Caspian Sea Region: Diplomacy Adrift," in Lane (1999), pp. 189–212.

48. Bruce Kellison, "Tuimen, Decentralization, and Center-Periphery Tension," in Lane (1999), pp. 127–140.

49. See David M. Woodruff, "It's Value That's Virtual: Bartles, Rubles, and the Place of Gazprom in the Russian Economy," *Post-Soviet Affairs* 15:2 (1999), pp. 135–142.

50. Alexander L. George and William E. Simons, eds. *The Limits of Coercive Diplomacy,* 2nd ed. (Boulder, CO: Westview Press, 1994), p. 2.

51. On the distinction between brute force and coercion, see Robert A. Pape, *Bombing to Win: Air Power and Coercion in War* (Ithaca: Cornell University Press, 1996), pp. 13–19; and Daniel Byman and Matthew Waxman, *The Dynamics of Coercion: America's Foreign Policy and the Limits of Military Might* (New York: Cambridge University Press, 2002), pp. 3–6. Because the wide range of credible threats and measurable cost of using

nonnuclear force, the distinction between deterrence and compellence is less significant for the practice of nonnuclear statecraft than it is for nuclear diplomacy. See Gary Schab, "Compellence: Resuscitating the Concept," in Lawrence Freedman, ed., *Strategic Coercion: Concepts and Cases* (Oxford: Oxford University Press, 1998), pp. 40–44.

52. Pape (1996), p. 16.

53. Ibid., pp. 18–38.

54. David Baldwin, "The Power of Postive Sanctions," *World Politics* 24:1 (October 1971), pp. 19–38; and William J. Long, *Economic Incentives and Bilateral Cooperation* (Ann Arbor: The University of Michigan Press, 1996). On tactical or structural linkage, see Michael Mastanduno, *Economic Containment: CoCom and the Politics of East-West Trade* (Ithaca: Cornell University Press, 1992), pp. 52–57.

55. Stephen R. Rock, *Appeasement in International Politics* (Lexington: University of Kentucky Press, 2000), pp. 12–15.

56. Long (1996); and Eileen M. Crumm, "The Value of Economic Incentives in International Politics," *Journal of Peace Research* 32:3 (1995), pp. 321–322.

57. Long (1996).

58. Baldwin (1971), p. 28.

59. Daniel W. Drezner, "The Trouble with Carrots: Transaction Costs, Conflict Expectation, and Economic Inducements," *Security Studies* 9:1/2 (Autumn 1999–Winter 2000), pp. 188–218.

60. On the diffusion of political power and changing role of states amid globalization, see Strange (1996), pp. 66–87. Others are more optimistic about the resilience of states but sympathetic to the variable effects of globalization on the different dimensions of national sovereignty and governance. See discussion in Stephen D. Krasner, "Problematic Sovereignty," in Stephen D. Krasner, ed., *Problematic Sovereignty: Contested Rules and Political Possibilities* (New York: Columbia University Press, 2001), pp. 1–23; and Robert O. Keohane and Joseph S. Nye, "Governance in a Globalizing World," in Robert O. Keohane, ed., *Power and Governance in a Partially Globalized World* (London: Routledge, 2002), pp. 193–215. On the need for analytical updating, see Byman and Waxman (2002), pp. 14–18; and Peter Viggo Jakobsen, "The Strategy of Coercive Diplomacy: Refining Existing Theory to Post-Cold War Realities," in Freedman (1998), pp. 61–85.

61. Joseph S. Nye, *The Paradox of American Power* (New York: Oxford University Press, 2002), pp. 8–12. See also Susan Strange, "The Persistent Myth of Lost Hegemony," *International Organization* 41:4 (Autumn 1987), pp. 564–565; and Peter Bacharach and Morton Baratz, *Power and Poverty: Theory and Practice* (New York: Oxford University Press, 1970).

62. Rodman (2001). On Russia's multidimensional forms of soft power, see Hill (2004).

63. Pape (1996), p. 16.

64. On fixed preferences versus outcomes, and the strategies for realizing these preferences, see James D. Morrow, *Game Theory for Political Scientists* (Princeton: Princeton University Press, 1994), pp. 18–20.

65. Jonathan Kirshner, *Currency and Coercion* (Princeton: Princeton University Press, 1995), pp. 115–148.

66. Rowe (2001).

67. Ibrahim A. Karawan, "Identity and Foreign Policy: The Case of Egypt," in Shibley Telhami and Michael Barnett, eds., *Identity and Foreign Policy in the Middle East* (Ithaca: Cornell University Press, 2002), pp. 160–167.

68. Randall E. Newnham, "More Flies with Honey: Positive Economic Linkage in German Ostpolitik from Bismarck to Kohl," *International Studies Quarterly* 44 (2000), p. 75; and Paul Milgrom and John Roberts, "Predation, Reputation, and Entry Deterrence," *Journal of Economic Theory* 27:2 (1982), pp. 280–312.

69. Rock (2000), p. 4.

70. Ibid., pp. 5–9; 155–177; Jack Hirshsliefer, "Appeasement: Can It Work?" *American Economic Review* 91:2 (2001), pp. 342–346; and Daniel Triesman, "Rational Appeasement," *International Organization* 58 (Spring 2004), pp. 345–373. The balance sheet is much more ambiguous when considering the tremendous defense expenditures and open-ended foreign commitments that states incur bolstering the credibility of threats and demonstrating their resolve.

71. Michael Mastanduno, "Economic Statecraft, Interdependence, and National Security," in Jean-Marc F. Blanchard, Edward S. Mansfield, and Norrin M. Ripsman, eds., *Power and the Purse* (London: Frank Cass, 2000), pp. 299–300; and Randall E. Newnham, *Deutsche Mark Diplomacy: Economic Linkage in German-Russian Relations* (University Park: Penn State University Press, 2000).

72. Byman and Waxman (2002), p. 38. See also discussion in Blanchard and Ripsman (1999–2000), pp. 223–224; and Rowe (2001).

73. Stephen J. Cimbala, *Russia and Armed Persuasion* (New York: Rowman & Littlefield Publishers, 2001), pp. 4–5.

74. Byman and Waxman (2002), p. 33.

75. See especially Richard Ned Lebow, "Conclusions," in Robert Jervis, Richard Ned Lebow, and Janice Gross Stein, eds., *Psychology and Deterrence: Perspectives on Security* (Baltimore: Johns Hopkins University Press, 1985), pp. 203–232.

76. David M. Rowe, "Economic Sanctions Do Work: Economic Statecraft and the Oil Embargo of Rhodesia," in Blanchard, Mansfield, and Ripsman (2000), pp. 254–287.

77. Baldwin (1985), p. 132.

78. Coercive strategies can backfire by provoking an adversary to choose a more undesirable option. This, however, is ultimately a form of failure that can lead the initiator to back down or escalate coercive pressure. Byman and Waxman (2002), pp. 35–37.

79. Schultz (2001); Shambaugh (1999); and Rodman (2001).
80. George E. Shambaugh, IV, "Globalization, Sovereign Authority and Sovereign Control over Economic Activity," *International Politics* 37 (December 2000), p. 405.
81. James D. Fearon, "Rationalist Explanations for War," *International Organization* 49 (1995), pp. 379–414.
82. Fearon (1994), pp. 577–592; and Schultz (2001).

Chapter 2

1. Richard Ned Lebow and Janice Gross Stein, "Rational Deterrence Theory: I Think, Therefore Deter," *World Politics* 41 (1989), pp. 208–224; and John Steinbrunner, *The Cybernetic Theory of Decision* (Princeton: Princeton University Press, 1976).
2. Daniel Kahneman and Amos Tversky, "Prospect Theory: An Analysis of Decisionmaking Under Risk," *Econometrica* 47 (1979), p. 277. For applications to international bargaining, see Barbara Farnham, ed., *Avoiding Losses/Taking Risks* (Ann Arbor: University of Michigan Press, 1994); Jack S. Levy, "Prospect Theory, Rational Choice, and International Relations," *International Studies Quarterly* 41:1 (1997), p. 90; Victor D. Cha, "Hawk Engagement and Preventive Defense," *International Security* 27:1 (Summer 2002), pp. 40–78; Jeffrey W. Taliaferro, *Balancing Risks: Great Power Intervention in the Periphery* (Ithaca: Cornell University Press, 2004); and Jeffrey D. Berejikian, *International Relations Under Risk: Framing State Choice* (Albany: State University of New York Press, 2004).
3. Kahneman and Tversky (1979), pp. 263–291. This can occur even though the choice may have no effect on the expected value of the alternatives.
4. Preliminary tests of prospect theory reveal that decision makers tend to be more compliant in the face of threats that make uncertain gains from noncompliance more risky, while they are more responsive to inducements that ameliorate certain losses. James W. Davis, Jr., *Threats and Promises: The Pursuit of International Influence* (Baltimore: Johns Hopkins University Press, 2000), pp. 26–43; Gitty Madeline Amini, *Sanctions and Reinforcement in Strategic Relationships: Carrots and Sticks, Compellence and Deterrence* (Los Angeles, CA: UCLA Dissertation, 2001); and Berejikian (2004), pp. 34–35.
5. Jack S. Levy, "Prospect Theory and International Relations: Theoretical Applications and Analytical Problems," in Farnham (1994), p. 136.
6. Daniel W. Drezner, *The Sanctions Paradox: Economic Statecraft and International Relations* (Cambridge: Cambridge University Press, 1999); and Dale C. Copeland, "Economic Interdependence and War: A Theory of Trade Expectations," *International Security* 20:4 (1996), pp. 5–41. This builds on the insight that decision makers must make tradeoffs between present and future values and risks. Alexander L. George, "The 'Opera-

tional Code': A Neglected Approach to the Study of Political Leaders and Decisionmaking," *International Studies Quarterly* 13 (1969), pp. 190–222. Trade expectations are consistent with frames around aspirations rather than the status quo. Accordingly, the extent to which the status quo deviates from the future reference point predisposes a target to take more risks. Alternatively, a decision maker should be risk-averse in a situation where the status quo exceeds expectations concerning a future reference point. See discussion in Jack S. Levy, "Prospect Theory and the Cognitive-Rational Debate," in Nehemia Geva and Alex Mintz, eds., *Decisionmaking on War and Peace: The Cognitive-Rational Debate* (Boulder: Lynn Rienner Publishers, 1997). p. 37; and Taliaffero (2004), pp. 37–51.

7. Kenneth A. Schultz, *Democracy and Coercive Diplomacy* (Cambridge: Cambridge University Press, 2001), pp. 23–115.
8. The literature on risk versus uncertainty is vast and arcane. Formal mathematical and economic accounts typically distinguish the two in terms of the viability of assigning numerical probabilities (risk). This becomes problematic once we accept that decision makers use subjective interpretations of probabilities. See discussion in Jack Hirshliefer, *Time, Uncertainty, and Information* (New York: Basil Blackwell, 1998), pp. 152–165; and Christian Schmidt, "Risk and Uncertainty: A Knightian Distinction Revisited," in Christian Schmidt, ed., *Uncertainty in Economic Thought* (Cheltenham, UK: Edward Elgar, 1996), pp. 65–84. Instead, I distinguish the two in terms of asymmetries of information (uncertainty) and assessments of prospective positive and negative values of specific outcomes. Risk, therefore, is inherent in a policymaker's choice, regardless of the information between rivals or probabilities assigned to specific options. Instead, it is a function of the values that decision makers assign to specific outcomes. See discussion of risk in Levy (1997), pp. 135–138; Rose McDermott, *Risk-Taking in International Politics: Prospect Theory in American Foreign Policy* (Ann Arbor: University of Michigan Press, 1998), pp. 1–6; and Taliaferro (2004), p. 26. Finally, some scholars suggest that risk is captured in the utility function of preferred and less preferred outcomes. Given that risking-taking propensities differ for actors in the domain of gains versus losses, and that this distinction has a measurable effect on compliance, I choose to separate it from considerations of utility and uncertainty. See James D. Morrow, *Game Theory for Political Scientists* (Princeton: Princeton University Press, 1994), p. 29.
9. Taliaferro (2004), p. 39.
10. Davis (2000), pp. 26–43; and Amini (2001), pp. 36–39.
11. Copeland (1996), pp. 5–41.
12. Taliaferro (2004), pp. 36–37.
13. George A. Quattrone and Amos Tversky, "Contrasting Rational and Psychological Analyses of Political Choice," *American Political Science Review* 82:3 (September 1988), p. 726.
14. Robert A. Pape, *Bombing to Win: Air Power and Coercion in War* (Ithaca: Cornell University Press, 1996).

15. Jack S. Levy (1997), pp. 135–138; Berejikian (2004), pp. 26–29; and Kenneth A. Schultz, "The Politics of Risking Peace: Do Hawks or Doves Deliver the Olive Branch," *International Organization* 59 (Winter 2005), pp. 1–38.

16. Raymond Vernon, *Storm over the Multinationals: The Real Issues* (Cambridge: Harvard University Press, 1977), p. 105; and Kenneth A. Rodman, *Sanctions Beyond Borders: Multinational Corporations and U.S. Economic Statecraft* (New York: Rowman & Littlefield Publishers, 2001), p. 5.

17. On the importance of specifying microfoundations for theory-building, see David A. Lake and Robert Powell, "International Relations: A Strategic-Choice Approach," in David A. Lake and Robert Powell, eds., *Strategic Choice and International Relations* (Princeton: Princeton University Press, 1999), pp. 21–25.

18. Robert Jervis, "Cooperation Under the Security Dilemma," *World Politics* 30 (January 1978), pp. 168–214.

19. Gideon Rose, "Neoclassical Realism and Theories of Foreign Policy," *World Politics* 51 (October 1998), p. 152; and Fareed Zakaria, *From Wealth to Power: The Unusual Origins of America's World Role* (Princeton: Princeton University Press, 1998).

20. Zakaria (1998), pp. 9–11.

21. Ibid.

22. This is best depicted by the "two-level game" metaphor in international relations theory. See Robert D. Putnam, "Diplomacy and Domestic Politics: The Logic of Two-Level Games," *International Organization* 42:3 (Summer 1988): 427–460. See also critical reflections in Andrew Moravcsik, "Introduction," in Peter B. Evans, Harold K. Jacobson, and Robert D. Putnam, eds., *Double-Edged Diplomacy* (Berkeley: University of California Press, 1993), pp. 15–17.

23. The notion of "situational rationality" subsumes both concepts of "conservative rationality" and "gambling rationality" discussed in Arthur A. Stein, *Why Nations Cooperate: Circumstance and Choice in International Relations* (Ithaca: Cornell University Press, 1990), pp. 88–95.

24. On the distinction between "outcome (substantive) rationality" and "procedural rationality," see especially Herbert A. Simon, "From Substantive to Procedural Rationality," in S.J. Latsis, ed., *Method and Appraisal in Economics* (Cambridge: Cambridge University Press, 1976); Jack Levy (1997), pp. 101–102; and Zeev Maoz, "Framing the National Interest: The Manipulation of Foreign Policy Decisions in Group Settings," *World Politics* 43 (October 1990), pp. 82–83.

25. On the violation of invariance that assumes that preference orders among prospects do not depend on how their outcomes and probabilities are described, see discussion in Quattrone and Tversky (1988), pp. 727–730; and McDermott (1998), p. 17.

26. On the critical elements of transitive and completeness for rational decision making, see especially James D. Morrow, *Game Theory for Political Scientists,* pp. 17–19. This discussion is derived loosely from James D. Morrow,

"A Rational Choice Approach to International Conflict," in Geva and Mintz (1997), pp. 14–15; and Jack S. Levy (1997), pp. 100–102.

27. This is akin to the claims by structural realists that the international system matters for foreign policy because one state's perceptions of threat are partly shaped by its relative material power.

28. On the aggregation problem, see Jack S. Levy (1997), pp. 102–104. This discussion, however, does not imply that all domestic actors respond in the same way to a frame, as they operate with different institutional authority and interests.

29. Ibid., p. 89.

30. This is derived from insights on agenda control in John W. Kingdon, *Agendas, Alternatives, and Public Policies* (Boston: Little, Brown, and Company, 1984), pp. 206–218; and Helen Milner, *Interests, Institutions, and Information: Domestic Politics and International Relations* (Princeton: Princeton University Press, 1997), pp. 101–103.

31. McDermott (1998), pp. 15–35; and Jack S. Levy, "Introduction to Prospect Theory," in Farnham (1996), pp. 15–20.

32. Robert Keohane and Joseph S. Nye, Jr., *Power and Interdependence: World Politics in Transition* (Boston: Little, Brown, 1977), pp. 44–45. See also discussion of political leverage derived from "vulnerability," in Albert O. Hirschman, *National Power and the Structure of Foreign Trade* (Berkeley: University of California Press, 1980), pp. 18–20; David A. Baldwin, "Interdependence and Power: A Conceptual Analysis," *International Organization*, Vol. 34, No. 4 (Autumn 1989), p. 487; James A. Caporaso, "Dependence, Dependency, and Power in the Global System: A Structural and Behavioral Analysis," p. 21; Beverly Crawford, *Economic Vulnerability in International Relations: The Case of East-West Trade, Investment, and Finance* (New York: Columbia University Press, 1993), pp. 165–171; George E. Shambaugh, *States, Firms, and Power: Successful Sanctions in United States Foreign Policy* (Albany: State University of New York Press, 1999), pp. 14–24; Walter Carlsnaes, *Energy Vulnerability and National Security*, pp. 9–25; and Paul J. D'Anieri, *Economic Interdependence in Ukrainian-Russian Relations* (Albany: State University of New York Press, 1999), pp. 45–46.

33. On the opportunity costs of exercising power, see John C. Harsanyi, "Measurement of Social Power, Opportunity Costs, and the Theory of Two-Person Bargaining Games," *Behavioral Science* VII (January 1962); ibid., "Measurement of Social Power in n-Person Reciprocal Power Situations," *Behavioral Science* VII (January 1962); Hirschman (1980), pp. 29–34; D'Anieri (1999), pp. 47–56; and Jean-Marc F. Blanchard and Norrin M. Ripsman, "Asking the Right Questions: When Do Economic Sanctions Work?" *Security Studies* 1, 2 (Autumn 1999–Winter 2000), pp. 219–253.

34. McDermott (1998), p. 25.

35. Zeev Maoz, *Paradoxes of War: On the Art of National Self-Entrapment* (Boston: Unwin Hyman, 1989), pp. 279–282.

36. Ibid. Maoz argues that decision makers sustain commitments and strive to avoid sharp breaks in policy in order to avert the domestic political costs of being challenged for pursuing fluid policies. They are cognizant that they are in a policy trap, but because of the domestic costs of adjustment they cannot afford to extract themselves. Conversely, I argue that a target state can become entrapped by an initiator's "salami tactics" within the domain of gains. The inclination to maintain consistency with a previous decision is a function of the risk aversion that accompanies decision making in the domain of gains, as opposed to either cognitive imperfections or domestic political pressure.
37. On cancelation as an editing strategy, see especially McDermott (1998), p. 23; and Levy (1996), p. 16
38. McDermott 1998), p. 23. The greater likelihood of a pipeline route materializing is not trivial to a target's choice.
39. Scott C. James and David A. Lake, "The Second Face of Hegemony: Britain's Repeal of the Corn Laws and America's Walker Tariff of 1846," *International Organization* 43:1 (1989), pp. 1–29; and Shambaugh (1999), pp. 6–30.
40. R. Harrison Wagner, "Economic Interdependence, Bargaining Power, and Political Influence," *International Organization* 42:3 (Summer 1988), pp. 462–472.
41. Rodman (2001), p. 15.
42. McDermott (1998), p. 39.
43. Ibid., pp. 29–33; and Jack S. Levy (1997), pp. 135–138.
44. McDermott (1998), p. 29–33.
45. This is loosely derived from the Stolper-Samuelson theorem that demonstrates that domestic incentives for international openness turn on the relative abundance of factors of production and the capacity of states to alter domestic demand. See Jeffry Frieden and Ronald Rogowski, "The Impact of the International Economy on National Policies," in Robert O. Keohane and Helen V. Milner, eds., *Internationalization of Domestic Politics* (New York: Cambridge University Press, 1996), pp. 25–47; Ronald Rogowski, *Commerce and Coalitions: How Trade Affects Domestic Political Alignments* (Princeton: Princeton University Press, 1989); and Jeffry A. Frieden, "Sectoral Conflict and US Foreign Economic Policy, 1914–1940," *International Organization* 42 (1988), pp. 59–90. For application to national security issues, see Etel Solingen, "The Political Economy of Nuclear Restraint," *International Security* 19:2 (Fall 1994), pp. 136–142.
46. For a general discussion of the systematic impact of domestic institutions on the extension of credible foreign commitments, see John S. Odell, "International Threats and Internal Politics," in Evans, Jacobson, and Putnam (1993), pp. 233–264; Peter Cowhey, "Domestic Institutions and the Credibility of Commitments: Japan and the United States," *International Organization* 47:2 (Spring 1993), pp. 299–326; Helen Milner, *Interests, Institutions, and Interests: Domestic Politics and International Relations*

(Princeton: Princeton University Press, 1997), pp. 99–127; and Ronald Rogowski, "Institutions as Constraints on Strategic Choice," in Lake and Powell (1999), p. 117.

47. This is akin to Putnam's argument that central executives are constrained by domestic "win-sets," defined as the set of potential policies that would be acceptable to the range of domestic constituencies charged with either approving or implementing the state's efforts at altering the status quo in relations with a foreign target. The fundamental domestic constraint on statesmen is the size of the win-set that narrows the range of foreign policies that can be effectively approved and implemented. Putnam (1988).

48. This discussion is based on the new-institutional literature on agenda control. See especially Mathew D. McCubbins, Roger G. Noll, and Barry R. Weingast, "Administrative Procedures as Instruments of Political Control," *Journal of Law, Economics, and Organization* 3:2 (Fall 1987), pp. 243–277; Ibid., "Structure and Process, Politics and Public Policy: Administrative Arrangements and the Political Control of Agencies," *Virginia Law Review* 75:2 (March 1989), pp. 431–507; and Kenneth A. Shepsle and Barry R. Weingast, "The Institutional Foundations of Committee Power," *American Political Science Review* 81:1 (1987), pp. 85–104. For application to foreign policy decision making, see Mark A. Pollack, "Delegation, Agency, and Agenda Setting in the European Community," *International Organization* 51:1 (Winter 1997), pp. 99–134; David P. Auerswald, "Inward Bound: Domestic Institutions and Military Conflicts," *International Organization* 53:3 (Summer 1999), pp. 469–504; and Kenneth A. Rodman, "Sanctions at Bay? Hegemonic Decline, Multinational Corporations, and U.S. Economic Sanctions Since the Pipeline Case," *International Organization* 49:1 (Winter 1995), pp. 105–137.

49. Oliver E. Williamson, *The Economic Institutions of Capitalism* (New York: Free Press, 1985); Armen A. Alchian and Harold Demsetz, "Property Rights Paradigm," *Journal of Economic History* 33 (1973), pp. 16–27; Benjamin Klien, Robert G. Crawford, and Armen A. Alichan, "Vertical Integration, Appropriable Rents, and the Competitive Contracting Process," *Journal of Law and Economics* 21 (October 1979), pp. 297–326; Mancur Olson, *The Logic of Collective Action* (Cambridge: Harvard University Press, 1965); and James Buchanan and Gordon Tullock, *The Calculus of Consent* (Ann Arbor: University of Michigan Press, 1962). See applications to post-Soviet policy making in Maxim Boyco, Andrei Shileffer, and Robert Vishny, *Privatizing Russia* (Cambridge, MIT Press, 1994), pp. 26–29; and Alexander Cooley, "Imperial Wreckage," *International Security*, 25:3 (Winter 2000/01), pp. 107–109.

50. This argument derives from the attenuation and partitioning of property rights. A.A. Alchian, "Corporate Management and Property Rights," in H.G. Manne, ed., *Economic Policy and the Regulation of Corporate Securities* (Washington, DC: American Enterprise Institute, 1969); Gary Libecap, *Contracting for Property Rights* (New York: Cambridge University Press,

1989); Douglas C. North, *Institutions, Institutional Change, and Economic Performance* (New York: Cambridge University Press, 1990); Thrainn Eggertsson, *Economic Behavior and Institutions* (New York: Cambridge University Press, 1990); and Michael Jensen and William C. Meckling, "Theory of the Firm: Managerial Behavior, Agency Costs, and Ownership Structure," *Journal of Financial Economics,* 3 (1976), pp. 305–360. On the relative importance of "clear" decision rules for extending credible commitments and crafting stable foreign policies, see Rogowski (1999), pp. 119–136.

51. Ibid.
52. Levy (1997), pp. 132–135.
53. According to research on international trade, a sender responsible for 50 percent of the supply of a good with a high substitution elasticity is poised to influence the terms of trade of that good. Drezner (1999), p. 15.
54. Jean-Marc F. Blanchard and Norrin Ripsman, "Measuring Economic Interdependence: A Geopolitical Perspective," *Geopolitics* 1:3 (Winter 1996), pp. 229–231.
55. Thomas R. Stauffer, "Caspian Fantasy: The Economics of Political Pipelines," *Brown Journal of World Affairs* VII: 2 (Summer/Fall 2000).
56. This comes from estimates of an initiator's proven and unproven energy reserves, as percentages of global supply and exports. Although energy estimates are notoriously imprecise, I use median estimates provided by the International Energy Agency and Energy Information Administration of the U.S. Department of Energy. Unlike the case for assessing a target state's frame, where national energy estimates can be used as an indicator of a leadership's subjective assessment of the state's international standing, the emphasis placed here is on examining objective criteria that set limits or create opportunities for statecraft. While any discrepancy between what an initiator thinks it can do to a target and what its real market power allows it to do is relevant for understanding the quality of an initiator's decision making; it is less significant for understanding the ultimate success of energy statecraft.
57. McDermott (1998), pp. 38–40; Eldar Shafir, "Prospect Theory and Political Analysis: A Psychological Perspective," in Farnham (1996), pp. 147–158; and Victor Cha, *Nuclear North Korea: A Debate on Engagement Strategies* (New York: Columbia University Press, 2003), pp. 26–32.
58. McDermott (1998), pp. 39–40.
59. Drezner (1999), p. 107.
60. Andrew Bennett and Alexander L. George, "Case Studies and Process Tracing in History and Political Science: Similar Strokes for Different Foci," in Colin Elman and Miriam Elman, eds., *Bridges and Boundaries: Historians, Political Scientists, and the Study of International Relations* (Cambridge: MIT Press, 2001).
61. See discussion in Timothy W. Crawford, *Pivotal Deterrence: Third-Party Statecraft and Pursuit of Peace* (Ithaca: Cornell University Press, 2003), pp.

43–45; and Alexander L. George and Andrew Bennett, "The Role of Congruence Method in Case Study Research," (unpublished paper, 2002).

62. For critical assessments of the merits of using plausible counterfactual analysis, see especially Richard Ned Lebow, "What's So Different about a Counterfactual?" *World Politics* 52 (July 2000), pp. 550–585; and Philip E. Tetlock and Aaron Belkin, eds., *Counterfactual Thought Experiments in World Politics: Logical, Methodological, and Psychological Perspectives* (Princeton: Princeton University Press, 1996).

63. On "strong" test cases, see especially Stephen Van Evera, *Guide to Methods for Students of Political Science* (Ithaca: Cornell University Press, 1997), pp. 74–88.

64. Ibid.

65. D'Anieri (1999).

66. Drezner (1999), pp. 153–247.

Chapter 3

1. In 1991, Russia possessed approximately 86 percent of the Soviet Union's total reserves. Russia's dominance was magnified by the fact that it was both commercially and technically infeasible to access approximately half of the world's natural gas reserves. Moreover, the magnitude and accessibility of Russia's supergiant Siberian and European fields made it possible to bring new fields on line as reserves were depleted for at least 40 years, despite questions concerning the profitability of domestic sales. See Matthew J. Sagers, "The Energy Industries of the former USSR: A Mid-Year Survey," *Post-Soviet Geography* 34:6 (1993), p. 377; and International Energy Agency (IEA), *Russian Energy Survey 2002* (March 2002), pp. 111–116.

2. IEA (2002), pp. 109–116; 133–137; U.S. Energy Information Administration (EIA), "Russia Country Analysis," November 2002, <www.eia.doe.gov/cabs/russia.html; EIA, "Caspian Sea Region: Reserves and Pipelines Tables," July 2002, <www.eia.doe.gov/emeu/cabs/caspgrph.html#TAB1; IEA, *Caspian Oil and Gas: The Supply Potential of Central Asia and Transcaucasia* (Paris: OECD, 1998), p. 112; and Fiona Hill and Florence Fee, "Fueling the Future: The Prospects for Russian Oil and Gas," *Demokratizatsiya*, 10:4 (Fall 2002), p. 468. Proven reserves are defined as natural gas deposits that are 90 percent probable. The comparison to Caspian states includes proven reserves in Azerbaijan, Kazakhstan, Turkmenistan, and only those regions in Iran adjacent to the Caspian Sea.

3. Gas markets are essentially regional due to the high costs of pipelines and storage. Natural gas delivery systems are very inflexible, supply and prices in one market do not directly affect those in other markets, contracts are typically long-term, and a premium is placed on maintaining a strict match between supply and demand at both ends of the pipelines. Although some

of these factors are attenuated by monetizing liquified natural gas, the high capital investments in ports and liquefaction facilities introduce additional expenditures and rigidities into international gas market transactions. See IEA, *The IEA Natural Gas Security Study* (Paris: OECD, 1995); and Leslie Dienes, Istvan Dobozi, and Marian Radetzki *Energy and Economic Reform in the Former Soviet Union* (New York: St. Martin's Press, 1994).

4. European Commission, "Towards a European Strategy for the Security of Energy Supply," *EU Green Paper* (2001), p. 24; and IEA (1998), pp. 104–108.

5. European Commission (2001). On the "tyranny of distance" and critical importance of economies of scale for cost-effective delivery of gas to regional markets, see Thomas R. Stauffer, "Caspian Fantasy: The Economics of Political Pipelines," *Brown Journal of World Affairs* VII: 2 (Summer/Fall 2000).

6. Fiona Hill, "Energy Integration and Cooperation in Northeast Asia," *NIRA Policy Research Paper* 15:2 (February 2002), available at www/brookings.edu; and Selig S. Harrison, "Gas and Geopolitics in Northeast Asia," *World Policy Journal* (Winter 2002/2003), pp. 23–36.

7. Gazprom was converted to a private state concern during the first phase of voucher privatization in November 1992, with shares divided between the government (40 percent), company employees (15 percent), private owners (28.7 percent), ethnic minorities (5.2 percent), and downstream distributors (one percent). In February 1993, the company was transformed into a joint stock company. Of the 37 enterprises owned by Gazprom, 24 were gas producing, transmission, or processing enterprises; the remainder were specialized engineering, research, or scientific enterprises. Matthew J. Sagers, "The Russian National Gas Industry in the Mid-1990s," *Post-Soviet Geography* 36:9 (1995), p. 523; ibid., "Joint Management of Oil and Gas Resources in Russia," *Post-Soviet Geography* 39:7 (1998), p. 597; and *IEA (2002),* p. 111. Although Gazprom maintained an effective monopoly over domestic production throughout the 1991–2002, it showed signs of relaxing its grip over the domestic industry. By 2000, several additional domestic suppliers—including private companies, oil companies, and regional companies—captured approximately 10 percent of annual production. Chief among them was Itera, that was responsible for producing 3 percent of Russia's total annual production, and had plans for increasing its share to 8 percent in 2005 and 12 percent in 2010. IEA (2002), pp. 112, 116–118.

8. IEA (2002), pp. 118–119.

9. Arild Moe and Valeriy A. Kryukov, "Joint Management of Oil and Gas Resources in Russia," *Post-Soviet Geography* 39:7 (1998), p. 597.

10. The prices served primarily as internal accounting measures that were not subject to direct government oversight. Sagers (1995), pp. 522–525; and IEA (2002), pp. 121–124.

11. Moe and Kryukov (1998), p. 601; and IEA (2002), pp. 121, 145.

12. Itera raised its profile in 1994 with the delivery of Central Asian gas, operating as a private transit middleman and under the watchful eye of Gazprom. By 2000, Itera was the only large-scale user of the Gazprom's transmission system, and was assumed to receive special dispensation and insider privileges. That same year, the company proposed to construct a pipeline link between Georgia and Turkey that reflected ambitions to export gas outside of the CIS in excess of the 50 bcm delivered to Western Europe in 1999. The latter raised the specter of future competition with Gazprom, something that the Russian gas giant was loath to consider. By the end of 2002, Gazprom took steps to reclaim business with Turkmenistan and Ukraine, directly intervening to control gas traffic across the NIS. On Itera's suspicious relationship with Gazprom, see *Moskovskaya pravda,* 1 February 1999; *Nezavisimaya gazeta,* 23 March 1999; and IEA (2002), pp. 112, 116–118; and I. Makarov, "Itera, One of the Biggest Gas Suppliers in the CIS," *International Affairs* 46:2 (2000), pp. 69–76.

13. Deepak Gopinath, "Face-off in Moscow," *Infrastructure Finance* 6:6 (July/August 1997), pp. 36–39. On the government's attempt at reforming Gazprom's policies related to third-party access and payment of tax arrears, see *Izvestiya,* 20 June 1997.

14. *Kommersant,* 24 August 2001. In July 2001, the Russian government presented a draft plan for restructuring the gas industry by 2010. Although not implemented by 2003, the program called for breaking up Gazprom's upstream operations, creating separate producing companies, and introducing state control of the pipeline system to ensure nondiscriminatory access to both domestic and export markets. Despite provisions for raising domestic wholesale and retaining prices, the state was not committed to liberalizing gas prices for the foreseeable future. For critique of the government's restructuring program, see IEA (2002), pp. 145–148; EIA, *Russia: Energy Sector Restructuring,* April 2002, www.eia.doe.gov/cabs/russrest.html#OIL.

15. Sagers (1995), pp. 522–529.

16. *Ekspert,* 1 September 1997. Due to the devaluation of the ruble, receipts from gas sales to Europe were higher than those received from Russian customers in 1998 and 1999, even though volumes delivered to Europe were 40 percent lower than those to the domestic market. IEA (2002), p. 125.

17. Between 1991–1999, aggregate domestic demand declined from an initial high of 469.6 bcm to a low of 380.9 bcm in 1997, and closed at 392.4 bcm. From 1995 to 1999, demand remained relatively constant. IEA (2002), p. 125.

18. Sagers (1995), pp. 526–528; and IEA (2002), pp. 126–130. Domestic industrial gas prices in Russia fluctuated throughout the period but never reached more than 60 percent of international prices. From 1992–1994, industrial gas prices climbed from a low of 3 percent to 26 percent of international prices; they peaked at 60 percent of international prices in 1995; and from 1996 to 2001 steadily declined to 10 percent of international prices. Resi-

dential prices were significantly lower throughout the period. Although the situation began to improve between 2000 and 2002, as cash payments increased to 70 percent of receivables, the 30 percent nonpayments crisis continued to mar the effectiveness of market mechanisms in the domestic gas sector. On the centrality of gas price subsidies to Russia's value-subtracting "virtual economy," see especially Clifford C. Gaddy and Barry W. Ickes, "Russia's Virtual Economy," *Foreign Affairs* 77:5 (1998), pp. 53–67.

19. Matthew J. Sagers, "Resource Rent from the Oil and Gas Sector and the Russian Economy," *Post-Soviet Geography* 36:7 (1995), pp. 410–411; 418–419.

20. The World Bank, *Russia Oil Transport and Export Study: Strategic Export Expansion Options and Legal, Contractual, and Regulatory Framework* (August 1997); Matthew J. Sagers, "Developments in Russian Crude Oil Production in 2000," *Post-Soviet Geography and Economics* 42:3 (2001), pp. 153–155; IEA (2002), p. 88; EIA, *Russia: Oil and Natural Gas Exports* (November 2002), www.eia.doe.gov/cabs/russexp.html.

21. EIA (2002); and IEA (2002), p. 91. The revised Russian Energy Strategy in 2002 set a favorable target of growing crude oil exports from 192 million tons in 2000 to 208.5 million tons by 2020. More optimistic scenarios projected average annual rates of growth a 4–5 percent through 2027, with production nearing 575 million tons per annum by 2012. See discussion in Tsnetr Strategicheskikh Razrabotok, *O vozmozhnyi napravleniyakh razvitiya infrastruktury po transportirovke Rossiiskoi nefti*, 12 November 2004, available at www.csr.ru.

22. EIA (2002); and IEA (2002), p. 91.

23. Russia's proven reserves in 1999 were the seventh largest at 48.6 billion barrels barrels, and constituted only 66 percent of the next largest reserves in Venezuela. By comparison, proven reserves of Kazakhstan and Azerbaijan stood at 8 and 7 billion barrels, respectively. In contrast, Saudi Arabia reserves (263.5 billion barrels) provided 10 percent of the world's daily supply of oil and almost 25 percent of global exports, as well as was considerably cheaper to extract than either Russian or Caspian oil. John Roberts, *Caspian Pipelines* (London: Royal Institute of International Affairs, 1996), p. 2; John V. Mitchell with Peter Beck and Michael Grubb, *The New Geopolitics of Energy* (London: the Royal Institute of International Affairs, 1996), pp. 62–63; Robert Ebel, *Energy Choices in the Near Abroad* (Washington, DC: Center for Strategic and International Studies, April 1997), pp. 11–13; and Andrei Belopolsky and Manik Talwani, "Geological Basins and Oil and Gas Reserves of the Greater Caspian Region," in Yelena Kalyuzhnova, Amy Myers Jaffe, Dov Lynch, and Robin C. Sickles, eds., Energy in the Caspian Region (Hampshire, UK: Palgrave, 2002), p. 28.

24. Sagers (2001), p. 159. One expert attributes the post-1999 surge in production to the extraction of 400 million tons of oil bypassed during the previous slowdown, and to the unsustainable "creaming" of easy exploitable

reserves in rapidly aging fields from the Soviet-era. Moreover, the 110 new oil fields opened from 1999 to 2002 produced only 1.3 million tons in their first year on line. Leslie Dienes, "The Present Oil Boom," *Johnson's Russia List #8399*, 6 October 2004, www.cdi.com/jrl.

25. "The actual route traveled from well-head to final customer by any particular barrel is determined by knife-edge differences in transportation costs, as well as by vagaries in viscosity, sulfur content, and other factors that affect how crude oil is refined into useful products." David G. Victor and Nadejda M. Victor, "Axis of Oil?" *Foreign Affairs* 82:2 (March/April 2003), p. 51. See also Steve A. Yetiv, *Crude Awakenings: Global Oil Security and American Foreign Policy* (Ithaca: Cornell University Press, 2004), pp. 188–190.

26. The World Bank (1997), pp. 136–138.

27. Russian and international industry experts estimate that the existing western pipeline infrastructure could handle at most 23 million tons per year of the projected 70–90 million tons of Russian production available for export by 2015 (with shortages projected at 15 million tons and 35 million tons per year in 2005 and 2007, respectively). The problem was compounded by the fact that export projects outside of Russia required additional transit operations with significantly higher overhead costs (approximately $20–25 per ton) due to transfer delays and environmental risks. Tsnetr Strategicheskikh Razrabotok (2004).

28. The World Bank (1997), pp. 133–148.

29. The projected 25 million tons of oil production per year from these fields was dwarfed by the 80 million tons per year that China was expected to demand by 2010. Tsnetr Strategicheskikh Razrabotok (2004); Hill (2002); Sagers (2001), pp. 185–188.

30. Dag Harald Claes, *The Politics of Oil Producer Cooperation* (Boulder: Westview Press, 2001), p. 45

31. Riyadh first played the role of benevolent hegemon to stabilize world prices by lowering production in the mid-1980s, and then acted as a market predator by overproducing to counter supply increases by the other cartel members and to improve its market share through the mid-1990s. Saudi Arabia's hegemony persisted through 2002, as Riyadh alone maintained at least a 3 million bbl/d spare capacity that was used to stabilize international prices and restrain competition among rival suppliers. Riyadh wielded the technical capacity to expand rapidly production both to defend global consumers against price spikes and to compensate for "shock" surges in demand. By accounting for 10 percent of daily deliveries, the kingdom was well positioned to take action with an immediate impact on international customers. Claes (2001), pp. 225–237.

32. Edward L. Morse and James Richard, "The Battle for Energy Dominance," *Foreign Affairs* (March/April 2002), p. 20. See also discussion in John Mitchell with Koji Morita, Norman Selly, and Jonathan Stern, *The New Economy of Oil: Impact on Business, Geopolitics, and Society* (London: The Royal Institute of International Affairs, 2001), pp. 162–165.

33. On the relationship between price and supply controls, see Mitchell with Morita, Selly, and Stern (2001), pp. 173–175; Claes (2001), pp. 45–82; J.E. Hartshorn, *Oil Trade: Politics and Prospects* (Cambridge: Cambridge University Press, 1993), pp. 195–224; and Yetiv (2004), pp. 5–6.

34. On the defensive role played by non-OPEC suppliers at the end of the 1990s, see Mitchell with Morita, Selly, and Stern (2001), pp. 167–171; and Yetiv (2004).

35. Relaxing these assumptions does not alter the picture, given that access to oil reserves on the Caspian shelf was beyond the financial and technological frontiers of both states, and depended on the commercial considerations of leading international oil companies. Moderate projections of annual increases in the demand for oil, combined with forecasts for stable annual increases in supply—driven by advances in oil drilling, producing and refining technologies; the marketization and deregulation of much of the world's energy sectors; and the discovery of new oil reserves outside of OPEC and the former Soviet Union—augur well for an increasingly competitive and transparent global market in which no one supplier can exercise effective control over world prices. These are rough estimates based on the data presented in Ebel (1997), pp. 12–13; and Shane S. Streifel, *Review and Outlook for the World Oil Market* (Washington, DC: World Bank Discussion Paper, 1995), pp. 96, 104. In 1994–2001, Russia steadily reduced oil imports from 4.6 to 4 million metric tons. Under a free market scenario, Russia was projected to import at most only 13 percent of the crude oil exported by Kazakhstan and Azerbaijan. See The World Bank (1997), pp. 107, 129, and 131.

36. The bulk of these independent production entities were small, regional companies that accounted for less than 3 percent of the country's production. Eugene Khartukov, "Incomplete Privatization Mixes Ownership of Russia's Oil Industry," *Oil & Gas Journal* (18 August 1997), p. 39; Leslie Dienes, "The Russian Oil and Gas Sector: Implications of the New Property System," *National Bureau of Asian Research* No. 4 (March 1996), pp. 8–10; and Sagers (2001), p. 167.

37. Ownership of the subsidiaries was divided between the state and private sector, with the government holding a 38 percent stake, 17 percent sold to citizens for vouchers, and the remainder sold to enterprise workers, management, and local administrations. Peter Rutland, "Lost Opportunities: Energy and Politics in Russia," *National Bureau of Asian Research Analysis* 8:5 (1997), p. 9; David Lane and Iskander Seifulmulukov, "Structure and Ownership," in David Lane, ed., *The Political Economy of Russian Oil* (Lanham, MD: Rowman & Littlefield, 1999), p. 24; and Dienes (1996), p. 9.

38. Khartukov (1997), pp. 36–40. This situation persisted until the end of 2004, when the government began to show interest in reclaiming strategic ownership of the industry. The latter included a proposed merger of the state-owned Rosneft oil firm with Gazprom, and the subsequent forced auc-

tioning of the prized YuKOS subsidiary, Yuganskneftgaz, to Rosneft to settle the private company's tax debt. This was seemingly abandoned in Spring 2005, however, in favor of a plan for the Kremlin to purchase outright additional (controlling) shares in Gazprom via sales of minority stakes in Rosneft. See discussion in Martha Brill Olcott, *The Energy Dimension in Russia's Global Strategy: Vladimir Putin and the Politics of Oil* (Houson: James A. Baer, III Institute for Public Policy of Rice University, 2005); and *New York Times*, 18 May 2005.

39. Lane and Seifulmulukov (1999), pp. 23–26; and Dienes (1996), pp. 8–11. In 1993, nearly all of Russia's vertically integrated oil companies were 100 percent owned by the state. By 2000, there was only one major state oil company (Rosneft), as well as two firms in which the state owned more than a 75 percent stake, and several companies where the state held stakes of 36.8 percent, 14.1 percent, and 1.1 percent, respectively. The remaining companies were completely privatized. IEA (2002), p. 69.

40. Lane and Seifulmulukov (1999), pp. 19–23. The Transneft system, which transported over 95 percent of all Russian crude oil, was initially comprised of nearly 40,000 miles of trunk lines, 70 percent located solely on Russian territory. The pipeline system connected 17 countries including Russia, Ukraine, Kazakhstan, Belarus, Lithuania, Latvia, Uzbekistan, Turkmenistan, Azerbaijan in the FSU, and Germany, Poland, the Czech Republic, Slovakia, Hungary, Slovenia, Croatia, and Serbia. In addition, the Transneft system pumped Russian and NIS oil to marine terminals on the Black Sea (Novorossiysk, Tuapse, and Odessa), and the Baltic Sea (Ventspils). The trunk line network in Russia shrunk slightly from 30,000 miles in 1991 to 28,000 miles by 1999. The World Bank (1997), pp. 32–35; and IEA (2002), p. 88.

41. Stephen D. Krasner, *Defending the National Interest: Raw Materials Investments and U.S. Foreign Policy* (Princeton: Princeton University Press 1978), pp. 245–273.

42. C. Locatelli, "The Russian Oil Industry Restructuration: Towards the Emergence of Western Type Enterprises," *Energy Policy* 27 (1999), p. 441; Khartukov (1997), pp. 36–40; and Dienes (1996), pp. 15–29.

43. IEA (2002), p. 78; Laura Wakefield, "The Need for Comprehensive Legislation in the Russian Oil and Gas Industries," *Case Western Reserve Journal of International Law* 29:1 (Winter 1997), pp. 149–163; and Moe and Kryukov (1998), pp. 588–605.

44. Sagers (2001), pp. 178–181.

45. Wakefield (1997), pp. 149–163; Mathew J. Sagers, Igor A. Didenko, and Valeriy A. Kryukov, "Distribution of Refined Petroleum Products in Russia," *Post-Soviet Geography and Economics* 40:5 (1999), pp. 407–439; Matthew J. Sagers, "Russian Crude Oil Production in 1996: Conditions and Prospects," *Post-Soviet Geography and Economics* 37:9 (1996), pp. 532–535; Sagers (2001), pp. 169–182; Moe and Kryukov (1998), pp. 588–605; and IEA (2002), pp. 79–83.

46. Wakefield (1997), pp. 149–153; IEA (2002), pp. 83–87; and James Watson, "Foreign Investment in Russia: The Case of the Oil Industry," *Europe-Asia Studies* 48:3 (May 1996), pp. 429–456;

47. IEA (2002), p. 87. Foreign direct investment in the Russian oil sector totaled $4 billion, with the annual average ranging between $200–$400 million from 1994–2000. Investment peaked in 1999 at $1.2 billion but declined in 2000 due to continued problems with implementing PSAs.

48. Eugene M. Khartukov, "Low Oil Prices, Economic Woes Threaten Russian Oil Exports," *Oil & Gas Journal* (8 June 1998), pp. 25–30; Matthew J. Sagers, Valeriy A. Kryukov, and Vladimir V. Shmat, "Resource Rent from the Oil and Gas Sector and the Russian Economy," *Post-Soviet Geography and Economics* 36:7 (1995), pp. 401–414; and Sagers (1996), pp. 532–537. The same diffusion of responsibility characterized regulation of tax, legal, and foreign direct investment in the Russian oil sector. See Charles P. McPherson, "Policy Reform in Russia's Oil Sector," *The World Bank* (June 1996).

49. IEA (2002), p. 77.

50. Ibid., pp. 92–93.

51. Ibid, 89–94; The World Bank (1997), pp. 220–226; and Watson (1996).

52. LUKoil planned for at least 20 percent of its production to come from outside of Russia. *Novaya Gazeta,* 13 May 1996; *Segodnya,* 19 June 1996; and *Nezavisimaya gazeta,* 14 January 1997. See also Yakov Pappe, "Neftyanaya i gazovaya diplomatiya Rossii," *Pro et Contra* 2:3 (Summer 1997), pp. 61–69; Igor Khripunov and Mary M. Matthews, "Russia's Oil and Gas Interest Group and Its Foreign Policy Agenda," *Problems of Post-Communism* (May/June 1996), pp. 404–447; Isabel Gorst and Nina Poussenkova, *Petroleum Ambassadors of Russia: State Versus Corporate Policy in the Caspian* (Houston: James A. Baker III Institute for Public Policy, Rice University, 1998); and David Lane and Iskander Seifulmulukov, "Company Profiles: LUKoil, YuKOS, Surgutneftgaz, and Sidanko," Lane (1999), p. 124.

53. Over 70 percent of Russia's proven oil reserves were concentrated in West Siberian fields. In addition, 14 percent were located in the Volga region, 7 percent in the underdeveloped Tyuman-Pechora Basin, 4 percent in East Siberia, 3 percent in Sakhalin, with the remainder located in the North Caucasus and Kaliningrad. The Tyuman oblast in West Siberia accounted for nearly 66 percent of national production Sagers (2001), p. 159.

54. Bruce Kellison, "Tyuman, Decentralization, and Center-Periphery Tension," in Lane (1999), pp. 127–142.

55. Tsnetr Strategicheskikh Razrabotok (2004).

56. Cited in Michael Lelyveld, "Russia: Pipeline Projects to Remain Under State Control," *RFE/RL,* 13 January 2003. Transneft undermined private initiatives to construct an oil pipeline from Irkutsk to China, as well as subverted private proposals to ship exports through the Latvian port of Ventspils and to construct an export pipeline and terminal for the Artic port of Murmansk. Russian oil executives claimed that administrative bottlenecks in the

Transneft system presented them with the difficult choice of curbing production during winter months or suffering huge commercial losses with
over 134 billion barrels shut-in from higher priced export markets. Tsnetr
Strategicheskikh Razrabotok (2004).
57. IEA (2004), p. 93.
58. At independence, Russia provided nuclear fuel service to commercial plants
located in Ukraine, Lithuania, Kazakhstan, Armenia, Finland, Hungary,
Slovakia, the Czech Republic, and Bulgaria.
59. IEA (2002), pp. 171, 175–177.
60. There were 16 NPPs either under construction or planned for operation by
2015, with 9 NPPs slated for decommissioning.
61. Nuclear Energy Agency (NEA), *Uranium 1999: Resources, Production, and
Demand* (Paris: OECD, 1999).
62. Canada and Australia together captured over 50 percent of the global
market of natural uranium supply by 2000. "DATA FEATURE: 2000
World Uranium Production," *NUKEM Market Report* (April 2001).
Russia's uranium production was slightly higher than both Kazakh and
Uzbek annual production, but constituted only 60 percent of their combined output by the end of the decade. World Nuclear Association (WNA),
"Uranium Production Figures, 1995–2003," *Information and Issue Briefs*
(April 2004), www.world-nuclear.org/info/uprod.htm.
63. "Minatom Rising Updated," NUKEM (December 2002). World natural
uranium production satisfied only 60 percent of global reactor requirements
for the period. NEA (1999), p. 63.
64. As of 2000, uranium production in both Uzbekistan and Kazakhstan
exceeded Russian production, constituting 5.8 and 5.1 percent, respectively,
of the global share. By the end of 2002, the two countries surpassed Russian production, accounting for 6.9 and 5.5 percent, respectively. It also was
estimated that Russia's secondary sources of blended down LEU would be
sufficient over the short-term to meet nearly 40 percent of the world uranium demand outside of the FSU. "Data Feature" (2001); WNA (2004);
Gerard Pauluis, "The Global Nuclear Fuel Market to 2020," *The Uranium
Institute Twenty-third Annual International Symposium 1998*, www.uilon-
don.org/uilondon/sym/1998/pauluis.htm; and NEA (1999), pp.
23–24, 190, 241–242.
65. Soizick Martin, "EC's 'Nuclear Package' to Harmonize Atomic Energy in
Expanded EU—but Environmentalists Cry Foul," *Bellona Report* (17 October 2003), www.bellona.no. This issue intensified during deliberations over
EU enlargement in 2004. In response, Russian officials pledged to preserve
the European niche without increasing its overall supply quota to Europe.
Soizick Martin and Charles Digges, "Upcoming EU Enlargement Revives
Long-Standing Nuclear Battle," *Bellona Working Paper* (10 February
2004), www.bellona.no.
66. Under the terms of the agreement, an American utility entering into a
matched sales deal paid an aggregate price, with U.S producers receiving a

higher price than the Russian supplier. Amendments were attached that imposed and then lowered quotas for matched sales, and that placed enriched uranium imported from Russia under similar "matched deal" restrictions. Although Russian officials refused to acknowledge that sales were adversely affected by "matching," they complained that the 4.2 metric tons sold in 2002 were below its commercial potential. Charles Digges, "Tenex Might be Stuck with Tonnes of Uranium it Produces for U.S. Consumption," *Bellona Report* (11 November 2002), www.bellona.no; and "Uranium Imports to the USA from CIS Countries," *Uranium Institute Trade Briefing* 1 (1 August 1999).

67. As early as 1997, Russia threatened to pull out of the deal unless it received assurances that the private Western selling parties agreed to honor the pricing terms of the 1993 government-to-government agreement. "The Elusive Peace Dividend," *The Ux Weekly* (18 October 1999); and Rashid Alimov, "Misinformation about the Uranium Deal," *Bellona Report* (2 August 2001), www.bellona.no.

68. The three stages constituted the critical link between the nuclear fuel cycle and nuclear power plant, representing approximately 10 percent of total costs for nuclear electricity generation. NEA, *Trends in the Nuclear Fuel Cycle: Economic, Environmental, and Social Aspects* (Paris: Nuclear Energy Agency, 2001), p. 34.

69. NEA (2001), pp. 31–32; and Peter Drasdo, "Impacts of Power Market Deregulation on the Nuclear Fuel Cycle Markets and Industries," *EWI-Working Paper* (February 1999), pp. 5–6.

70. Russia's traditional customers in the Czech Republic, Finland, and Slovenia purchased enriched uranium from other suppliers. NEA (2001), pp. 32–34; Drasdo (1999), pp. 6–8; and "Country Nuclear Fuel Cycle Profiles," *International Atomic Energy Agency Technical Reports Series* 404 (2001), p. 12.

71. Nuclear fuel fabrication services are less homogenous than other front-end services. Consequently, fuel fabricators are typically fuel service and reactor vendors. In addition, finished fuel assemblies are usually tied to specific customers, given economies of scale of established suppliers and transportation licensing and import taxes. NEA (2001), pp. 34–37; Drasdo (1999), pp. 9–11; Country Nuclear Fuel Cycle Profiles (2001), p. 13; and Igor Kudrik, "Nuclear Fuel Producer Gets Nervous," *Bellona Report*, 23 June 2000, www.bellona.no.

72. Renaissance Capital, "UESR Appeals Anti-Monopoly Ministry Ruling," *Morning Monitor*, 19 September 2001, cited in Michael T. Maloney and Oana Diaconu, *Analysis of Privatization of Russia's Nuclear Ministry on the Nuclear Nonproliferation Objectives of the United States* (The Strom Thurmond Institute, Clemson University Special Report, 2001), p. 34, www.strom.clemson.edu.

73. Sales of Russian nuclear energy to CIS customers at prices that were slightly higher than domestic wholesale prices would not have generated cash flows sufficient to "support expansion of nuclear generation capacity." The sales

provided supplemental energy for primary electricity consumption, selling for "less than the full costs of baseload power." Maloney and Diaconu (2001), pp. 33–37.

74. By the end of 2002, Russia retained a monopoly of nuclear fuel supply for local power generation in only three states: Armenia, Bulgaria, and Hungary. "Russian Nuclear Ministry to Lose Monopoly Over Nuclear Fuel delivery in Ukraine," *Bellona Report* (9 June 2000), www.bellona.no.; and Country Nuclear Fuel Cycle Profiles (2001).

75. The nuclear sector accounted for 21 percent of electricity production in the European part of Russia in 2002, and was projected to supply 45 percent of that demand by 2030. Maloney and Diaconu (2001), p. 32. See also Oana Diaconu and Michael T. Maloney, "Is Nuclear Power Viable in Russia?" *The Electricity Journal* (January/February 2003), pp. 80–87. The authors cite pessimism concerning the commercial viability of shifting to greater reliance on nuclear power among senior government officials both inside and outside of the nuclear sector. Senior Minatom officials acknowledged explicitly that nuclear power could not compete against natural gas-fired power generation under "normal cost assumptions." Oana C. Diaconu and Michael T. Maloney, "Russian Commercial Nuclear Initiatives and U.S. Nuclear Nonproliferation Interests," *The Nonproliferation Review* (Spring 2003), p. 7.

76. *Radio Mayak,* 22 December 2000, as translated in *FBIS-SOV CEP 20010102000111,* 22 December 2000. See also discussion in Charles E. Zeigler and Henry B. Lyon, "The Politics of Nuclear Waste in Russia," *Problems of Post-Communism* 49:4 (July/August 2002), pp. 33–42.

77. *Vedomosti,* 21 November 2000, p. 2, as translated in *FBIS-SOV CEP 20001121000385,* 21 November 2000. The supply of Russian-origin spent nuclear fuel was projected to expand to 35,000 tons by 2025.

78. Unable to cover the cost of securing 230,000 tons of its own nuclear waste, Kazatomprom (Minatom's equivalent in Kazakhstan) began in 2002 to set the stage for luring former Soviet and foreign spent fuel contracts as a panacea. Marat Yermukanov, "Nuclear Waste Disposal Scheme Sets Off Wave of Protests in Kazakhstan," *Central Asia-Caucasus Analyst Field Report* (15 January 2003), www.cacianalyst.org/2003–01–15/20030115_Kazakhstan_Nuclear_Waste.htm.

79. Michael Knapik, Mark Hibbs, and Ann MacLachlan, "U.S. Open to Nuclear Pact With Conditions on International Spent Fuel Storage in Russia," *NuclearFuel* 26:12 (11 June 2001), www.mhenergy.com.

80. Minister Rumyantsev stated explicitly that Washington's reticence in approving the Russian plan cost Moscow a very lucrative contract with Slovenia in 2003. *Kommersant,* 4 November 2003.

81. Sam Vaknin, "Surviving on Nuclear Waste," *United Press International* (22 November 2002).

82. Russian law prohibited private ownership of nuclear power plants.

83. In 1998, there was a proposal to combine all three entities into a single state-owned joint stock company, Atomprom. This entity was modeled on

Gazprom, with the exception that exclusive ownership would reside with the federal government.

84. Sonia Ben Ougrahm, *Study of Minatom's Organizational Structure* (unpublished paper, September 2000).

85. Ibid., p. 19. Unlike the "power ministries" (i.e., the Ministry of Defense, Ministry of Foreign Affairs, Interior Ministry, and Federal Security Services) that reported directly to the president, and other ministries that reported to the prime minister via a designated deputy prime minister, Minatom reported directly to the prime minister.

86. Acquiring licensing-evaluation authority by the end of the decade, Minatom also secured greater discretion to relax safety standards to accommodate narrow commercial and technological objectives. Igor Khripunov, "MINATOM: Time for Crucial Decisions," *Problems of Post-Communism* 48:4 (July/August 2001), pp. 56–57.

87. Sonia Ben Ougrahm, "Minatom's Regional Strategy," in James Clay Moltz, Vladimir A. Orlov, and Adam N. Stulberg, eds., *Preventing Nuclear Meltdown: Managing the Decentralization of Russia's Nuclear Complex* (London: Ashgate Academic Publishers, 2004), pp. 69–74.

88. Ibid. Minatom was represented on three Security Council commissions.

89. *Kommersant,* 4 July 2001.

90. *Nezavisimaya gazeta,* 8 August 2001. Minatom consistently complained that the domestic pricing system orchestrated by UES deprived the nuclear industry of access to coveted cash-paying customers, as well as created opportunities to mask per unit savings of nuclear power generation via "accounting techinques."

91. Russian Federal Law, "On the Use of Atomic Energy," *Sobraniye zakonodatel'stva Rossiskoy Federatsii,* 27 November 1995, No. 48, Articles 10–12.

92. In 1997–1998, Minatom covered only 20 percent of the actual costs of operating the plutonium complex. Oleg Bukharin, "The Future of Russia's Plutonium Cities," *International Security* 21:4 (Spring 1997), p. 132; and Igor Khripunov and Maria Katsva, "Russia's Nuclear Industry: The Next Generation," *The Bulletin of Atomic Scientists* (March/April 2002), pp. 51–57.

93. Ben Ougrahm (2004), pp. 74–78.

94. See detailed discussion in Adam N. Stulberg, "Nuclear Regionalism in Russia: Decentralization and Control in the Nuclear Complex," *The Nonproliferation Review* (Fall/Winter 2002), pp. 31–46.

Chapter 4

1. Peter Rutland, "Lost Opportunities: Energy and Politics in Russia," *National Bureau of Asian Research Analysis (NBR)* 8:5 (1997), pp. 17–21.

2. *Ekspert,* 4 October 1999; E.A. Telegina, "Geopoliticheskie i ekonomicheskie interesy Rossii v global'nom energeticheskom prostranstve," *Energiya: ekonomika, tekhnika, ekologiya* 1 (2001), pp. 11–17; and *Izvestiya,* 22

June 1999. On Gazprom's regional strategy, see Yakov Pappe, "Neftyanaya i gazovaya diplomatiya Rossii," *Pro et Contra* 2:3 (Summer 1997), pp. 60–61; Isabel Gorst and Nina Poussenkova, *Petroleum Ambassadors of Russia: State versus Corporate Policy in the Caspian Basin* (Houston, TX: Center for International Political Economy, James A. Baker III Institute for Public Policy, Rice University, 1998), pp. 12–14; Igor Khripunov and Mary M. Matthews, "Russia's Oil and Gas Interest Group and Its Foreign Policy Agenda," *Problems of Post-Communism* (May/June 1996), pp. 38–48; Astrid Sahm and Kirsten Westphal, "Power and the Yamal Pipeline," in Margarita M. Balmaceda, James I. Clem, and Lisbeth L. Tarlow, eds., *Independent Belarus: Domestic Determinants, Regional Dynamics, and Implications for the West* (Cambridge: Distributed by Harvard University Press for the Ukraine Research Institute and Davis Center for Russian Studies, Harvard University, 2002), pp. 277–285; and Matthew J. Sagers, "The Russian Natural Gas Industry in the Mid-1990s," *Post-Soviet Geography* 36:9 (1995), pp. 551–557.

3. *Nezavisimaya gazeta,* 19 February 1999.
4. David M. Woodruff, "It's Value That's Virtual: Bartles, Rubles, and the Place of Gazprom in the Russian Economy," *Post-Soviet Economy* 15:2 (1999), pp. 136–139.
5. Domestic receipts dropped below foreign revenues in absolute terms following the ruble devaluation in 1998. International Energy Agency (IEA), *Russia's Energy Survey, 2002* (Paris, OECD 2003), pp. 131–132.
6. The push toward market liberalization within the European Union promised to drive gas prices downward and generate smaller profits for long-distance gas exports from Russia. European Commission, "Towards a European Strategy for the Security of Energy Supply," *EU Green Paper,* 2001; and *Ekspert* 4 October 1999.
7. Pappe (1997), pp. 59–61; and *Kommersant,* 24 August 2001.
8. Woodruff (1999), pp. 135–142; and Jonathan P. Stern, *Russian Natural Gas Bubble: Consequences for European Gas Markets* (London: The Royal Institute of International Affairs, 1993), pp. 34–38.
9. IEA (2002), pp. 127–130. Confirmed in interviews with Gazprom officials conducted by author in November 1997.
10. Energy Information Agency (EIA), "Russia: Energy Sector Restructuring" (April 2002); www.eia.doe.gov/cabs/russrest.html#OIL.
11. IEA (2002), pp. 302–306.
12. Ibid., p. 306. By the end of 2004, the Russian government officially acknowledged the strategic purpose of folding Central Asian gas suppliers into its overall energy blueprint. See especially "Development Strategies for Leading Russian Economic Sectors," Ministry of Economic Development and Trade, 14 December 2004, as translated in *Foreign Broadcast Information Service* (FBIS) CEP20050114000257, 14 January 2005.
13. "Text of Putin's Speech at the Foreign Ministry 12 July," *Moscow RTV,* 12 July 2002, as translated in *FBIS-SOV,* 15 July 2002. The goal was to direct

the strategic behavior of the gas industry to reinforce the country's global position via a combination of administrative and market mechanisms. On Putin's strategic energy vision, see especially Vladimir Putin, "Mineral Raw Materials in the Strategy for Development of the Russian Economy," *Notes of the Mining Industry* (January 1999), cited in Martha Brill Olcott, *The Energy Dimension in Russian Global Strategy: Vladimir Putin and the Politics of Oil* (Houston, TX: James A. Baker III Institute for Public Policy, Rice University, 2005).

14. Jan S. Adams, "Russia Gas Diplomacy," *Problems of Post-Communism* 49:3 (May/June 2002), p. 21

15. Statement by Gazprom official, cited in Ibid., p. 21. The point was affirmed in author's interview with senior Gazprom official, November 2002.

16. Sergei Kamenev, "Ekonomika Turkmenistana na sovremennom etape," *Tsentral'naya aziya i kavkaz* 3:21 (2002), pp. 194–205; and Igor Torbakov, "Russia-Turkmen Pacts Mark Strategic Shift for Moscow in Central Asia," *EurasiaNet/Eurasia Insight,* 15 April 2003.

17. Russian officials pressured President Niyazov to reinstate the Turkmenrosgaz joint-stock company that was 45 percent owned by Gazprom and that had a record of mismanaging transport and sales of Ashgabat's natural gas to CIS customers. Akira Miyamoto, *Natural Gas in Central Asia: Industries, Markets, and Export Options of Kazakhstan, Turkmenistan, and Uzbekistan* (London: Royal Institute of International Studies, 1997), pp. 44–48; and Daniel W. Drezner, *The Sanctions Paradox: Economic Statecraft and International Relations* (Cambridge: Cambridge University Press, 1999), pp. 169–173.

18. International Monetary Fund (IMF), *Turkmenistan: Recent Economic Developments,* Country Report No. 99/140 (December 1999), p. 119.

19. *Nezavisimaya gazeta,* 25 April 2002; and *Interfax,* 24 April 2002. In the 2002–2003 long-term contract with Russia, Turkmenistan agreed to accept a $22 per thousand cubic meters (cm) cash price for its exports to Russia until 2006. This contrasted with the $90–$120 per thousand cm that Russia charged in the European market.

20. Miyamoto (1997), pp. 42–43. On the costs of Turkmenistan's continued failure to realize its gas export potential, see Nancy Lubin, "Turkmenistan's Energy: A Source of Wealth or Instability?" in Robert Ebel and Rajan Menon, eds., *Energy and Conflict in Central Asia and the Caucasus* (Lanham: Rowman & Littlefield Publishers, 2000), pp. 107–121. Until 1994, Russia allowed Turkmenistan to export 11–14 bcm of gas per annum to Europe. With access cutoff in 1995, Turkmenistan's gas production dropped by nearly 45 percent, causing a 30 percent drop in GDP. The resumption of gas exports to Russia in 1999 produced a 10 percent increase in Turkmenistan's GDP during the first half of 2000. See IEA, *Caspian Oil and Gas: The Supply Potential of Central Asia and Transcaucasia* (Paris: OECD, 1998), p. 243.

21. *Nezavisimaya gazeta,* 27 October 1999. See also Sergei Kamenev, "Toplivno-energeticheskii kompleks Turkmenistana: sovremennoe sostoyanie i perspektivy razvitiya," *Tsenral'naya aziya i kavkaz* 6:18 (2001), pp. 190–191; and Richard M. Auty, "Sustainable Mineral Driven Development in Turkmenistan," in Shiroz Akiner, Sader Tideman, and John Hay, eds., *Sustainable Development in Central Asia* (Surrey, UK: Curzon Press, 1998), p. 170.

22. Cited in Naz Nazar, "Turkmenistan: U.S. Official Discusses Trans-Caspian Pipeline," *RFE/RL,* 14 August 2000.

23. Sergei Kamenev, "Vneshnyaya politika Turkmenistana," *Tsentral'naya aziya i kavkaza* 4:22 (2002), pp. 91–92. "With construction costs of $2.2 billion, of which 50 percent would be spent in Turkmenistan, the size of this project in relation to macroeconomic data from 1998 was roughly 33 percent of GDP, with full shipments of gas projected to nearly double national exports. IMF (1999), p. 17. However, Ashgabat was willing to relax this standard for delivering gas to Iran and Russia. With only 50 percent of the TCP pipeline capacity dedicated to Turkmen gas, Ashgabat was presented with a transit option that could deliver more gas than it had planned to deliver via the Iranian pipeline and onto Russia prior to 2006. "Turkmenistan Ponders Russian-initiated Gas Alliance," *Ashgabat Ministry of Foreign Affairs Press Service,* 23 July 2002, as translated in *FBIS-SOV CEP20020724000328,* 24 July 2002.

24. Murad Esenov, "Vneshnyaya politika Turkmenistana i ee vliyanie na sistemu regional'noi bezopasnosti," *Tsentral'naya aziya i kavkaz,* 1:13 (2001), p. 62.

25. Author's interviews with U.S. State Department officials, July-November, 2002. Turkmenistan's proposal for the pipeline link to Afghanistan and Pakistan was under consideration at the UN in May 2002.

26. U.S. proven gas reserves were estimated at 4.6 tcm, while Russia was estimated to possess 49.3 tcm during the period. In contrast, the Turkmen government optimistically claimed that the country possessed up to 24 tcm in proven and unproven natural gas reserves, and that national production could reach 230 bcm per year by 2010. Robert Ebel, *Energy Choices in the Near Abroad* (Washington, DC: Center for Strategic and International Studies, April 1997), p. 130; Miyamoto (1997), p. 41; Ottar Skagen, *Caspian Gas* (London: The Royal Institute of International Affairs, 1997), pp. 35–38; Andrei V. Belopolsky and Manik Talwani, "Geological Basins and the Oil and Gas Reserves of the Greater Caspian Region," in Yelena Kalyuzhnova, Amy Myers Jaffe, Dov Lynch, and Robin Sickles, eds. *Energy in the Caspian Region* (Hampshire, UK: Palgrave, 2002), pp. 24–33; and IEA (1998), pp. 32–34.

27. EIA, "Central Asia: Turkmenistan Energy Sector," May 2002, www.eia.doe. gov/emeu/cabs/turkmen/html; and Stuart S. Brown and Misha V. Belkindas, "Who's Feeding Whom? An Analysis of Soviet Interrepublic Trade," in

Richard F. Kaufman and John P. Hardt, eds., *The Former Soviet Union in Transition* (Armonk, NY: M.E. Sharpe, 1993), pp. 34–35.

28. Sally N. Cummings and Michael Ochs, "Turkmenistan: Saparmurat Niyazov's Inglorious Isolation," in Sally N. Cummings, ed., *Power and Change in Central Asia* (London: Routeledge, 2002), p. 123.

29. Kamenev (2002), pp. 199–200. See also Neil MacFarlane, *Western Engagement in the Caucasus and Central Asia* (London: Royal Institute of International Affairs, 1999), pp. 35–39.

30. For a discussion of Ashgabat's independent internal gas distribution system and optimistism on the development of its fuel and energy complex, see especially IMF (1999); Kamenev (2001), pp. 190–192; and Andrei Kalyuzhnov, Jilian Lee, and Julia Nanay, "Domestic Use of Energy: Oil Refineries and Gas Processing," in Kalyuzhnova, Jaffe, Lynch, and Sickles (2002), p. 157.

31. As of 2000, the country consisted of 77 percent Turkmen. According to official Turkmen sources, the Russian minority dropped from 15 percent to 2 percent of the population by 1999. To the extent that an ethnic minority presented a threat to the regime's stability, it came from the Uzbek minority that was predominantly situated along the Uzbek-Turkmen border. Cummings and Ochs (2002), p. 126.

32. The regime was characterized by the concentration of arbitrary power in the hands of President Niyazov, fusion of state and regime interests with the leader's personality cult, and absence of a credible political opposition. See Cummings and Ochs (2002), pp. 116–124; and Steven Sabol, "Turkmenbashi: Going it Alone," *Problems of Post-Communism* 50:5 (September/October 2003), pp. 48–57.

33. Drezner (1999), p. 170. See also discussion in Lubin (2000), pp. 107–121; and Michael Ochs, "Turkmenistan: The Quest for Stability and Control," in Karen Dawisha and Bruce Parrott, eds., *Conflict, Cleavage and Change in Central Asia and the Caucasus* (Cambridge: Cambridge University Press, 1997), p. 314. Drezner claims that for Turkmenistan, "the short-term costs of substituting away from the Russian pipelines were greater than the costs of conceding." While this was the case in terms of short-term costs, Ashgabat's long-term prospects were potentially more favorable. Moreover, this does not explain why a strong Turkmen leadership would be preoccupied with a short-term strategic calculus.

34. Paul Kubicek, "Regionalism, Nationalism, and *Realpolitik*," *Europe-Asia Studies* 49:4 (1997), pp. 643–644; and Lena Jonson, *Russia and Central Asia: A New Web of Relations* (London: Royal Institute of International Affairs, 1998), pp. 41–47.

35. Cummings and Ochs (2002), pp. 116–124.

36. Esenov (2001), pp. 56–64.

37. *Nezavisimaya gazeta,* 19 February 1997.

38. Andrei P.Tsygankov, *Pathways after Empire* (Boulder: Rowman & Littlefield Publishers, 2001); pp. 176–181; and R. Frietag-Wirminghaus, "Turk-

menistan's Place in Central Asia and the World," in Mehdi Mozaffari, ed., *Security Politics in the Commonwealth of Independent States: The Southern Belt* (London: St. Martin's Press, 1997), pp. 77.
39. Interview with former U.S. State Department official, 20 November 2002.
40. Lubin (2000), pp. 110–113.
41. Cummings and Ochs (2002), p. 122.
42. Miyamoto (1997), p. 49. See also IMF (1999), p. 118; and IEA (1998), p. 245.
43. Belopolsky and Talwani (2002), pp. 13–33.
44. Lubin (2000), pp. 108–110.
45. IMF (1999), p. 123. After suffering a precipitous 25.9 percent decline in GDP in 1998, Turkmenistan experienced four successive years of growth; yet, GDP in 2001 remained only 70 percent of its 1990 level. See discussion in EIA (May 2002).
46. EIA, *Russia: Oil and Natural Gas Exports* (November 2002); www.eia.doe.gov/cabs/russexp.html
47. European Commission (2001), pp. 24, 41.
48. The Yamal Pipeline was constructed in 1999 and was initially scaled to deliver 28 bcm at full compression to Germany via Belarus and Poland. During the first year of operation, 10–14 bcm of natural gas was exported to Germany via this pipeline. Russia's competitive advantages in the EU market were augmented by the depletion of North Sea reserves, and the high costs of building undersea pipeline from Algeria. IEA (2002), p. 138.
49. Russia's gas reserves are as much as 16 times larger than those of the Caspian region, and 24 times larger than those of Turkmenistan. Sagers (1995), p. 529. Turkmenistan's main hope for breaking into the crowded European gas market over the ensuing 30–40 years depended on decisions by foreign consumers to opt for supply diversity over lower prices (i.e., canceling commercially favorable long-term "take-or-pay contracts" with traditional suppliers), as well as on Ashgabat's willingness to sell gas at bargain basement prices while paying higher-transit costs than its competitors. Ira Joseph, *Caspian Gas Exports: Stranded Reserves in a Unique Predicament* (Houston, TX: Center for International Political Economy and the James A. Baker III Institute for Public Policy, Rice University, April 1998). Russia's success in November 2000 at locking in an agreement with the EU to supply two-thirds of the European market by 2020 suggests that such a shift was unlikely.
50. Skagen (1997), pp. 54–81. The cost of supplying Turkmen gas to the West European market was estimated between $127 and $152/1000 cm. In comparison, Algeria was projected to deliver gas for $65/1000 cm, while Russia could have delivered it for $113 to $131/1000 cm. Accordingly, it was expected to require the delivery of more than 200 bcm for it to be cost-effective for Western Europe to import gas from Turkmenistan.
51. *Ekspert,* 4 October 1999. The Blue Stream project, which came on line at the end 2002, promised to expand Russia's two-thirds share of the Turkish

market previously serviced by two existing pipelines. The Turkish government was severely criticized for caving into Moscow's pressures and overpaying for Russian gas. In return for hastening the construction of the project, Turkey initially pledged to pay Russia $114–$125 per 1,000 cm for the delivery of 16 bcm per annum, while both Turkmenistan and Iran offered the same amount for roughly $70 per 1,000 cm.

52. Thomas Stauffer, "Caspian Fantasy: The Economics of Political Pipelines," *Brown Journal of World Affairs* VII:2 (Summer/Fall 2000). The uncertainty associated with Turkey's economic downturn and inflated demand for natural gas imports was born out in April 2003 when Ankara faced a domestic gas glut and stopped the flow of Russian gas from the Blue Stream pipeline. Turkish parliamentarians claimed that by Spring 2003 the country had already lost $196 million in excess payments to Gazprom and had become precariously dependent on Russia for 70 percent of its gas imports. Relations seemed back on track by 2004, as Gazprom purchased shares in Turkey's internal gas distribution and storage system, as well as acquired rights to transit deliveries across Turkey to third countries. *Reuters,* 28 April 2003; Michael Lelyveld, "Turkey: Ankara Cuts Russian Gas, Courts Iran," *RFE/RL,* 29 April 2003; and *Gazeta.ru,* 12 January 2005.

53. This would have been the case even with the lifting of U.S. sanctions on Iran and the easing of American pressure on the Turkmen-Iran-Turkey project. Miyamoto (1997) pp. 64–71; IEA (1998), pp. 254–260; and Skagen 1997), pp. 83–85.

54. Robert Cutler, "The Indo-Iranian Rapproachment: Not Just Natural Gas Anymore," *Central Asian Caucasus Analyst* (9 May 2001).

55. "Niyazov, Turkish Deputy Premier Discusses Pipeline," *Ashgabat Turkmen Television,* 6 October 1999, as translated in *FBIS-FTS19991011001397,* 11 October 1999.

56. *Kommersant,* 18 October 2000; and Michael Lelyveld, "Turkmenistan: Niyazov Rejects Long-term Gas Deal with Russia," *RFE/RL,* 26 September 2000.

57. Deliveries prior to 2000 were restricted to less than 2 bcm due to limited Iranian demand. There were plans to link this pipeline to the existing 600–mile pipeline running west across Northern Iran to Turkey.

58. Lubin (2000), p. 114.

59. "Turkmenistan Ponders Russian Initiated Gas Alliance," *Turkmen Foreign Ministry Press Service,* 23 July 2002, as translated in *FBIS CEP20020724-000328,* 24 July 2002.

60. Turkmen gas deliveries to Iran via the Korpedzhe-Kordkuy line were 1.8 bcm in 1998; 2 bcm in 1999; 1 bcm in 2000, 4.5 bcm in 2001; and 4.5 bcm in 2002. The price was generally $2 per cm more than what Russia paid. "Turkmen President Says Iran Best Route for Pipeline to Turkey," *Tehran IRNA,* 22 September 2000, as translated in *FBIS-IAP20000922000023,* 22 September 2000.

61. *Wall Street Journal*, 25–27 April 2003. Pessimism intensified in 2005, as India began to expand direct cooperation with Gazprom on gas extraction and LNG projects in both countries.
62. *Nezavisimaya gazeta*, 8 June 2000.
63. *Nezavisimaya gazeta*, 7 August 1997; and *Segodnya*, 8 August 1997. According to Russian sources, the 1997 deal did not provide for the construction of alternative pipelines. Furthermore, Gazprom was able to use new standing in Turkmenistan to participate as a "strategic partner" in the development of the southern pipeline route via Pakistan. Interests in directing the flow of Turkmenistan's gas to Southwest Asia and in restricting Ashgabat's access to Turkish and West European markets were reiterated by the chairman of Gazprom in February 2000 during a meeting with President Niyazov. See *Russian Oil & Gas Report*, 25 February–2 March 2000.
64. *Nezavisimaya gazeta*, 7 August 1997; *Segodnya*, 8 August 1997; *Nezavisimaya gazeta (sodruzhestvo)*, 28 April 1999; *Izvestiya*, 22 June 1999; and *Ekspert*, 31 January 2000.
65. Adams (2002), p. 16.
66. *Delovie lyudi*, May 1996, pp. 146–48. By mid-1996, Gazprom received only 26 percent of the payments for gas delivered to consumers in Russia, and was prevented by presidential decree from cutting-off domestic supplies.
67. The chairman of Gazprom, Rem Vyakhirev warned that failure to lower domestic gas prices would ensure that "whatever is still showing signs of life will shutdown once and for all." *Kommersant* 10 April 1997; and Woodruff (1999), pp. 138–139.
68. *Ekspert*, 13 January 1997. See also discussion in David Woodruff, *Money Unmade: Barter and the Fate of Russian Capitalism* (Ithaca: Cornell University Press, 1999), pp. 192–193; and Yu. I. Koryakin, "Gde i kak gorit Rossiiskii gaz," *Energiya: Ekonomika, tekhnika, ekologiya* 12 (2000), p. 19.
69. IEA (2002), pp. 131–132.
70. Ibid., p. 132.
71. *Izvestiya*, 22 April 1999; and *Ekspert*, 18 January 2000.
72. *Nezavisimaya gazeta*, 22 December 1999. To sweeten the deal for Gazprom, Kiev was charged $80 per 1,000 cm for the delivery of Turkmen gas by Russia. Russia insisted that Turkmenistan pay a transit fee of $1.75 per 1,000 cm of gas per 100 km, which was 20 percent more than the international rate. By contrast, in 1998, Russia paid $32 per 1,000 cm for 20 bcm, of which only 30 percent was paid in hard currency. *ITAR-Tass*, 22 April 1998; and *Interfax*, 12 March 1998.
73. *Segodnya*, 19 May 2000; and *Izvestiya*, 22 August 2000. The final terms contrasted with Gazprom's initial unwillingness to pay more than $32 per 1,000 cm, with 70 percent covered in goods and equipment. Turkmenistan was demanding at the time a sales price of $40 per 1,000 cm, half of which would be covered in hard currency. Ultimately, however, Ashgabat had to

standby while Russia reexported Turkmen gas to Azerbaijan at a markup. *Interfax,* 24 January 2000. According to the President of Turkmenistan, this deal was neither "profitable" for Ashgabat, nor was it an acceptable precedent for future transactions with Russia or other states. See *Interfax,* 22 September 2000.

74. *ITAR-Tass,* 16 February 2001; and *Nezavisimaya gazeta,* 17 May 2001. Moscow subsequently charged $42 per 1,000 cm for Turkmenistan gas that passed through the Russian pipeline system. This deal left Turkmenistan in the precarious position of subsidizing gas sales to Russia while doubling gas exports to an unreliable customer, Ukraine, via pipelines controlled by Russia.

75. Torbakov (2003); and *Nezavisimaya gazeta,* 22 January 2002.

76. This echoed Niyazov's long-standing concerns about signing long-term contracts with Russia that, in his own words "can lead to arguments." Cited in Michael Lelyveld, "Turkmenistan: Niyazov Rejects Long-term Gas Deal with Russia," *RFE/RL,* 26 September 2000. These demands contrasted with previous deliveries to Russia that included 30 bcm in 2000 and 10 bcm in 2001, of which only 40 percent of the sales were paid in cash at $36–$40 per thousand cm. See also *Nezavisimaya gazeta,* 23 January 2002.

77. Although this presented Ashgabat with an alternative transit option that could deliver more gas than via the Iranian pipeline, it was a far cry from meeting the 30 bcm reference point prior to 2006.

78. *Moscow News,* 16 April 2003; and *Wall Street Journal,* 25–27 April 2003. Although Russia agreed to pay cash for deliveries after 2006, it reserved the right to negotiate annual prices and volumes. In 2005, Turkmenistan disrupted supply to force a price hike to $58 per cm in order to "cover the real costs" of delivery. Putin and Gazprom held firm, however, forcing Ashgabat to concede to "unacceptable levels" in the hopes of raising future deliveries and terms for cash payments in 2007–2008. *ITAR-Tass,* 11 February 2005.

79. *Izvestiya,* 7 August 1997.

80. Cited in Mikhail Alexandrov, *Uneasy Alliance: Relations between Russia and Kazakhstan in the Post-Soviet Era, 1992–1997* (Westport, CT: Greenwood Press, 1999), p. 292.

81. This shift in policy notably drew harsher criticism from Tehran than Moscow.

82. For a review of Turkmenistan's prolonged waffling and reluctance to outpace shifts in Russia's policy on the legal status of the Caspian Sea, see especially *Nezavisimaya Gazeta,* 13 April 2001; *Financial Times,* 29 May 2001; and Vladimir Mesamed, "Turkmenistan: Oil, Gas, and Caspian Politics," in Michael P. Croissant and Bulent Aras, eds. *Oil and Geopolitics in the Caspian Sea Region* (Westport, CN: Praeger, 1999), pp. 213–219.

83. Turkmenistan's claims to national sectoral ownership of contested undersea deposits were manifest in an "on-again-off-again" battle with Azerbaijan over the legal status of specific oil fields (not gas). Moreover, Niyazov's idiosyncrasies impeded international exploration of the TCP alternative. Yet

with seemingly nothing to lose, given Russia's stranglehold over Turkmenistan's Caspian options. Niyazov's erratic behavior both toward Baku
and potential investors in the TCP does not seem irrational.

84. Tsygankov (2001), p. 177; and Sabol (2003), pp. 51–53.

85. *Delovoi mir,* 8 August 1997. President Niyazov's refusal to sign the Ankara
Declaration (for "commercial reasons") that endorsed the Baku-Ceyhan
main oil pipeline route, suggests that Moscow's leverage extended to issues
beyond the gas sector. *Moscow News,* 26 November 1998.

86. Kamenev (2001), p. 93.

87. Esenov (2001), pp. 56–64.

88. *Izvestiya,* 22 January 2002. Kamenev (2001), pp. 95–98; and Kamenev
(2002), p. 201. Ukraine was Ashgabat's largest gas customer by the end of
2001.

89. Martha Brill Olcott, *Kazakhstan: Unfulfilled Promise* (Washington, DC:
Carnegie Endowment for International Peace, 2002), pp. 49–50.

90. EIA, *Kazakhstan,* July 2003; www.eia.doe.gov/emeu/cabs/kazak.html.

91. IMF, *Republic of Kazkhastan: Selected Issues and Statistical Appendix,*
County Report, No. 00/29 (March 2000); IMF, *Republic of Kazakhstan:
Recent Economic Developments,* Country Report No. 98/84 (July 1998);
and IMF, *Russian Federation Statistical Appendix,* Country Report, No.
03/145 (May 2003).

92. IMF (March 2000). See also discussion in Skagen (1997), pp. 18–19; and
Miyamoto (1997), pp. 18–24. For discussion of the generic and energy
sector-specific problems confronted by foreign investors in Kazakhstan, see
Olcott (2002), pp. 144–171.

93. Proven reserves were comparable to only four percent and 64 percent of
those of Russia and Turkmenistan, respectively.

94. EIA (2003); and Belopolsky and Talwani (2002), p. 29.

95. The country received support from international lending institutions to
explore these plans, and to improve the utilization of associated gas by
domestic oil producers. Kalyuzhnov, Lee, and Nanay (2002), pp. 161–162.

96. MacFarlane (1999), p. 42; and Murat Asipov, "Kazakhstan: Living Standards During the Transition," *World Bank Report,* No. 17520–KZ (23
March 1998).

97. EIA (2003); and Olcott (2002), p. 168.

98. Nazarbaev's authority was secured by adroitly manipulating formal institutions to: dissolve the first parliament in 1995 and reconfigure its successor;
subordinate the government to the presidential administration and enhance
its administrative authority; postpone presidential elections and then pro-
long his personal term in office; vest the president with direct authority to
appoint regional political officials; and orchestrate the relocation of the
capital in order to secure federal control over the northern regions. These
efforts were enhanced by successive political campaigns to assert centralized control over the media and independent civic outlets. See discussion in
Sally N. Cummings, "Kazakhstan: An Uneasy Relationship—Power and

Authority in the Nazarbaev Regime," in Cummings (2002); and Olcott (2002), pp. 87–127.

99. The Kazakh leadership's preoccupation with diffusing social unrest and collecting rents via the early privatization of the energy sector and attraction of foreign investment was counterproductive over the long haul, as it inflated expectations of future energy wealth, crowded out investment in other sectors, and fostered large-scale discretionary government spending and corruption. "In sum, Kazakhstan's approach towards its energy sector is aimed at promoting political acquiescence and providing social and economic relief in the short-term, yet it is actually increasing the likelihood for political instability and socioeconomic decay over the long-term." Pauline Jones Luong, "Kazakhstan: The Long-term Costs of Short-term Gains," in Ebel and Menon (2000), p. 80.

100. The field was estimated to hold as much as 500 bcm of natural gas, constituting roughly 40 percent of the country's proven natural gas reserves. Miyamoto (1997), p. 37.

101. *Kazakhstanskaya pravda,* 11 October 1997.

102. Ibid., pp. 279–280.

103. Gaye Christoffersen, "China's Intentions for Russian and Central Asian Oil and Gas," *NBR Analysis,* 9:2 (March 1998), p. 26; and Kent Calder, "Japan's Energy Angst and the Caspian Great Game," *NBR Analysis,* 12:1 (March 2001), pp. 30–31.

104. "Turkmen-Kazakh Heads Hold Talks," *Ashgabat Turkmen Television,* 9 April 1999, as translated in *FBIS-FTS19990409001755, 9 April 1999.*

105. Olcott (2002), p. 45.

106. *Nezavisimaya gazeta,* 28 December 2001.

107. Olcott (2002), pp. 24–50.

108. On the early strategy toward Asia, see especially *Ekspert,* 4 October 1999; and Christoffersen (1998). In an effort to woo foreign investment and to accommodate the prospects for extending the pipeline to South Korea and Japan, the proposed pipeline was scaled to deliver 32 bcm of gas per year and was configured to deliver gas to an LNG terminal to ease exports to Japan. The proven reserves of the Kovytka field totaled at least 1.6 tcm. According to analysts this is sufficient under optimal commercial conditions to provide 20 bcm of gas annually to China for 25 years, plus an additional 10 bcm to South Korea and 10 bcm to meet the growing gas demand in Siberia. Selig S. Harrison, "Gas and Geopolitics in Northeast Asia," *World Policy Journal* (Winter 2002/03), p. 27.

109. Fiona Hill, "Energy Integration and Cooperation in Northeast Asia," *NIRA Policy Research (Japan)* 15:2 (25 February 2002), www.brook.edu/views/articles/hillf/20020225.htm.

110. Ibid. The international project operators estimated that the 1.4 tcm in unproven gas reserves located in new Sakhalin concession areas would be sufficient to meet the future gas demand in the Russian Far East and the neighboring northeast regions of China. See discussion in Harrison (2002/03), pp. 29–30.

111. Calder (2001).
112. IEA (2002), p. 112.
113. Fiona Hill and Florence Fee, "Fueling the Future: The Prospects for Russian Oil and Gas," *Demokratizatsiya* 10:4 (Fall 2002), pp. 470–471.
114. Christofferson (1998); Calder (1998); and Harrison (2002/03), p. 27. Throughout the period, feasibility studies were conducted on the construction of a gas liquification plant proximate to the Karachaganak field, an internal pipeline linking western oil fields to all regions of Kazakhstan, an 8,000 km export pipeline that could bring gas from Central Asia to China, and an undersea export pipeline to Turkey and other European markets. None of these projects were approved, and the commercial viability of each project remained highly suspect. See also Alexandrov (1999), pp. 261–262; and EIA (1999).
115. *Interfax*, 21 July 1999; ibid , 18 August 1999; *Interfax Oil, Gas, and Coal Report*, 29 (393), 23–29 July 1999, as translated in *FBIS-FTS199907210-01076*, 21 July 1999. See also Hill and Fee (2002), p. 474.
116. See review in Naz Nazar, "Kazakhstan/Turkmenistan: President's Agree on Gas and Oil Exports to China, Japan," *RFE/RL*, 12 April 1999.
117. Hill and Fee (2002), pp. 476–477.
118. Alexandrov (1999), p. 281.
119. *Interfax*, 28 June 1997
120. See discussion in Hill and Fee (2002), p. 477
121. Ibid. See also Miyamoto (1997), pp. 17–39; and Joseph (1998), pp. 16–17.
122. "Kazakhstan: Company President Explains Kazmunaygaz Structure, Purpose, Goals," *Almaty Ekspress*, 22 May 2002, as translated in *FBIS CEP20020607000241*, 7 June 2002.
123. Alexandrov (1999), p. 281.
124. The fact that Gazprom was not able to coerce a preferred 26 percent stake in the venture, and acquired an equity share only after the Russian government negotiated the agreement with Astana, speaks to the decisiveness of Moscow's intervention. Alexandrov (1999), pp. 280–282.
125. Miyamoto (1997), pp. 17–39; and Joseph (1998), pp. 16–17.
126. Hill (2002).
127. "Putin, Kazakh President Discuss Gas Project, Need for New Look at CIS," *Russian Public Television ORT*, 3 March 2003, as translated in *FBIS CEP 200203020C0009*, 2 March 2002.
128. *ITAR-Tass*, 1 March 2002. See critique in Michael Denison, "Putin Aims to Lock up Caspian Gas with or without Niyazov, *Central Asia Caucasus Analyst*, 27 February 2002; and Hill and Fee (2002), p. 475.
129. *Panorama*, 27 Jan 2002; and Marat Yermukanov, "CIS Leaders Meet at Summit in Kazakhstan," *Central Asia Caucasus Analyst*, 27 February 2002.
130. *Kommersant*, 19 February 2002.
131. Ibid.
132. *Wall Street Journal*, 1 February 2002.

133. *Kommersant,* 19 February 2002.
134. *Interfax,* 21 May 2002. The Kazakh state oil and gas company, KazMunaiGaz, owned the remaining 50 percent of the joint venture.
135. *Interfax-Kazakhstan,* 5 December 2002.
136. Ibid., and *Interfax,* 7 June 2002.

Chapter 5

1. Caspian oil reserves include offshore and onshore deposits in Azerbaijan, Kazakhstan, Uzbekistan, and Turkmenistan. Energy Informational Administration (EIA), "Caspian Sea Region" (February 2002); EIA, "Caspian Sea Region" (August 2003); and International Energy Agency (IEA), *Caspian Oil and Gas: The Supply Potential of Central Asia and Transcaucasia* (Paris: OECD, 1998).
2. Robert Ebel, *Energy Choices in the Near Abroad* (Washington, DC: Center for Strategic and International Studies, April 1997), pp. 11–19; Amy Myers Jaffe and Robert A. Manning, "The Myth of the Caspian Great 'Game': The Real Geopolitics of Energy," *Survival* 40:4 (Winter 1998–1999), pp. 112–131; Andrei V. Belopolsky and Minik Talwani, "Geopolitical Basins and Oil and Gas Reserves of the Greater Caspian Region," in Yelena Kalyuzhnova, Amy Myers Jaffe, Dov Lynch, and Robin C. Sickles, eds., *Energy in the Caspian Region: Present and Future* (Hampshire, England: Palgrave, 2002), pp. 13–33; and Terry Adams, "Caspian Hydrocarbons, the Politicization of Regional Pipelines, and the Destabilization of the Caucasus," Center for European Policy Studies, Brussels, Belgium (27–28 January 2000). On the impact of size, location, and geology on the economic rents of oil exploration, see especially Oystein Noreng, *Crude Power: Politics and the Oil Market* (London: I.B. Taurus, 2002), pp. 156–161.
3. S. Frederick Starr, "Power Failure: American Policy in the Caspian," *The National Interest* (Spring 1997), pp. 20–31; Stephen J. Blank, "Every Shark East of Suez: Great Power Interests, Policies, and Tactics in the Transcaspian Energy Wars," *Central Asian Survey* 18:2 (1999), pp. 149–184; and Terry Karl, "Crude Calculations: OPEC Lessons for the Caspian Region," in Robert Ebel and Rajan Menon, eds. *Energy and Conflict in Central Asia and the Caucasus* (Lanham, MD: Rowman and Littlefield Publishers, 2000), p. 54.
4. *Nezavisimaya gazeta,* 4 November 1995; and ibid., 26 March 1997.
5. Russia's strategic interests in the Caspian Basin were codified in the 1997 and 2000 versions of Russia's National Security Concept, as well as in Yeltsin's 1994 presidential directive "On Protecting the Interests of the Russian Federation in the Caspian Sea." See discussion in Andrei Konoplyanik, *Kaspiiskaya neft': na evraziiskom perekrestke* (Moscow: IGiRGI, 1998), pp. 68–71; Yuri Fedorov, "Kaspiiskaya politika Rossii: k konsenssusu elit," *Pro et Contra* 2:3 (Spring 1997), pp. 72–89; Andrei

Shoumikhin, "Russia: Developing Cooperation on the Caspian," in Michael P. Croissant and Bulent Aras, eds., *Oil and Geopolitics in the Caspian Region* (Westport, CN: Praeger 1999), pp. 144–150; and Timothy L. Thomas, "Russian National Interests in the Caspian Sea," *Perceptions* 4:4 (December 1999–February 2000), pp. 75–96. This range of interests was corroborated by the author's interviews with Russian legislators and foreign ministry officials in 1997–1998.

6. On Russia's legal claims, see Will Raczka, "A Sea or a Lake? The Caspian's Long Odyssey," *Central Asian Survey* 19:2 (2000), pp. 189–221.
7. Fedorov (1997), pp. 72–89; P. Shultse, "Kaspiiskaya neft': podkhody k probleme," *Kaspiiskaya neft' i mezhdunarodnaya bezopastnost'* 2 (1996), pp. 8–9; and Abraham S. Becker, "Russia and Caspian Oil: Moscow Loses Control," *Post-Soviet Affairs* 16:2 (2000), pp. 91–132.
8. Konoplyanik (1998), pp. 71–73; Douglas W. Blum, "Domestic Politics and Russia's Caspian Policy," *Post-Soviet Affairs* 14:2 (1998), pp. 137–164; and Becker (2000), pp. 92–104. This was affirmed in a secret directive signed by President Yeltsin in July 1994, as discussed in *Kazakhstanskaya pravda,* 15 November 1994.
9. Blum, (1998), pp. 138–140; and Fedorov (1997).
10. Pavel Baev, "Russia's Policies in the Southern Caucasus and Caspian Region," *European Security* 10:2 (Summer 2001), pp. 95–110; and Adam N. Stulberg, "Moving Beyond the Great Game: The Geoeconomics of Russia's Influence in the Caspian Energy Bonanza," *Geopolitics* 10:1 (2005), pp. 1–25.
11. *Nezavisimaya gazeta,* 11 April 2002.
12. According Putin's Caspian envoy, "[t]he concept of a 'single package,' by which all agreements must enter into force simultaneously, appears to us [Russia] unproductive, because it blocks the possibility of implementation of understandings already reached and holds them hostages to less successful negotiation processes." See statement by Viktor Kalyuzhniy, Deputy Minister of Foreign Affairs and Special Representative of the Russian President for Issues of Settlement of the Status of the Caspian Sea in, "Caspian: Legal Problems," Moscow, Russia, 26 February 2002, *Ministry of Foreign Affairs of the Russian Federation Bulletin,* as translated in *Foreign Broadcast Information Service (FBIS)-SOV 344–27–02–2002,* 28 February 2002.
13. *Komsomolskaya pravda,* 2 June 2001.
14. *Nezavisimaya gazeta,* 13 March 2001. Presidents Putin and Khatami reiterated these positions at the April 2002 Caspian Summit.
15. Russia staged slightly smaller naval exercises in January 2001. On the swelling Caspian naval imbalance of power, see *Nezavisimaya gazeta,* 1 August 2003; ibid., 13 August 2002; and Mehrdad Haghayeghi, "The Coming of Conflict to the Caspian Sea," *Problems of Post-Communism* 50:3 (May-June 2003), pp. 32–41.
16. Baev (2001), p. 6. For more on Russia's strategy, see statement by "Caspian: Legal Problems" (2002).

17. Yukos began monthly shipments of 270,000 metric tons of crude in July 2002.

18. Baev (2001); and Vladimir Putin, "Mineral Raw Materials in the Strategy for Development of the Russian Economy," cited in Martha Brill Olcott, *The Energy Dimension in Russian Global Strategy: Vladimir Putin and the Politics of Oil* (Houston, TX: James A. Baker III Institute for Public Policy, Rice University, 2005), pp. 16–22.

19. *Nezavisimaya gazeta*, 24 April 2002; and Putin (2005). However, Putin was adamant about retaining government control of the national pipeline system. *ITAR-Tass*, 29 April 2004.

20. In May 2002, the Russian government lifted oil-export limits that it imposed jointly with OPEC in January 2002. On Russia's challenge to OPEC, see Edward L. Morse and James Richard, "The Battle for Energy Dominance," *Foreign Affairs* 81:2 (March/April 2002), pp. 28–31; and Lynne Kiesling and Joseph Baker, "Russia's Role in the Shifting World Oil Market," *Caspian Studies Policy Brief,* Harvard University (8 May 2002).

21. Cited in Nasib Nassibli, "Azerbaijan: Oil and Politics in the Country's Future," in Croissant and Aras (1999), pp. 111–112.

22. Blum (1998), p. 138; Nassibli (1999), pp. 110–119; and *Kazakhstanskaya pravda,* 15 November 1994.

23. Robert Ebel and Rajan Menon, "Introduction: Energy, Conflict, and Development in the Caspian Sea Region," in Ebel and Menon (2000), p. 5.

24. Ronald Soligo and Amy Myers Jaffee, "The Economics of Pipeline Routes: The Conundrum of Oil Exports from the Caspian Basin," in Kalyuzhnova, Jaffee, Lynch, and Sickles (2002), pp. 126–127.

25. It cost nearly $274 million per year to pump Azerbaijani's early oil through two pipeline systems rather than consolidating shipments in a single larger diameter pipeline via Georgia or Russia. Ronald Soligo and Amy Jaffee, *The Economics of Pipeline Routes: The Conundrum of Oil Exports from the Caspian Basin* (Houston, TX: James A. Baker III Institute for Public Policy, Rice University, April 1998), p. 15.

26. Soligo and Jaffee (2002), pp. 113–125; and Thomas R. Stauffer, "Caspian Fantasy: The Economics of Political Pipelines," *Brown Journal of World Affairs* VII:2 (Summer/Fall 2000). The estimated construction cost of the BTC varied between $2.5 and $4 billion. This compared roughly to $2.5 billion for a new Russian pipeline to Novorossiysk; $1.8 billion for construction of the segment through Georgia (Supsa); and $1 billion for boosting oil swaps with Iran to 50,000 barrels.

27. This comment was made by a senior Vice President of Conaco, quoted in Robert Lyle, "Views Differ on Viability of Oil Pipelines," *RFE/RL,* 4 March 1999.

28. Marina Ottaway, *Democracy Challenged: The Rise of Semi-Authoritarianism* (Washington, DC: Carnegie Endowment for International Peace, 2002), pp. 51–70; and Shireen T. Hunter, "Azerbaijan: Searching for New Neighbors," in Ian Bremmer and Ray Taras, eds., *New States New Politics: Building the Post-Soviet Nations* (Cambridge: Cambridge University Press,

1997), pp. 446–461. For an insider's account, see especially Rasul Gouliev, *Oil and Politics: New Relationship among the Oil-Producing States: Azerbaijan, Kazakhstan, and Turkmenistan* (New York: Liberty Publishing House, 1997), pp. 116–137.

29. This dovetailed with the rise in foreign direct investment from $30 to $250 million in the nonoil sector from 1994 to 1997. International Monetary Fund (IMF), *Azerbaijan Republic: Recent Economic Developments* Report No. 98/83 (August 1998), pp. 40, 64.

30. EIA, *Azerbaijan Country Analysis Brief* (June 2003), www.eia.doe.gov/emeu/cabs/azerbjan.html; and *RIA-Novosti,* 19 August 1997. The parent companies of the AIOC consortium originally included State Oil Companies of Azerbaijan (20 percent); BP (17 percent); Amoco (17 percent); LUKoil (10 percent); Penzoil (9.8 percent); Unocal (9.5 percent), Statoil (8.6 percent); McDermott International (2.4 percent); Turkish TPAO (1.7 percent); ITOCHU (3.9 percent); and Delta Hess (2.7 percent). The consortium operated the Azeri, Chirag-1, and Deepwater Gunashli oil field that possessed roughly 40 percent of Azerbaijan's oil reserves. The original plan called for delivering 400,000 bbl/d by 2004 and 1 million bb/d by 2008.

31. *ITAR-Tass,* 30 September 1997.

32. *Nezavisimaya gazeta,* 29 July 1997. See also Nassibli (1999), pp. 119–120; and Hendrik Spruyt and Laurent Ruseckas, "Economics and Energy in the South: Liberal Expectations versus Likely Realities," in Rajan Menon, Yuri E. Federov, and Ghia Nodia, eds., *Russia, the Caucasus, and Central Asia* (New York: M.E. Sharpe, 1999), pp. 103–106.

33. EIA (2003); and Michael Wyzan, "Azerbaijan: Economy Grows, But Problems Persist," *RFE/RL,* 6 May 1999.

34. *Nezavisimaya gazeta,* 1 February 2000; and *Financial Times,* 9 April 2002.

35 Alec Rasizade, "Azerbaijan after a Decade of Independence: Less Oil, More Graft and Poverty," *Central Asian Survey* 21:4 (2002), pp. 354–360; and Arif Yunusov, "Azerbaijani Security Problems and Policies," in Dov Lynch, ed. *The South Caucasus: A Challenge for the EU,* Chaillot Papers, 65 (December 2003), p. 150.

36. IMF, *Azerbaijan Republic-Selected Issues and Statistical Appendix,* Report 03/130 (May 2003).

37. Rasizade (2002), pp. 357–364; and Audrey L. Altstadt, "Azerbaijan and Aliev: A Long History and Uncertain Future," *Problems of Post-Communism,* 50: 5 (September–October 2003), pp. 9–11.

38. *Financial Times,* 31 May 1994.

39. *UN Document* A/49/475, 5 October 1994.

40. Gouliev (1997), p. 140; and *Kazakhstanskaya pravda,* 15 November 1994.

41. Yuri Federov, "Russia's Policies Towards Caspian Region Oil: Neo-imperial or Pragmatic?" *Perspectives on Central Asia* I:6 (September 1996); Shoumikhin (1999), p. 138; and Racza (2000), pp. 189–221.

42. It was not until Washington pressured Baku and the AIOC to preclude Iran from participating in the "contract of the century" that Tehran adopted a strict legal interpretation. Nassibli (1999), p. 115.

43. *Kommersant,* 23 August 1995.
44. Becker (2000), p. 95. The proposal called for Baku to sacrifice claims to large reserves in its self-described sector, and predictably complicated competing claims with Turkmenistan by legitimatizing Moscow's new claims to the same resources. According to Aliev, Moscow's interests in developing the contested Kyapaz/Sedar field provoked Ashgabat's harsh rejoinder. *Bakinskiy Rabochiy,* 26 July 1997, as translated in *FBIS-SOV-97–209,* 28 July 1997.
45. Einar Bergh, "AIOC Current Developments," *Azerbaijan International* 4:1 (Spring 1996), p. 44.
46. *United Press International,* 13 November 1996; and Andrei Shoumikhin, "Developing Caspian Oil: Between Conflict and Cooperation," *Comparative Strategy* 16:4 (Fall 1997), p. 342.
47. *Kommersant,* 2 August 2000; and "Azerbaijan, Iran Sign Memorandum on Geological Cooperation," *Baku Turan,* 2 August 2000, as translated in *FBIS CEP20000802000190,* 2 August 2000. Turkmenistan also harshly denounced Moscow's modified median proposal for "deepening divisions" that confounded resolution of the Caspian Sea's legal status. *Interfax,* 5 October 2000.
48. *Interfax,* 22 February 2002.
49. On the comparative economic costs of oil pipeline options, see Soligo and Jaffe (2002). The study concludes that the most cost-effective route for delivering Azerbaijan's main oil to Western markets, factoring in political and security risks of disruption and economies of scale, was the proposed Baku-Supsa-Turkish Thrace Bypass. The least cost-effective alternative was the Baku-Ceyhan that was projected at one dollar per barrel more than the bypass through Thrace or the Russian route. Yet even this route was potentially more attractive than the Russian route, if connected to large oil shipments from Kazakhstan's Kashagan offshore field, and if synergies were created with gas pipelines from Baku's Shah Deniz field. The shorter Baku-Supsa-Thrace route more than offset initial savings linked to upgrading the Russian route (as opposed to repairing and upgrading Georgian pipelines and ports).
50. Cited in Michael Lelyveld, "Azerbaijan: Caspian Sea Pipeline Get Major Backing," *RFE/RL,* 20 October 1999.
51. The Sponsor Group was comprised of companies that overlapped the AIOC, including SOCAR, BP-Amoco, Unocal, Statoil, Ithochu, Turkish Petroleum, Ramco, and Delta-Hess. The group initially committed $25 million to cover the basic engineering study, and $125 million for follow on engineering and construction studies. The AIOC later upgraded estimates from 4.6 to 5.3 billion barrels. BP pledged to invest $5.2 billion in the next stage of development of Azerbaijan's resources to fill the BTC with one million barrels per day.
52. Michael Lelyveld, "Azerbaijan: Progress Seen on Baku-Ceyhan Oil Pipeline," *RFE/RL,* 5 June 2002.

53. "Sezer, Aliyev, and Shevardnadze Hold News Conference at End of Trabzon Summit," *Ankara Anatolia*, 30 April 2002, as translated in *FBIS GMP20020403000085,* 30 April 2002; and Jean-Christophe Peuch, "Caucasus: Energy Projects Given Impetus by Regional Summit, Arrival of U.S. Soldiers," *RFE/RL,* 7 May 2002.

54. Soligo and Jaffe (1998), pp. 13–15. The economies of scale linked to building new large diameter pipelines also reduced the commercial viability of investing in multiple pipelines, notwithstanding political and security risks of relying on one main route. This offset the initial appeal of upgrading the Russian variant as a secondary route. As late as December 1998, there were serious misgivings among AIOC partners as to the cost-effectiveness of the BTC, absent these concessions. According to the outgoing president of the AIOC, "whoever wants to waste money on building the BTC pipeline may waste it." *Moscow News,* 10 December 1998.

55. European Commission: "Towards a European Strategy for the Security of Energy Supply," *EU Green Paper* (2001), pp. 37, 67, 73–74. The demand for oil among EU member states was expected to grow 60–90 percent by 2030, depending on respective enlargement scenarios. However, the long-term profitability of this market was jeopardized by: the EU's commitment to diversifying supply, competition by rival producers, costly delivery discounts, and festering infrastructure problems. See discussion in chapter 3, as well as in Tsnetr Strategicheskikh Razrabotok, *O vozmozhnyi napravleni-yakh razvitiya infrastruktury po transportirovke Rossiiskoi hefti,* 12 November 2004, available at www.csr.ru.

56. See especially comments by U.S. Special Envoy to the Caspian Region, Steve Mann, cited in "Georgia, U.S. Deny Russian Allegations About Caspian Oil Pipeline," *Georgian TV1,* 20 September 2002, as translated in *FBIS CEP20020921000110,* 20 September 2002.

57. Jeanne Whalen, "U.S., Russia Vie to Dominate Georgia Gate to Caspian Oil," *RFE/RL,* 8 October 2002.

58. On competing claims to the Sedar/Kyapaz deposit, see discussion in Vladimir Mesamed, "Turkmenistan: Oil, Gas, and Caspian Politics," in Croissant and Aras (1999), pp. 214–215.

59. Geoffrey Kemp, "U.S.-Iranian Relations: Competition or Cooperation in the Caspian Basin," in Ebel and Menon (2000), pp. 150–157.

60. Jennifer Delay, "The Caspian Oil Pipeline Tangle: A Steel Web of Confusion," in Croissant and Aras (1999), pp. 54–57.

61. Jaffe and Manning (1998–1999).

62. It was shorter than the BTC and Russian main pipeline, as well as involved cheaper tanker and bypass pipeline costs than the BTC. Soligo and Jaffe (1998), pp. 16–23.

63. Gareth Winrow, "Turkish National Interests," in Kalyuzhnova, Jaffe, and Lynch, and Sickles (2002), pp. 240–242.

64. Ibid.

65. Delay (1999), p. 52.

66. Russian energy officials intimated that Baku's concessions on pipeline transportation would have "an impact on solving the Karabkah conflict." *Segodnya,* 25 November 1993.

67. Sheila Heslin, *Key Constraints to Caspian Energy Development: Status, Significance, and Outlook* (Houston, TX: The Center for International Political Economy and the James A. Baker III Institute for Public Policy, Rice University, April 1998).

68. Author's interview with LUKoil officials, 20 March 1997. LUKoil's strategy at the time called for 25 percent of the company's production by 2005 to come from foreign oil fields, approximately 80 percent of which would derive from the Caspian Basin. See discussion in *Segodnya,* 30 September 1996; and David Lane and Iskander Seifulmulkov, "Company Profiles: LUKoil, YuKOS, Surgutneftgaz, and Sidanko," in David Lane, *The Political Economy of Russian Oil* (Lanham, MD: Rowman & Littlefield, 1999), pp. 113–114.

69. Author's interview with LUKoil officials, 20 March 1997; *Rabochaya tribuna,* 18 July 1995; and Lev Klepatsky and Valery Pospelov, "Maneuvering Round the Caspian Sea," *International Affairs* 11:12 (1995).

70. While Rosneft, a state company, subsequently withdrew from the contract, LUKoil remained party to the agreement while it was put on hold. *Segodnya* 7 August 1997. Yet even Rosneft officials stated that the company "is not planning to leave the Caspian region and is developing a host of projects to open up promising fields on the Caspian shelf, including projects with Azerbaijani and Turkmen participation." *Russia Today,* 7 August 1997.

71. *Interfax,* 19 February 1998; and ibid., 9 December 1997.

72. *Nezavisimaya gazeta,* 9 July 1997.

73. *RIA-Novosti,* 3 September 1997; and Igor Rotar, "Will Caspian Oil Flow over the Northern Variant?" *Prism* (Jamestown Foundation) 3:12; part 3 (25 July 1997).

74. *RIA-Novosti,* 3 September 1997.

75. Delay (1999), pp. 50–51.

76. *Obshchaya gazeta,* 20–26 August 1998.

77. Cited in Michael Lelyveld, "Azerbaijan: Oil and Gas Problems: A Contentious Issue with Russia," *RFE/RL,* 16 June 2000. See also *Kommersant,* 22 April 2000.

78. Transneft was under pressure by the government to lower tariffs from $15.67 to $8–$10 per ton. This was still approximately double the transit fees charged by Georgia. See discussion in *Kommersant,* 8 July 2000.

79. *Kommersant,* 19 September 2002.

80. *Interfax,* 18 January 2001.

81. *Kommersant,* 8 July 2000; and *Interfax,* 30 June 2000.

82. *Rossiyskaya gazeta,* 19 January 2001.

83. *Interfax,* 14 March 2000.

84. Ibid., 26 November 2001; and Michael Lelyveld, "Azerbaijan: Cooperation with Russia Seems to be Paying Off," *RFE/RL,* 20 November 2001.

85. *Nezavisimaya gazeta,* 29 July 1997. See also *Moskovskiy komsomolets,* 20
August 1997.

86. Author's interview with a senior Azerbaijani national security staff
member, 24 July 1997.

87. *Interfax,* 15 April 1998.

88. LUKoil joined development (with a 57.7 percent share together with Agip)
of Azerbaijan's offshore Karabakh field in November 1995 (a project subsequently shutdown due to dry holes), and pursued a 20–40 percent stake
in the international consortium formed to develop the Sha Deniz oil and
gas field. Other Russian participants in Azerbaijani ventures included Rosneft and the Central Oil Company.

89. *Interfax,* 11 July 1997.

90. Nassibli (1999), pp. 113–115.

91. *Baku Zaman,* 20 January 2001, as translated in *FBIS CEP2001012400-
0279,* 20 January 2001.

92. *ITAR-Tass,* 9 January 2001. Putin's Caspian envoy interpreted Aliev's declaration as representing a "180 degree change" regarding delimitation of
the Caspian and rapprochement with Russia. *Nezavisimaya gazeta,* 16
January 2001.

93. *Baku Zaman,* 20 January 2001, as translated in *FBIS CEP2001012400-
0279,* 20 January 2001.

94. *ITAR-Tass,* 26 January 2002; and *Interfax,* 24 September 2002.

95. Aliev also announced plans to deploy an Iranian-built destroyer. Haghayeghi (2003), pp. 36–37.

96. *Novaya vremya,* 13 January 2001; *Vedomosti,* 28 January 2002; and
Kommersant, 25 January 2002.

97. *Nezavisimaya gazeta,* 16 September 2002;

98. EIA, *Azerbaijan: Production-Sharing Agreements* (June 2002), www.eia.
doe.gov/emeu/cabs/azerproj.html. LUKoil acquired a 12.5 percent stake
and shared a 45 percent stake (LUKoil-Agip) in the Caspian International
Oil Consortium formed in 1995 to develop the Karabakh offshore oil field
(80–120 million tons of estimated reserves). This project was canceled in
1998. LUKoil also shared a 60 percent operator stake with Arco (LukArco)
to develop the Yamala/D-222 offshore oil field (750 million barrels of estimated reserves) in 1997. Both SOCAR and LukArco considered abandon-
ing the project in 2001 due to concerns about seismic activity.

99. *Nezavisimaya gazeta,* 2 February 2002. The first section of the pipeline
officially opened in spring 2005.

100. *Interfax,* 27 May 2002. Although LUKoil disposed of its shares in the
AIOC in fall 2002, it did so as part of a corporate retrenchment strategy
and after Moscow had acknowledged the inevitability of the BTC. See discussion in Douglas Blum, "Why Did LUKoil Really Pull Out of the Azer-

Chirag-Guneshli Oilfield," *PONARS Policy Memo 286* (January 2003); and Michael Lelyveld, "Azerbaijan: LUKoil's Plan to Sell Azerbaijani Assets Raises Questions," *RFE/RL*, 29 October 2002.

101. *Vremya novostey,* 27 September 2002, as translated in *FBIS CEP2002-0927000187,* 27 September 2002.

102. EIA, *Kazakhstan Country Analysis Brief* (Washington, DC: Department of Energy, July 2003).

103. Cited in Alexandrov (1999), p. 261. See also Vladimir Babak, "Kazakhstan: Big Politics around Oil," in Croissant and Aras (1999), pp. 184–185; and Andrei Kalyuzhnov, Julian Lee, and Julia Nanay, "Domestic Use of Energy: Oil Refineries and Gas Processing," in Kayuzhnova, Jaffe, Lynch, and Sickles (1999), pp. 142–146.

104. IMF *Republic of Kazakhstan: Recent Economic Developments,* Country Report No. 98/84 (August 1998), pp. 5, 30–31, 37–39; and The Word Bank Country Study, *Kazakhstan: Transition of the State* (Washington, DC: World Bank 1997), p. 228.

105. IMF (1998), p. 85. On Kazakhstan's reserve base see Belopolsky and Manik (2002), pp. 13–33.

106. EIA, *Kazakhstan Country Analysis Brief* (Washington, DC: Department of Energy, July 2003); and IMF, *Republic of Kazakhstan: Selected Issues and Statistical Appendix,* Country Report No. 03/211 (July 2003), p. 8. Kazakhstan's net wealth from the oil industry was expected to skyrocket during 2022 to 2046, generating $150 billion. IMF *Republic of Kazakhstan: Selected Issues and Statistical Appendix,* Country Report No. 02/64 (February 2002), p. 19.

107. There were plans for constructing two additional refineries. On Kazakhstan's domestic oil development, see especially Kalyuzhnov, Lee, and Nanay (2002), pp. 142–146; and *Interfax-Kazakhstan,* 21 February 2002, as translated in *FBIS CEP20020221000180,* 21 February 2002.

108. During 1998 to 2002, oil exports jumped from less than 30 percent to over 50 percent of Kazakhstan's exports, exceeding nonoil exports for the first time and accounting for nearly 25 percent of GDP. IMF (2003), pp. 5–6 and 38.

109. Ibid., pp. 97, 103, 104. On official corruption, foreign investor frustration and "unnecessary obstacles" that marred the investment climate in the Kazakh oil industry, see Martha Brill Olcott, *Kazakhstan: Unfulfilled Promise* (Washington, DC: Carnegie Endowment for International Peace, 2002), pp. 159–171; and *Financial Times,* 16 July 2001.

110. Segei Smirnov, "Neftyanaya strategiya Kazakhstana: mirazhi i real'nost'," *Tsentral'naya aziya i kavkaz* 6:12 (2000), pp. 186–192.

111. *Nezavisimaya gazeta,* 28 April 1998.

112. Regine A. Spector, "The Caspian Basin and Asian Energy Markets," *The Brookings Institution Background Paper* (24 May 2001).

113. On the cost-effectiveness of Kazakhstan's alternative pipeline routes, see Ebel (1997), pp. 88–89; and Soligo and Jaffe (1998), p. 17.

114. Babak (1999), p. 188.
115. See discussion in Raczka (2000), pp. 205–212; and Alexandrov (1999), pp. 285–289.
116. *Reuters,* 12 July 1994; and *Kommersant,* 23 August 1995.
117. Mesamed (1999), pp. 215–219; and *Interfax,* 20 September 1999.
118. Alexandrov (1999), p. 289.
119. *Kazakhstanskaya pravda,* 11 October 1997.
120. *Reuters,* 22 January 1993. See also Alexandrov (1999), pp. 265–266.
121. Alexandrov (1999), pp. 265–266.
122. Ibid.
123. LUKoil acquired a 5 percent stake in the Tengizchevoil joint venture for $30 million "bonus payment" in April 1996.
124. Comment made by the first vice president of LUKoil, cited in Carol Saivetz, "Putin's Caspian Policy," Caspian Studies Program, Harvard University (October 2000).
125. The United States was the principal source of foreign direct investment in Kazakhstan, accounting for 28.5 percent from 1993 to 1996, and 36 percent from 1998 to 2002. China, Germany, Britain, and Turkey followed with approximately 7–9 percent per annum. Russia did not rank among the top ten sources of foreign direct investment, nor did it account for more than 0.3 percent per annum. IMF (2003), p. 103.
126. In November 1997, Kazakhstan transferred rights to develop 10 blocks in the Kashagan field to Western firms in return for a pledge to invest over 30 billion in the Kazakh oil sector by 2010 and to split profits 80:20 on related projects.
127. Babak (1999), p. 195.
128. *Jamestown Monitor,* 5 June 1998.
129. Ibid., 24 January 2001; and *Interfax,* 5 February 2001. In 2001 TotalFinaElf purchased BP's 9.5 percent stake and Agip became the sole operator of the consortium.
130. *Almaty Interfax-Kazakhstan,* 26 March 2002, as translated in *FBIS CEP20020326000228,* 26 March 2002.
131. China National Petroleum Company (CNPC) outbid foreign competition in 1997 to purchase ($9.5 billion) a 60 percent stake in Aktyubinsk oil fields (1 billion barrels) and exclusive rights to develop the large Uzen onshore deposits (1.5 billion barrels), as well as agreed to construct a 1,920-mile pipeline to Chinese markets. The CNPC also promised in 1998 to finance construction of a 150-mile pipeline link to the Iranian border, and a consortium of Chinese oil firms pledged in 1999 to build a 236 mile pipeline from Iran's Caspian port of Neka to its major refinery at Tehran to handle the oil swaps from Kazakhstan. Gaye Christoffersen, "China's Intentions for Russian and Central Asian Oil and Gas," The *National Bureau of Asian Research Analysis,* 9:2 (1998), pp. 24–28. Although Kazakhstan and the CNPC remained deadlocked by the end of 2002 on commitments to construct the 1,920 mile pipeline to China, Beijing reiter-

ated a strong interest in bringing the project to fruition. Conversely, the CNPC officially withdrew in June 2000 from the project to construct a pipeline from Iran's Caspian port of Neka to refineries in Tehran. *Interfax,* 16 April 2002.

132. *Interfax,* 16 April 2002. China's CNPC claimed that a feasibility study on the 1,920-mile pipeline revealed that the project would not be commercially viable as long as world prices remained low and the pipeline would be filled by 20 to 25 million tons of oil per year. Michael Lelyveld, "Kazakhstan: Oil Pipeline to China a Victim of Dimplomatic Dispute," *RFE/RL,* 19 September 2001. One expert study concluded: "If Saudi Arabia insists on maintaining a significant presence in the European market, the option for Kazakhstan to export its oil to Pakistan by pipeline might provide a higher new revenue to Europe." Soligo and Jaffe (2002), pp. 119–125; and *Almaty-MN-Movosti Nedeli,* No. 9, 6 March 2002, as translated in *FBIS-CEP20020306000201,* 6 March 2002.

133. Olcott (2002), pp. 152–153; and *Wall Street Journal,* 10 July 2001.

134. *Russian TV RTR,* 22 November 2000, as translated in *FBIS CEP 20001122000404,* 22 November 2000. Kalyuzhniy repeatedly claimed that Russian, Iranian, and eastern pipeline routes offered more feasible options than the highly uncertain and costly BTC. *Interfax,* 2 October 2002; and *Kommersant,* 19 February 2002.

135. Statement by Kazakh Deputy Foreign Minister, V. Gizzatov in 1995, cited in Alexandrov (1999), p. 287.

136. Ibid., p. 289.

137. The joint Russian-Kazakh statement in April 1996 acknowledged mutual rights to exploit seabed resources in national waters, as well as mutual commitments to working toward an eventual consensus agreement on the sea's legal status.

138. Michael Lelyveld, "Iran: Tehran Puts Off Oil Swaps with Astana," *RFE/RL,* 21 March 2001. Mercaptan content is a sulfur derivative that can corrode pipelines and storage tanks.

139. Established in 1992, the CPC originally allocated 25 percent stakes to the Kazakh and Russian governments, as well as a 50 percent to Oman for arranging international financing. After successive delays, internal intrigue, and financial difficulties, the CPC was reorganized in 1996 leaving the Russian and Kazakh governments with 24 and 19 percent stakes, respectively, Oman with a 7 percent stake, and the remaining 50 percent divided among leading oil firms—including Lukoil, Rosneft, Arco, Chevron, Mobil, Agip, and British Gas—that were responsible for financing construction. The consortium was split into two entities- CPC Russia and CPC Kazakhstan—due to the tax implications of the separate value streams across the two countries. The pipeline was initially projected to deliver 560,000 barrels per day (28 million barrels per year), with construction costs that spiraled to $2.6 billion. On the early internal history of the CPC, see Alexandrov (1999), pp. 263–278.

140. *Almaty Karavan,* 20 April 2001, as translated in *FBIS CEP20010511-000391,* 20 April 2001; and Segodnya, 1 April 1994.
141. *Reuters,* 12 March 2002.
142. This entailed constructing a $1.6 billion pipeline from Kazakhstan to Iran, as well as building a 235-mile internal link between Iran's Caspian port facilities in Neka to its major refinery in Tehran.
143. Soligo and Jaffe (1998), p. 31.
144. Delay (1999), pp. 64–65.
145. *RusEnergy.com,* 7 May 2001, as translated in *FBIS CEP2001050800-0293,* 7 May 2001.
146. Soligo and Jaffe (2002), pp. 119–125.
147. Jaewoo Choo, "The Geopolitics of Central Asian Energy," in James Sperling, Sean Kay, and S. Victor Papacosma, eds., *Limiting Institutions?: The Challenge of Eurasian Security Governance* (Manchester: Manchester University Press, 2003), pp. 105–121.
148. Yakov Pappe, "Neftyanaya i gazovaya diplomatiya Rossii," *Pro et Contra* 2:3 (Summer 1997), pp. 64–67; Peter Rutland, "Oil, Politics, and Foreign Policy," in Lane (1999), pp. 172–173; and Blum (1998), pp. 145–149.
149. Babak (1999), pp 190–192.
150. *Interfax,* 9 October 2000. In 2001, Russia raised Kazakhstan's transit quota to 17.3 million tons, of which 12.3 million tons were delivered to international markets and 5 million tons were slated for delivery to CIS markets. IEA (2002), p. 99.
151. *Interfax,* 15 February 2001. Ibid., 2 May 2001.
152. "Marina Dracheva, "Held Hostage by CPC Interests, Chevron Agrees to Kazakh Export Demands," *Economist Intelligence Unit* 51:26 (7 February 2001). See also "Tengiz Field Operation Marks Second Anniversary," *Oil and Gas Journal* (17 April 1995), p. 28.
153. Cited in Alexandrov (1999), p. 267.
154. *Kommersant,* 28 July 2001; and *Interfax,* 13 June 2001.
155. Michael Lelyveld, "Caspian: Russia Seeking to Build Oil Pipeline to Iran," RFE/RL, 9 July 2001.
156. *Izvestiya,* 15 June 2000.
157. *Almaty Khabar Television,* 19 May 2001, as translated in *FBIS CEP200-105119000114,* 19 May 2001. See also *Wall Street Journal,* 8 June 2001. Although the CPC project ultimately started commercial operation in September 2001, it continued to suffer repeated delays in both Kazakh and Russian routes.
158. *Prime-TASS,* 10 October 2002, as translated in *FBIS CEP20021011000-377,* 10 October 2002.
159. *Izvestiya,* 4 May 2000.
160. Ibid., 15 June 2000. Between 1999 and 2000, the international price for Russia's Urals blend jumped from $10.09 per barrel to 24.71 per barrel. EIA, *International Energy Annual 2001: Selected Crude Oil Prices,* 1992–2002.

161. Chelyabinsk, Orenburg, Sverdlovsk, Omsk, and Novosibirsk relied on partners in eastern Kazakhstan for as much as 36–80 percent of respective industrial trade. On the regional dimension to Russian-Kazak trade and the problems posed for respective federal relations, see Sergei Golunov, "Rossiya-Kazakhstan: dilemma 'bezopasnost'-sotrudnichestvo' v trans-granichnykh vzaimootnosheniyakh," *Tsentral'naya Aziya i Kavkaz* 4:22 (2002), pp. 66–78.

162. *Vek,* 4 August 2000.

163. Author's interviews with a senior western oil executive in Moscow, May 1997.

164. Ibid.

165. *Izvestiya,* 12 April 2001.

166. These interventions not only repeatedly delayed Moscow's preferred start-up date for the CPC, but bolstered the relative appeal of an alternate non-Russian pipeline system for Kazakhstan's oil that would link up to the delivery of Azerbaijan's main oil. *Nezavisimaya gazeta,* 7 October 1997; and *Vek,* 4 August 2000.

167. In two of these projects, Russian firms controlled less than a 15 percent stake. EIA, *Kazakhstan* (2002).

168. *Kazakhstanskaya pravda,* 12 November 2002; and *Interfax,* 29 November 2002. See also Michael Lelyveld, "Kazakhstan: Astana Plans to Boost Energy Exports to West," *RFE/RL,* 6 December 2002.

169. *Interfax,* 19 August 1995. See also "Itogi vtorogo soveschaniya rukovoditeley pravovykh sluzhb MID prikaspiyskikh gosudarstv po voprosam pravovogo status kaspiyskogo moray," *Diplomaticheskiy kurier* 1 (1996), p. 93.

170. *Almaty Interfax-Kazakhstan,* 22 September 2000, as translated in *FBIS CEP20000922000321,* 22 September 2000.

171. Babak (1999), p. 192.

172. Becker (2000), p. 123.

173. *ITAR-Tass,* 5 October 2000; and ibid., 11 October 2000.

174. *Interfax,* 13 May 2002. The Kurmangazy field was assessed by Astana at holding reserves at least as large as the combined total of the other two fields under Russian authority. Vladimir Socor, "Kazakhstan and Russia to Share Northern Caspian Oilfields," *The Jamestown Monitor,* VII:94 (14 May 2002).

175. *ITAR-Tass,* 24 May 2001.

176. *Interfax,* 7 June 2002.

177. ITAR-Tass, 24 May

178. Cited in Michael Lelyveld, "Kazakhstan: Nazarbaev to Seek Balance between U.S., Russia on Oil," *RFE/RL,* 7 December 1999. See also *Nezavisimaya gazeta,* 21 December 1999.

179. Cited in Michael Lelyveld, "Kazakhstan: Controversy Persists on Oil Pipeline to Iran," *RFE/RL,* 15 December 2000. On Nazarabev's promises

to ship Kazakh exports via the BTC, see *Interfax,* 2 March 2001; and Alec Rasizade, "The Bush Administration and the Caspian Oil Pipeline," *Contemporary Review* 279: 1626 (July 2001), pp. 21–26.

180. This posture persisted beyond the period of study, as Nazarbayev told Putin in January 2004 that his commitment to the BTC "will begin very soon." Putin openly expressed his frustration with the failure of Russian policy to derail Astana's preoccupation with this commercially suspect project that he regarded as symptomatic of a complicating pro-Western infatuation. *Kommersant,* 10 January 2004.

181. *Almaty Turkestan,* 26 September 2002; as translated in *FBIS CEP200-21008000344,* 26 September 2002.

182. *Almaty Kazakh,* 9 April 2002, as translated in *FBIS CEP2002040-9000289,* 9 April 2002; and *Almaty Turkestan,* 26 September 2002, as translated in *FBIS CEP20021008000344,* 26 September 2002; and *Nezavisimaya gazeta,* 14 April 2001.

183. *Kommersant,* 27 February 2003. By mid-2003, Nazarbaev reiterated to Washington his determination to participate in the BTC. *Izvestiya,* 7 June 2003.

Chapter 6

1. *Strategy of Nuclear Power Development in Russia in the First Half of the 21st Century,* endorsed by the Russian Government on 25 May 2000 (Protocol No. 17). The optimistic "high growth" scenario envisioned nuclear power production to rise from 21GWe in 2000 to 32GWe, 50GWe, and 60GWe by 2010, 2020, and 2050, respectively. Alternatively, the "low-growth" scenario projected that nuclear energy production would increase by 21GWe by the end of the period. Specifically, the plan called for: developing third-generation commercial plant designs and constructing a pilot plutonium burning fast-breeder reactor by 2020; establishing a closed nuclear fuel cycle for domestic plants by 2030; and operating a pilot thermal reactor by 2050.

2. Ibid. See also Igor Khripunov, "MINATOM: Time for Crucial Decisions," *Problems of Post-Communism* 48:4 (July/August 2001), pp. 51–52; and Adrian Collins, "Nuclear Power in Russia," *CORE Issues* (Journal of the Uranium Institute) 2 (December 2000), available at: http://www.world-nuclear.org/coreissues/2000/no2/features/feature4.htm.

3. *Vek,* 14 April 2000.

4. *Nezavisimaya gazeta (nauka),* 18 October 2000; and *Segodnya,* 22 July 2000. Minatom officials were careful to exclude defense-related or basic scientific activities from this reform agenda.

5. *Nezavisimaya gazeta (nauka),* 18 October 2000. See also *Segodnya,* 22 July 2000; *Interfax,* 5 September 2001, *Byulletin po atomnoy energii,* 31 July 2003, pp. 6–12.

6. *Segodnya,* 22 July 2000. See also discussion in Ann MacLachlan, "Rosener-goatom Change Raises Hopes for More Nuclear Reforms," *Nucleonics Week* 42:37 (13 September 2001), http://www.mhenergy.com.

7. *Russian TV RTR,* 15 April 2001, as translated in *Foreign Broadcast Information Service (FBIS) CEP20010415000031,* 15 April 2001. See Also *Obshchaya gazeta,* 23 August 2000; *Moskovskiy komsomolets,* 33 March 1999; *Nezavisimaya gazeta,* 2 June 2000; and *Segodnya,* 16 March 2001.

8. *Nuclear.ru,* 4 December 2002; and *Moskovskiy byulletin po atomnoy energii,* 31 July 2003, pp. 6–12. See commentary on Minatom's post-Adamov commercial priorities in "Minatom Rising," *NUKEM* (December 2002).

9. Minatom also was intent on modernizing and supplying Soviet designed VVER-440 reactors built in Eastern Europe, and fleshing out the signed agreement with Syria to cooperate on "peaceful use" of nuclear energy. Khripunov (2001), pp. 53–54.

10. See discussion in Charles Zeigler and Henry B. Lyon, "The Politics of Nuclear Waste in Russia," *Problems of Post-Communism* 49:4 (July/August 2002), pp. 33–42; and Adam N. Stulberg, "The Federal Politics of Importing Spent Nuclear Fuel: Inter-Branch Bargaining and Oversight in the New Russia," *Europe-Asia Studies* 56:4 (June 2004), pp. 491–520.

11. *Strategy of Nuclear Power Development in Russia in the First Half of the 21st Century;* and *Rossiskaya federatsiya segodnya,* 31 May 2003.

12. *Vek,* 14 December 2001. See also *Polyarnaya pravda* (Murmansk), 31 January 2001; and *Vek,* 11 March 2000.

13. *Vek,* 27 December 2001.

14. *Parlamentskaya gazeta,* 11 January 2001.

15. *Sovetskaya Rossiya,* 5 June 2001.

16. The Zarechnoye deposit contained 14,500 proven metric tons of uranium and 4,500 probable metric tons of uranium.

17. *Parlamentskaya gazeta,* 11 January 2001.

18. *Kommersant,* 25 October 2000; and *Moscow News,* 1 November 2000.

19. "Kazakhstan: Ulba Metallurgy Plant," 15 February 2001, available at cns.miis.edu.

20. *Kazakhstanskaya pravda,* 29 November 2001.

21. Ulba also agreed to cooperate with TVEL by expanding services to include production lines for medium-enriched uranium. *UMZ-inform* 12:32 (11 May 2001), p. 2.

22. *Vek,* 3 November 2000; and

23. *Kommersant,* 29 February 2000.

24. "Economic Research to Assume Forefront in the Minatom," *Elektrostal Atom-Pressa,* 21 (429), January 2001, as translated in *FBIS CEP20010-221000227,* 31 January 2001.

25. *Vek,* 8 December 2000. On the geoeconomic significance of the deals, see *Vek,* 3 November 2001; and *Izvestiya,* 26 February 2003.

26. Ymircerik Kasenov, "Novaya 'bol'shaya' Tsentral'nou Asii," *Tsentral'naya aziya i kavkaz* 8 (2001); and *Izvestiya,* 26 February 2003.

27. On Minatom's international activism, see especially Igor Khripunov (2001), pp. 53–54; C.M. Johnson, *The Russian Federation's Ministry of Atomic Energy: Programs and Developments* (Paper Prepared by Pacific Northwest Laboratory for the U.S. Department of Energy, Contract DE-AC06-76RLO, February 2000); and Vladimir Orlov, "Chto bygodno Minatomu," *Pro et Contra* 2 (Summer 1997).

28. *Kommersant,* 25 October 2000. For a discussion of Kyrgyzstan's plans for the hydroelectric power, see Gregory Gleason, "Russia and the Politics of the Central Asian Electricity Grid," *Problems of Post-Communism* 50:3 (May/June 2003), pp. 47–48.

29. Yuriy Razgulyaev, "There Is Enough Gold for Everyone," *Delovy mir,* 16 July 1996, as translated in *FBIS-SOV-96-155–S,* 16 July 1996; *RFE/RL Newsline,* 23 June 2000; and *Vechernyi Bishkek,* 23 October 2000. On the significance of the gold industry to the Kyrgyz economy, see especially International Monetary Fund (IMF), *Kyrgyz Republic: Selected Issues and Statistical Appendix,* Country Report No. 03/53 (February 2003), p. 46.

30. From 1998 to 2002, purchases from the U.S., Germany, China, and Turkey comprised significantly larger shares of total imports than during the first half of the decade. Furthermore, the rise in gold exports allowed Kyrgyzstan's share in these non-CIS markets to hold steady while it slipped in other areas. IMF (2003), pp. 12, 31; and IMF, *Kyrgyz Republic: Recent Economic Developments,* Staff Country Report No. 98/8 (January 1998), p. 119. See also discussion of Kyrgyzstan's strategic trade diversification through the 1990s in Richard Pomfret, "Central Asian Regional Integration and New Trade Patterns," in Yelena Kalyuzhnova and Dov Lynch, eds., *The Euro-Asian World: A Period of Transition* (London: MacMillan Press, 2000), pp. 198–202.

31. Gleason (2003), pp. 47–48.

32. In 2001, the U.S. established an air base near Manas from which it staged operations into Afghanistan.

33. *Vek,* 15 December 2000.

34. Askar Akaev, *Pamyatnoe desyatiletie: trudnaya doroga k demokratii* (Moskva: Mezhdunarodhye otnosheniya, 2002), pp. 436–440.

35. See, for example, "Interview with Kyrgyz President on Policies," *Moscow Central Television First Program Network,* 2 July 1991, as translated in *FBIS-SOV-91-127,* 2 July 1991; and Akaev (2002), pp. 317, 321, 322.

36. Askar Akaev, *Kyrgyzstan: An Economy in Transition* (Sidney: Asia Pacific Press, Australian National University, 2000), pp. 50–53. Since independence foreign direct investment from non-CIS countries far outstripped that from members of the CIS. The main multilateral donors were the World Bank, the Asian Development Bank, and the EBRD; while the main bilateral donors were Japan, Germany, and the U.S. See especially IMF (1998), p. 28; and IMF (2003), p. 80.

37. From 1994 to 2003, Kyrgyzstan participated in over 100 NATO-directed exercises. *Krasnaya zvezda,* 15 July 2003.

38. *Krasnaya zvezda,* 25 September 2003. For a summary of Kyrgyzstan's conspicuous tilt to Moscow during 1999 to 2003, see Henry Plater-Zyberk,

Kyrgyzstan-Focusing on Security (Surrey, England: Conflict Studies Research Center, November 2003).

39. Plater-Zyberk (2003), pp. 6–9, 14–16. Border disputes with Uzbekistan were especially volatile, as Tashkent unilaterally mined both sides of the territory under dispute (which according to Bishkek killed numerous citizens inside Kyrgyz territory). Tashkent also frequently crossed the border to conduct counter-terrorist operations without notifying Bishkek.

40. _The Times of Central Asia,_ 2 May 2002. There were over 3000 young male recruits alone in the terrorist organization, Hizb al-Tahrir, operating in southern Kyrgyzstan from 1999 to 2001. See Plater-Zyberk (2003), pp. 7–8.

41. Eugene Huskey, "An Economy of Authoritarianism?: Askar Akaev and Presidential Leadership in Kyrgyzstan," in Sally N. Cummings, ed., _Power and Change in Central Asia_ (New York: Routledge, 2002), pp. 74–96; John Anderson, _Kyrgyzstan: Central Asia's Island of Democracy_ (Australia: Harwood Academic Publishers, 1999), pp. 23–63; and Pauline Jones Luong, _The Transformation of Central Asia: States and Societies from Soviet Rule to Independence_ (Ithaca: Cornell University Press, 2004).

42. Huskey (2002), pp. 74–96; and Anderson (1999), pp. 23–63.

43. Regine Spector, _The Transformation of Askar Akaev, President of Kyrgyzstan,_ Berkeley Program in Soviet and Post-Soviet Studies Working Paper Series (Spring 2004), p. 22. Parenthesis added.

44. IMF and the World Bank, "Country Notes: Debt Sustainability and Policy Priorities to Ensure Growth in the CIS7 Countries," February 6, 2001, pp. 14–20.

45. Akaev (2002), p. 238.

46. Akaev (2000), pp. 157–186.

47. _Project of Scientific and Research Investigations,_ Kyrgyz National Academy of Sciences, Institute of Rock Physics and Mechanics (1995), p. 2, cited in "Kyrgyzstan: Spent Fuel and Radioactive Waste," available at cns.miis.edu.

48. "Kyrgyzstan: Uranium Mining and Milling," available at cns.miis.edu.

49. The arrest of senior officials at the Kara-Balta mine, who were found guilty of diverting significant sales from the inventory, ignited protests against the revival of uranium operations. "Kyrgyz Security Minister on Uranium Corruption Case," _Kyrgyz Television Network,_ 1 July 1999, as translated in _FBIS FTS199907030003788,_ 1 July 1999.

50. "CIS Republics Sign U Suspension Agreement to End Anti-Dumping Agreement," _Nuclear Fuel,_ 21 October 1994, pp. 1–6. In addition, the Kara-Balta mine filed suit against customers in the West for failing to pay for uranium purchases due to the imposition of the 1992 antidumping agreement.

51. Margarita Sevcik, "Uranium Tailings in Kyrgyzstan: Catalyst for Cooperation and Confidence Building?" _The Nonproliferation Review_ (Spring 2003), pp. 147–154; and _Washington Times,_ 6 March 2004. In 1994, 1000 cm of radioactive waste was pushed into the Mayluu-Suu River and contaminated dozens of square kilometers of arable land. In 1999, the adminis-

tration of the southern Dzha_alabad region petitioned the Kyrgyz cabinet to adopt urgent meas_res to secure 23 storage sites for tailings and slag heaps in the vicinity of the Mayluu-Suu uranium deposit. *Interfax*, 25 February 1999.

52. "Kyrgyzstan," *Nuexco Review* (1994), p. 52, available at cns.miis.edu.
53. The agreement also include_ provisions for future partnership with Tajikistan's Leninabad Mining and Metallurgical Complex. "Kazakhstan, Kyrgyzstan Sign Mining/Processing Pact," *FreshFUEL*, 15 September 1997, pp. 1–2.
54. *BBC Monitoring Central Asia Unit*, 14 April 2000.
55. The United States proposed arrangements ranged from 260 thousand pounds of U_3O_8 at $13 per pound to 650 thousand pounds U_3O_8 at $20 per pound. "Suspension Agreements: The Status Quota," *The Nuclear Review* (January 1996), p. 19, available at cns.miis.edu.
56. *Nezavisimaya gazeta*, 4 June 2001.
57. *Interfax*, 21 October 2000. A small portion of the Kara-Balta facility was used to assemble South Korean refrigerators and microwave ovens from imported parts.
58. *Nuclear Fuel*, 2 January 1995, available at cns.miis.edu.
59. *Interfax*, 13 September 200_.
60. *Rossiskaya gazeta*, 23 December 2000; and *RIA Novosti*, 28 July 2000. The Russian enterprises proceeded to contract with Kyrgyz partners were the Eleron state company, the START production association, Uran, and Atomredmetzoloto. *The Times of Central Asia*, 31 January 2001.
61. *Rossiskaya gazeta*, 23 December 2000.
62. *Polit.ru*, 24 November 2000.
63. *The Times of Central Asia*, 31 January 2001.
64. *Nuclear.ru*, 13 November 2000. See also Sevcik (2003), p. 153; and *ITAR-Tass*, 20 January 2004.
65. *Rossiskaya gazeta*, 23 December 2000. See also *Nezavisimaya gazeta*, 11 April 2001.
66. *Parlamentskaya gazeta*, 11 January 2001.
67. *Interfax*, 20 October 2000.
68. *Vechernyi Bishkek*, 13 December 2000.
69. *ITAR-Tass*, 2 October 2003.
70. *Interfax*, 31 July 2000.
71. Ibid., 13 April 2001; and *Uranium Institute News Briefing*, 18 October 2000.
72. By 2004, Minatom officials publicly acknowledged that an initial 500 metric tons would be sold to both Russian and foreign customers. *ITAR-Tass*, 20 January 2004.
73. *Rossiskaya gazeta*, 23 December 2000; and *Bishkek Slovo Kyrgyzstana*, 20 April 2001, as translated in *FBIS CEP20010428000132*, 20 April 2001. Author thanks Kenley Butler for these citations and for clarifying the transaction.

74. *ITAR-Tass,* 20 January 2004; and "TVEL, TENEX Will Be Involved in Kazakh Uranium Venture," *Platts Nuclear Fuel* 27:25 (9 December 2002), p. 5. Responsibility for arranging financing was onerous, as it took Russia at least two years to devise a credible commercial plan for the production of the first 500 metric tons of uranium. By early 2004, Minatom arranged $1.35 million in loans, and it was expected that it would take an additional two years to secure the residual financing required to begin the first phase of operations.

75. *Nezavisimaya gazeta,* 11 April 2001.

76. *Interfax,* 31 July 2000. See also *Bishkek slovo Kyrgyzstana,* 20 April 2001, as translated in *FBIS-CEP20010428000132,* 20 April 2001.

77. *BBC Monitoring Central Asia Unit,* 15 October 2000.

78. *Vek,* 3 November 2000; and *Interfax,* 29 August 2001.

79. Vitaly V. Naumkin, "Russian Policy Toward Kazakhstan," in Robert Legvold, ed., *Thinking Strategically: The Major Powers, Kazakhstan, and the Central Asian Nexus* (Cambridge: The MIT Press, 2003), pp. 41–42.

80. Murat Laumulin, "Nonproliferation and Kazakhstani Security Policy," *The Nonproliferation Review* (Spring/Summer 1998), pp. 129–130.

81. *Elektrostal Atom-pressa,* 21 January 2001, as translated in *FBIS CEP20010221000227,* 31 January 2001.

82. "O Natsionalnoy atomnoy kompanii Kazatomprom," *Kazatomprom web site,* http://www.kazatomprom.kz/About/rus.asp; and "Verification of Initial LEU Inventory Under Way at Ust-Kamenogorsk Plant," *Nuclear Fuel,* 31 July 1995, p. 6, available at cns.miis.edu. In addition, Kazakhstan inherited modest capabilities to process uranium dioxide that supplemented services provided by plants in Kyrgyzstan and Tajikistan. See summary in *Country Nuclear Fuel Cycle Profiles,* Technical Report Series No. 404 (Vienna, Austria: International Atomic Energy Association, 2001), pp. 44–46.

83. Naumkin (2003), p. 42.

84. *Gazeta SNG.ru,* 9 June 2001.

85. *Kazakh Khabar TV,* 6 June 2001, as translated in *BBC Monitoring Service,* 7 June 2001.

86. *Pravda,* 2 November 1996. See also analysis of the sector's initial contraction in "It's 2007—Do You Know Who Your Uranium Suppliers Are?" *NUKEM* (May 2003), p. 11; and Nuclear Energy Agency (NEA) and the International Atomic Energy Agency (IAEA), *Uranium 1999: Resources, Production, and Demand* (Paris, OECD, 1999), p. 195.

87. Alexander Pavlov and Nigel Moe, "The Nuclear Fuel Market in Russia and the Former Soviet Union: The Dreams and the Reality," in *Uranium and Nuclear Energy 1997: Proceedings of the Twenty-Second Annual Symposium of the Uranium Institute* (London: Uranium Institute, September 1997), p. 198.

88. World Nuclear Association, "Supply of Uranium" and "Uranium Production Figures, 1995–2003," www.world-nuclear.org/info/uprod; NEA and IAEA (1999); and "1999 World Natural Uranium Production," *NUKEM* (April 2000), pp. 28–41.

89. "It's 2007—Do You Know Who Your Uranium Suppliers Are?" p. 16; *Interfax-Kazakhstan*, 31 May 2000, as translated in *FBIS-CEP200006-01000301*, 31 May 2000; World Nuclear Association, "Supply of Uranium" and "Uranium Production Figures, 1995–2003;" NEA and IAEA (1999); "1999 World Natural Uranium Production," pp. 28–41; and "2000 World Natural Uranium Production," *NUKEM* (April 2001).

90. *Almaty Yegemen Qazaqstan*, 5 January 2001, as translated in *FBIS, CEP20010115000001*, 5 January 2001.

91. *Interfax*, 22 April 1998. In 2000, the President of Kazatomprom pledged to invest \$50 million to increase uranium production. *Interfax-Kazakhstan*, 31 May 2000, as translated in *FBIS, CEP20000601000301*, 31 May 2000.

92. Daphne Biliouri, "Kazakhstan Set to Follow Russia into Nuclear Waste Business," *Jane's Intelligence Review* (1 May 2001).

93. *Panorama*, 28 February 1997.

94. *Almaty Panorama*, 19 May 2000, translated in *FBIS CEP20000607-000339*, 19 May 2000.

95. *Interfax*, 4 January 2001; and ibid., 17 May 2001.

96. *Interfax CIS Daily News Brief*, II:100 (1 June 2000); "On to Kazakhstan," *Nukem* (October 1995), p. 8; and "Cogema, KATEP Set Up Joint Venture to Develop U Deposits in Kazakhstan," *Nuclear Fuel* (26 August 1996), p. 8; and "It's 2007—Do You Know Who Your Uranium Suppliers Are?"

97. "Israeli Company Invests in Kazakh Uranium Plant," Nuclear Engineering International, 31 March 2000; and *Interfax*, 22 April 1998.

98. In August 2001, Kazatomprom initiated negotiations with the Chinese Atomic Corporation for creation a joint venture to develop deposits in southern Kazakhstan.

99. *Interfax*, 17 June 1999. The quota for Kazakh sales on the U.S. market ranged from 440,000 pounds of U_3O_8 in 1994; to 500,000 pounds of U_3O_8 in 1995; to 700,000 pounds of U_3O_8 in 1996. *Uranium Institute News Briefing*, NB96.41–12, 10 September 1996. http://www.ilondon.org/nb/nb96/nb9641.html.

100. *Interfax-Kazakhstan*, 14 July 1999, as translated in *FBIS FTS199907-16000552*, 14 July 1999. For more background discussion on the antidumping suit see especially Michael Knapik and Elaine Hiruo, "ITC Issues Decision in Kazakh Case," *Nuclear Fuel* (9 August 1999), pp. 2, 12.

101. *Kazakhstanskaya pravda*, 29 November 2001.

102. *Interfax*, 4 February 2002; ibid., 8 April 2002; *Ekpress-K*, 6 January 2000, as translated in *FBIS-CEP20000117000018*, 6 January 2000.

103. *Novoe pokolenie*, 21 February 2003, p. 4.

104. *Interfax*, 24 June 1999.

105. *Kommersant*, 29 February 2000.

106. *Vek*, 27 December 2001.

107. *Kazakh Commercial Television*, 17 December 1999, as translated in *FBIS CEP19991220000020*, 17 December 1999.

108. *Interfax,* 27 December 1999.
109. Ibid., 24 June 1999.
110. *Interfax-Kazakhstan,* 17 May 2001, as translated in *FBIS-CEP20010-517000224,* 17 May 2001.
111. *BBC Monitoring Service,* 15 July 2002.
112. *Interfax,* 18 February 2003.
113. *Vek,* 27 December 2001.
114. The Russian enterprises included the Chepetsk Mechanical Plant, Novosibirsk Chemical Concentrate Plant, and the Priargunskiy Mining and Chemical Combine. *Panorama,* 3 January 2000; and *Kommersant,* 29 February 2000. The author thanks Kenley Butler for these citations and for sharing the details of these agreements.
115. *Kazakhstanskaya pravda,* 29 November 2001, http://www.kazpravda.kz/archive/29_11_2001/e.html.
116. *UMZ-Inform,* 14 April 2000; and *Vek,* 3 November 2000.
117. *Interfax,* 24 June 1999.
118. *Kommersant,* 29 February 2000.
119. Ibid.
120. *Interfax,* 27 December 1999.
121. Ibid., 24 June 1999; *Kazakh Commercial Television,* 17 December 1999, as translated in *FBIS CEP19991220000020,* 17 December 1999; and "Kazatomprom-TVEL Equity Swap," *Nuclear Engineering International,* 31 March 2000.
122. *UMZ-Inform,* 8 May 2000. The author thanks Kenley Butler for this citation.
123. Alexei Breus, "Ukranian-Russian-Kazakh Fuel Venture May Be Formally Launched in August," *Nuclear-Fuel* 26:15 (23 July 2001).
124. *Kommersant,* 5 July 2001.
125. *Vek,* 27 December 2001.

Chapter 7

1. See discussion in Jeffrey W. Taliaffero, *Balancing Risks: Great Power Intervention in the Periphery* (Ithaca: Cornell University Press, 2004), pp. 232–233.
2. Ibid.; and Victor D. Cha, "Hawk Engagement and Preventive Defense on the Korean Peninsula," *International Security* 27:1 (Summer 2002), pp. 40–78.
3. Paul J. D'Anieri, *Economic Interdependence in Ukrainian-Russian Relations* (Albany, NY: State University Press of New York, 1999); Oles Smolansky, "Fuel, Credit, and Trade: Ukraine's Economic Dependence on Russia," *Problems of Post-Communism* 46:2 (March/April 1999), pp. 49–58; Rawi Abdelal, "Interpreting Interdependence: Energy and Security in Ukraine and Belarus," *Unpublished paper* (Harvard University, July

2002), available at www.csis.org/ruseura/ponars/workingpapers/020.PDF; Margarita M. Balmaceda, *Explaining the Management of Energy Dependency in Ukraine: Possibilities and Limits of a Domestic-Centered Perspective* (Mannheim, Germany: Working Paper 79 Mannheimer Zentrum fur Europaische Sozialforschung, 2004), available at www.mzes.uni-mannheim.de; and Corina Herron Linden, *Power and Uneven Globalization: Coalitions and Energy Trade Dependence in the Newly Independent States of Europe,* Ph.D. dissertation, University of Washington, 2000.

4. D'Anieri (1999), pp. 69–96.
5. *Nezavisimaya gazeta,* 28 January 2000. See also discussion in Balmaceda (2004): 22–34; and Abdelal (2002).
6. D'Anieri (1999), p. 82; Balmaceda (2004): and Arkady Toritsyn and Eric A. Miller, "From East to West and Back Again: Economic Reform and Ukrainian Foreign Policy," *European Security* 11:1 (Spring 2002), pp. 102–126.
7. *Nezavisimaya gazeta,* 16 January 2004. See also Balmaceda (2004), pp. 11–13; and Smolansky (1999), pp. 51–54.
8. Cited in Margarita Mercedes Balmaceda, "Gas, Oil, and the Linkage Between Domestic and Foreign Policies: The Case of Ukraine," *Europe-Asia Studies* 50:2 (1998), p. 266. See also Rosario Puglisi, "Clashing Agendas? Economic Interests, Elite Coalitions, and Prospects for Cooperation between Russia and Ukraine," *Europe-Asia Studies* 55:6 (2003), pp. 827–845. Russia's oil diplomacy toward the Baltic states was tempered by the availability of imports from Finland, and the incentives to accelerate domestic energy reforms and efficiency generated by the prospects for EU membership.
9. See discussion in Margarita M. Balmaceda, "Belarus as a Transit Route: Domestic and Foreign Policy Implications," pp. 167–169; Yuri Drakokhrust and Dmitri Furman, "Belarus and Russia: The Game of Virtual Integration," pp. 253–255; Astrid Sahm and Kirsten Westphal, "Power and the Yamal Pipeline," pp. 270–301 in Margarita M. Balmaceda, James I. Clem, and Lisbeth L. Tarlow, eds., *Independent Belarus: Domestic Determinants, Regional Dynamics, and Implications for the West* (Cambridge: HURI/Davis Center for Russian Studies: distributed by Harvard University Press, 2002).
10. Cited in Sergei Danilochkin, "Belarus: Moscow and Minsk Back Down from Gas Crisis as Temporary Supplies Resume," *RFE/RL,* 19 February 2004. Although the final terms of Gazprom's stakes in the joint venture with Belarus's gas transit company remained deadlocked over the value of the company, Minsk authorities continued to pay higher rates and acknowledged Gazprom's 50 percent debt equity stake. By contrast, Lukashenko did not cave to Russian pressure in the oil sector, as he clutched to national ownership of the country's sole refinery. Balmacedes (2002), pp. 172–176.
11. Kenneth A. Schultz, *Democracy and Coercive Diplomacy* (Cambridge: Cambridge University Press, 2001).
12. Frederick Starr, "Power Failure: American Policy in the Caspian," *The National Interest* 47 (Spring 1997), pp. 20–31; and Jan S. Adams, "The

U.S.-Russian Face-off in the Caspian Basin," *Problems of Post-Communism* 47:1 (January/February 2000), pp. 49–58. See also Laurent Ruseckas, "State of the Field Report: Energy and Politics in Central Asia and the Caucasus," *Access Asia Review* 1:2 (February 1998), available at: www.nbr.org/publications/review/vol1no2/essay2.html.

13. Douglas W. Blum, "Domestic Politics and Russia's Caspian Policy," *Post-Soviet Affairs* 14:2 (1998), pp. 137–164; and Henry Hale, "The Rise of Russian Anti-Imperialism," *Orbis* 43:1 (Winter 1999), pp. 111–125.

14. Anatol Lieven, "The (Not So) Great Game," *The National Interest* 58 (Winter 1999–2000), pp. 69–80; Abraham S. Becker, "Russia and Caspian Oil: Moscow Loses Control," *Post-Soviet Affairs* 16:2 (2000), pp. 91–132; Peter Rutland, "Paradigms for Russian Policy in the Caspian Region," in Robert Ebel and Rajan Menon, eds., *Energy and Conflict in Central Asia and the Caucasus* (New York: Rowman & Littlefield Publishers, 2000), pp. 163–188; and David G. Victor and Nadejda M. Victor, "Axis of Oil?" *Foreign Affairs* 82:2 (March/April 2003), pp. 47–61.

Selected Bibliography

Abdelal, Rawi. "Interpreting Interdependence: Energy and Security in Ukraine and Belarus." Unpublished paper. Harvard University, July 2002. www.csis.org/ruseura/ponars/workingpapers/020.PDF.

———. *National Purpose in the World Economy: Post-Soviet States in Comparative Perspective.* Ithaca: Cornell University Press, 2001.

Adams, Jan S. "Russia Gas Diplomacy." *Problems of Post-Communism*, 49, 3 (May/June 2002): 14–22.

———. "The U.S.-Russian Face-off in the Caspian Basin." *Problems of Post-Communism*, 47, 1 (January/February 2000): 49–58.

Adams, Terry. "Caspian Hydrocarbons, the Politicization of Regional Pipelines, and the Destabilization of the Caucasus." Center for European Policy Studies (CEPS), Brussels, Belgium, 27–28 January 2000.

Akaev, Askar. *Kyrgyzstan: An Economy in Transition.* Sidney: Asia Pacific Press, Australian National University, 2000.

———. *Pamyatnoe desyatiletie: trudnaya doroga k demokratii.* Moskva: Mezhdunarodnye otnosheniya, 2002.

Akiner, Shiron, and Sader Tideman, and John Hay, eds. *Sustainable Development in Central Asia.* Surrey, UK: Curzon Press, 1998.

Alchian, Armen A., and Harold Demsetz. "Property Rights Paradigm." *Journal of Economic History,* 33 (1973): 16–27.

Alexandrov, Mikhail. *Uneasy Alliance: Relations Between Russia and Kazakhstan in the Post-Soviet Era, 1992–1997.* Westport, CT: Greenwood Press, 1999.

Alimov, Rashid. "Misinformation about the Uranium Deal." *Bellona Report* (2 August 2001), www.bellona.no.

Allison, Roy, and Christoph Bluth, eds. *Security Dilemmas in Russia and Eurasia.* London: The Royal Institute of International Affairs, 1998.

Altstadt, Audrey L. "Azerbaijan and Aliev: A Long History and Uncertain Future." *Problems of Post-Communism* 50, 5 (September–October 2003): 3–13.

Amini, Gitty Madeline. *Sanctions and Reinforcement in Strategic Relationships: Carrots and Sticks, Compellence and Deterrence.* Los Angeles, CA: UCLA Dissertation, 2001.

Anderson, John. *Kyrgyzstan: Central Asia's Island of Democracy.* Australia: Harwood Academic Publishers, 1999.

Asipov, Murat. "Kazakhstan: Living Standards During the Transition." *World Bank Report,* no. 17520-KZ (23 March 1998).

Auerswald, David P. "Inward Bound: Domestic Institutions and Military Conflicts." *International Organization* 53, 3 (Summer 1999): 469–504.

Bacharach, Peter, and Morton Baratz. *Power and Poverty: Theory and Practice.* New York: Oxford University Press, 1970.

Baev, Pavel. "Russia's Policies in the Southern Caucasus and Caspian Region." *European Security* 10, 2 (Summer 2001): 95–110.

Baldwin, David A. "Interdependence and Power: A Conceptual Analysis." *International Organization* 34, 4 (Autumn 1980): 471–506.

———. *Economic Statecraft.* Princeton: Princeton University Press, 1985.

———. *Paradoxes of Power.* New York: Basil Blackwell, 1989.

———. "The Power of Positive Sanctions." *World Politics* 24, 1 (October 1971): 19–38

Balmaceda, Margarita M. *Explaining the Management of Energy Dependency in Ukraine: Possibilities and Limits of a Domestic-Centered Perspective.* Mannheim, Germany: Working Paper 79 Mannheimer Zentrum fur Europaische Sozialforschung, 2004, www.mzes.uni-mannheim.de.

———. "Gas, Oil, and the Linkage Between Domestic and Foreign Policies: The Case of Ukraine." *Europe-Asia Studies* 50, 2 (March 1998): 257–286.

———, James I. Clem, and Lisbeth L. Tarlow, eds. *Independent Belarus: Domestic Determinants, Regional Dynamics, and Implications for the West.* Cambridge: MA: Distributed by Harvard University Press for the Ukrainian Research Institute and Davis Center for Russian Studies, Harvard University, 2002.

Barbieri, Katherine. "Economic Interdependence: A Path to Peace or a Source of Interstate Conflict?" *Journal of Peace Research* 33, 1 (1996): 29–49.

Barnes, Andrew. "Russia's New Business Groups and State Power." *Post-Soviet Affairs* 19, 2 (2003): 154–186.

Becker, Abraham S. "Russia and Caspian Oil: Moscow Loses Control," *Post-Soviet Affairs* 16, 2 (2000): 91–132.

———. "Russia and Economic Integration." *Survival* 38, 4 (Winter 1996–1997): 117–36

Berejikian, Jeffrey D. *International Relations Under Risk: Framing State Choice.* Albany: State University of New York Press, 2004.

Bergh, Einar. "AIOC Current Developments." *Azerbaijan International* 4, 1 (Spring 1996).

Blanchard, Jean-Marc F., and Norrin M. Ripsman. "Asking the Right Question: When Do Economic Sanctions Work Best?" *Security Studies* 1, 2 (Autumn 1999–Winter 2000): 218–253.

——— and Norrin Ripsman, "Measuring Economic Interdependence: A Geopolitical Perspective." *Geopolitics,* 1, 3 (Winter 1996): 229–231.

———, Edward S. Mansfield, and Norrin M. Ripsman, eds. *Power and the Purse*. London: Frank Cass, 2000.

Blank, Stephen J. "Every Shark East of Suez: Great Power Interests, Policies, and Tactics in the Transcaspian Energy Wars." *Central Asian Survey* 18, 2 (1999): 149–184.

———. "Russia's Real Drive to the South." *Orbis* 39:3 (Summer 1995): 369–386.

———. *Towards the Failing State: The Structure of Russian Security Policy*. Camberley, UK: Conflict Studies Research Center, Royal Military Academy Sandhurst, 1996.

Blum, Douglas W. "Domestic Politics and Russia's Caspian Policy." *Post-Soviet Affairs* 14, 2 (1998): 137–164.

———. "Why Did LUKoil Really Pull Out of the Azer-Chirag-Guneshli Oilfield." *Program on New Approaches to Russian Security*. Memo 286 (January 2003).

Bohi, Douglas R., and Micahel A. Toman. *The Economics of Energy Security*. Boston: Kluwer Academic Publishers, 1996.

Boyco, Maxim, Andrei Shileffer, and Robert Vishny. *Privatizing Russia*. Cambridge: MIT Press, 1994.

Bremmer, Ian and Ray Taras, eds. *New States New Politics: Building the Post-Soviet Nations*. Cambricge: Cambridge University Press, 1997.

Breus, Alexei. "Ukranian-Russian-Kazakh Fuel Venture May Be Formally Launched in August." *Nuclear-Fuel* 26:15 (23 July 2001).

Buchanan, James, and Gordon Tullock. *The Calculus of Consent*. Ann Arbor: University of Michigan Press, 1962.

Bukharin, Oleg. "The Future of Russia's Plutonium Cities," *International Security* 21:4 (Spring 1997): 126–158.

Bunce, Valerie J. "The Political Economy of the Brezhnev Era." *British Journal of Political Science* 13 (January 1983): 129–158.

Byman, Daniel, and Matthew Waxman. *The Dynamics of Coercion: America's Foreign Policy and the Limits of Military Might*. New York: Cambridge University Press, 2002.

Calder, Kent. "Japan's Energy Angst and the Caspian Great Game." *National Bureau of Asian Research, Analysis* 12, 1 (March 2001).

Caporaso, James A. "Dependence, Dependency, and Power in the Global System: A Structural and Behavioral Analysis." *International Organization* 32, 1 (Winter 1978): 13–43.

Carlsnaes, Walter. *Energy Vulnerability and National Security: The Energy Crises, Domestic Policy Responses and the Logic of Swedish Neutrality*. New York: Pinter Publishers.

Cha, Victor D. "Hawk Engagement and Preventive Defense" *International Security* 27, 1 (Summer 2002): 40–78.

———. *Nuclear North Korea: A Debate on Engagement Strategies*. New York: Columbia University Press, 2003.

Chan, Steve, and A. Cooper Drury, eds. *Sanctions as Economic Statecraft: Theory and Practice.* London: St. Martin's Press, 2000.

Christoffersen, Gaye. "China's Intentions for Russian and Central Asian Oil and Gas." *The National Bureau of Asian Research Analysis* 9, 2 (March 1998).

Chufrin, Gennady. *Russia and Asia: The Emerging Security Agenda.* Oxford: Oxford University Press, 1999.

———. *The Security of the Caspian Region.* Oxford: Oxford University Press. 2001.

Cimbala, Stephen J. *Russia and Armed Persuasion.* New York: Rowman & Littlefield Publishers, 2001.

Claes, Dag Harald. *The Politics of Oil Producer Cooperation.* Boulder: Westview Press, 2001.

Cochran, Thomas B, Robert S. Norris, and Oleg A. Bukharin. *Making the Russian Bomb: From Stalin to Yeltsin.* Boulder: Westview Press, 1995.

Collins, Adrian. "Nuclear Power in Russia." *CORE Issues (Journal of the Uranium Institute)* 2 (December 2000), http://www.worldnuclear.org/core-issues/2000/no2/features/feature4.htm.

Collins, Kathleen. "The Logic of Clan Politics: Evidence from the Central Asian Trajectories." *World Politics* 56, 2 (January 2004): 171–190.

Cooley, Alexander. "Imperial Wreckage: Property Rights, Sovereignty, and Security in the Post-Soviet Space." *International Security* 25, 3 (Winter 2000/01): 100–127.

Copeland, Dale C. "Economic Interdependence and War: A Theory of Trade Expectations." *International Security* 20, 4 (1996): 5–41.

Cowhey, Peter. "Domestic Institutions and the Credibility of Commitments: Japan and the United States." *International Organization* 47, 2 (Spring 1993): 299–326.

Crawford, Beverly. *Economic Vulnerability in International Relations: The Case of East-West Trade, Investment, and Finance.* New York: Columbia University Press, 1993.

Crawford, Timothy W. *Pivotal Deterrence: Third-Party Statecraft and Pursuit of Peace.* Ithaca: Cornell University Press, 2003.

Croissant, Michael P. and Bulent Aras, eds. *Oil and Geopolitics in the Caspian Sea Region.* Westport, CT: Praeger, 1999.

Crumm, Eileen M. "The Value of Economic Incentives in International Politics." *Journal of Peace Research* 32, 3 (1995): 321–322.

Cummings, Sally N., ed. *Power and Change in Central Asia.* London: Routledge, 2002.

Cutler, Robert. "The Indo-Iranian Rapprochement: Not Just Natural Gas Anymore." *Central Asian Caucasus Analyst* (9 May 2001).

D'Anieri, Paul J. *Economic Interdependence in Ukrainian-Russian Relations.* Albany: State University of New York Press, 1999.

"DATA FEATURE: 2000 World Uranium Production." *NUKEM Market Report* (April 2001).

Davis, Jr., James W. *Threats and Promises: The Pursuit of International Influence*. Baltimore: Johns Hopkins University Press, 2000.

Dawisha, Karen and Bruce Parrott, eds. *Conflict, Cleavage and Change in Central Asia and the Caucasus*. Cambridge: Cambridge University Press, 1997.

Diaconu, Oana C., and Michael T. Maloney. "Russian Commercial Nuclear Initiatives and U.S. Nuclear Nonproliferation Interests." *The Nonproliferation Review* (Spring 2003):1–17.

———. "Is Nuclear Power Viable in Russia?" *The Electricity Journal* (January/February 2003): 80–87.

Dienes, Leslie. "The Russian Oil and Gas Sector: Implications of the New Property System." *National Bureau of Asian Research* 4 (March 1996).

———, Istvan Dobozi, and Marian Radetzki. *Energy and Economic Reform in the FormerSoviet Union*. New York: St. Martin's Press, 1994.

Digges, Charles. "Tenex Might be Stuck with Tonnes of Uranium it Produces for U.S. Consumption." *Bellona Report* (11 November 2002), www.bellona.no.

Drasdo, Peter. "Impacts of Power Market Deregulation on the Nuclear Fuel Cycle Markets and Industries." *EWI-Working Paper* (February 1999).

Drezner, Daniel W. "The Trouble with Carrots: Transaction Costs, Conflict Expectation, and Economic Inducements." *Security Studies* 9, 1–2 (Autumn 1999–Winter 2000): 188–218.

———. *The Sanctions Paradox: Economic Statecraft and International Relations*. Cambridge: Cambridge University Press, 1999.

Ebel, Robert, and Rajan Menon, eds. *Energy and Conflict in Central Asia and the Caucasus*. New York: Rowman & Littlefield Publishers, 2000.

———. *Energy Choices in the Near Abroad*. Washington, DC: Center for Strategic and International Studies, April 1997.

Ebinger, Charles K. *The Critical Link: Energy and National Security*. Cambridge: Ballinger Press, 1982.

Eggertsson, Thrainn. *Economic Behavior and Institutions*. New York: Cambridge University Press, 1990.

Energy Informational Agency. "Caspian Sea Region: Reserves and Pipelines Tables." July 2002, www.eia.doe.gov/emeu/cabs/caspgrph.html#TAB1.

———. "Central Asia: Turkmenistan Energy Sector." May 2002, www.eia.doe.gov/emeu/cabs/turkmen/html.

———. "Azerbaijan Country Analysis Brief." June 2003, www.eia.doe.gov/emeu/cabs/azerbjan.html.

———. "Kazakhstan." July 2003, www.eia.doe.gov/emeu/cabs/kazak.html.

———. "Russia Country Analysis." November 2002, www.eia.doe.gov/cabs/russia.html.

———. "Russia: Energy Sector Restructuring." April 2002, www.eia.doe.gov/cabs/russrest.html#OIL.

———. *Russia: Oil and Natural Gas Exports*. November 2002, www.eia.doe.gov/cabs/russexp.html.

Elman, Colin and Miriam Elman, eds. *Bridges and Boundaries: Historians, Political Scientists, and the Study of International Relations.* Cambridge: MIT Press, 2001.

Esenov, Murad. "Vneshnyaya politika Turkmenistana i ee vliyanie na sistemu regional'noi bezopasnosti." *Tsentral'naya i kavkaz* 1, 13 (2001): 56–64.

European Commission. "Towards a European Strategy for the Security of Energy Supply." *EU Green Paper,* 2001.

Evans, Peter B., Harold K. Jacobson, and Robert D. Putnam, eds. *Double-Edged Diplomacy.* Berkeley: University of California Press, 1993.

Evera, Stephen Van. *Guide to Methods for Students of Political Science.* Ithaca: Cornell University Press, 1997.

Farnham, Barbara, ed. *Avoiding Losses/Taking Risks.* Ann Arbor: University of Michigan Press, 1994.

Fearon, James D. "Rationalist Explanations for War." *International Organization* 49 (1995): 379–414.

———. "Domestic Political Audiences and the Escalation of International Disputes." *American Political Science Review* 88, 3 (September 1994): 577–592.

Federov, Yuri. "Kaspiiskaya politika Rossii: k konsenssusu elit." *Pro et Contra,* 2, 3 (Spring 1997): 72–89.

———. "Russia's Policies Towards Caspian Region Oil: Neo-imperial or Pragmatic?" *Perspectives on Central Asia* 1:6 (September 1996).

Ferguson, Yale H., and R.J. Barry Jones, eds. *Political Space: Frontiers of Change and Governance in a Globalizing World.* Albany: State University of New York Press, 2002.

Financial Times

Freedman, Lawrence, ed. *Strategic Coercion: Concepts and Cases.* Oxford: Oxford University Press, 1998.

———, ed. *Strategic Choice.* Oxford: Oxford University Press, 1998.

Frieden, Jeffry A. "Sectoral Conflict and US Foreign Economic Policy, 1914–1940." *International Organization* 42, 2 (1988): 59–90.

———. "International Investment and Colonial Control: A New Interpretation." *International Organization* 48, 4 (Autumn 1994): 559–593.

Gaddy, Clifford C., and Barry W. Ickes. "Russia's Virtual Economy." *Foreign Affairs* 77, 5 (1998): 53–67.

Gadziyev, K.S. *Vvedeniye v geopolitiku.* Moskva: Logos, 2000.

Garret, Geoffrey. "Global Markets and National Politics: Collision Course or Virtuous Circle." *International Organization* 52, 4 (Autumn 1998): 787–824.

Gartzke, Erik, Quan Li, and Charles Boehmer. "Investing in the Peace: Economic Interdependence and International Conflict." *International Organization* 55, 2 (2001): 391–438.

Gaubatz, Kurt Taylor. "Democracy and Commitment." *International Organization* 50, 1 (Winter 1996): 109–140.

George, Alexander L. "The 'Operational Code': A Neglected Approach to the

Study of Political Leaders and Decisionmaking." *International Studies Quarterly* 13 (1969): 190–222.

———, and William E. Simons, eds. *The Limits of Coercive Diplomacy*. 2nd ed. Boulder, CO: Westview Press, 1994.

Geva, Nehemia, and Alex Mintz, eds. *Decisionmaking on War and Peace: The Cognitive-Rational Debate*. Boulder: Lynn Rienner Publishers, 1997.

Gill, Stephen, and David Law. "The Global Hegemony and the Structural Power of Capital." *International Studies Quarterly* 35, 3 (September 1991): 313–336.

Gilpin, Robert. *Global Political Economy: Understanding the International Economic Order*. Princeton: Princeton University Press, 2001.

Gleason, Gregory. "Russia and the Politics of the Central Asian Electricity Grid." *Problems of Post-Communism* 50, 3 (May/June 2003): 42–52.

Golunov, Sergei. "Rossiya-Kazakhstan: dilemma 'bezopasnost'-sotrudnichestvo' v transgranichnykh vzaimootnosheniyakh." *Tsentral'naya aziya i kavkaz*, 4, 22 (2002): 66–78.

Gopinath, Deepak. "Face-off in Moscow." *Infrastructure Finance* 6, 6 (July/August 1997): 36–39.

Gorst, Isabel, and Nina Pousssenkova. *Petroleum Ambassadors of Russia: State Versus Corporate Policy in the Caspian Region*. Houston, TX; Center for International Political Economy and the James A. Baker III Institute for Public Policy, Rice University, 1998.

Gouliev, Rasul. *Oil and Politics: New Relationship Among the Oil-Producing States: Azerbaijan, Kazakhstan, and Turkmenistan*. New York: Liberty Publishing House, 1997.

Haghayeghi, Mehrdad. "The Coming of Conflict to the Caspian Sea." *Problems of Post-Communism* 50, 3 (May–June 2003): 32–41.

Hale, Henry. "The Rise of Russian Anti-Imperialism." *Orbis* 43, 1 (Winter 1999): 111–125.

Harrison, Selig S. "Gas and Geopolitics in Northeast Asia." *World Policy Journal* (Winter 2002/2003): 23–36.

Harsanyi, John C. "Measurement of Social Power, Opportunity Costs, and the Theory of Two-Person Bargaining Games." *Behavioral Science* VII (January 1962).

Hartshorn, J.E. *Oil Trade: Politics and Prospects*. Cambridge: Cambridge University Press, 1993.

Heslin, Sheila. *Key Constraints to Caspian Energy Development: Status, Significance, and Outlook*. Working Paper: The Center for International Political Economy and the James A. Baker III Institute for Public Policy, Rice University, April 1998.

Hill, Fiona. *Energy Empire: Oil, Gas, and Russia's Revival* (London: The Foreign Policy Center, September 2004).

———. "Energy Integration and Cooperation in Northeast Asia." *NIRA Policy Research* (Japan) 15, 2 (25 February 2002), www.brook.edu/views/articles/hillf/20020225.htm.

————, and Florence Fee. "Fueling the Future: The Prospects for Russian Oil and Gas."*Demokratizatsiya* 10, 4 (Fall 2002): 462–487.

Hirschman, Albert O. *National Power and the Structure of Foreign Trade*. Berkeley: University of California Press, 1980.

Hirshliefer, Jack. *Time, Uncertainty, and Information*. New York: Basil Blackwell, 1998.

————. "Appeasement: Can It Work?" *American Economic Review* 91, 2 (May 2001): 342–346.

Holsti, Ole R., Randolph M. Siverson, and Alexander George, eds. *Change in the International System*. Boulder: Westview Press, 1980.

Hopmann, P.T. "Asymmetrical Bargaining in the Conference on Security and Cooperation in Europe." *International Organization* 32, 1 (1978): 141–178.

Ikenberry, G. John. *Reasons of State: Oil Politics and the Capacities of American Government*. Ithaca: Cornell University Press, 1988.

————, and Charles Kupchan. "Socialization and Hegemonic Power." *International Organization* 44, 3 (1990): 283–315.

International Atomic Energy Agency. *Country Nuclear Fuel Cycle Profiles*. Technical Report Series No. 404. Vienna, Austria. 2001.

International Energy Agency. *Caspian Oil and Gas: The Supply Potential of Central Asia and Transcaucasia*. Paris: OECD. 1998.

————. *The IEA Natural Gas Security Study*. Paris: OECD. 1995.

————. *Russia Energy Survey, 2002*. Paris: OECD. March 2002.

International Monetary Fund. *Azerbaijan Republic: Recent Economic Developments* Report No. 98/83. August 1998.

————. *Azerbaijan Republic—Selected Issues and Statistical Appendix*, Report 03/130. May 2003.

————. *Direction of Trade Statistics: 1998 Year Book*. 1998.

————. *Kyrgyz Republic: Selected Issues and Statistical Appendix*. Country Report No. 03/53. February 2003.

————. *Republic of Kazakhstan: Recent Economic Development*. Country Report No. 98/84. July 1998.

————. *Republic of Kazakhstan: Selected Issues and Statistical Appendix*, Country Report, No. 00/29. March 2000.

————. *Russian Federation Statistical Appendix*. Country Report, No. 03/145. May 2003.

————. *Turkmenistan: Recent Economic Developments*. Country Report No. 99/140. December 1999.

———— and the World Bank. "Country Notes: Debt Sustainability and Policy Priorities to Ensure Growth in the CIS7 Countries." 6 February 2001.

Izvestiya

Jaffe, Amy Myers, and Robert A. Manning. "The Myth of the Caspian 'Great Game': The Real Geopolitics of Energy." *Survival* 40, 4 (Winter 1998–1999): 112–131.

James, Scott C., and David A. Lake. "The Second Face of Hegemony: Britain's

Repeal of the Corn Laws and America's Walker Tariff of 1846." *International Organization* 43, 1 (1989): 1–29.

Jensen, Michael, and William C. Meckling. "Theory of the Firm: Managerial Behavior, Agency Costs, and Ownership Structure." *Journal of Financial Economics* 3 (1976): 305–360.

Jervis, Robert. "Cooperation Under the Security Dilemma." *World Politics* 30 (Jan. 1978): 168–214.

———, Richard Ned Lebow, and Janice Gross Stein, eds. *Psychology and Deterrence: Perspectives on Security.* Baltimore: Johns Hopkins University Press, 1985.

Johnson, C.M. *The Russian Federation's Ministry of Atomic Energy: Programs and Developments.* Paper Prepared by Pacific Northwest Laboratory for the U.S. Department of Energy. Contract DE-AC06–76RLO. February 2000.

Jonson, Lena. *Russia and Central Asia: A New Web of Relations.* London: The Royal Institute of International Affairs, 1998.

Joseph, Ira. *Caspian Gas Exports: Stranded Reserves in a Unique Predicament.* Houston: Center for International Political Economy and the James A. Baker III Institute for Public Policy, Rice University, April 1998.

Josephson, Paul R. *Red Atom: Russia's Nuclear Power Program from Stalin to Today.* New York: W.H. Freeman and Company, 2000.

Kahneman, Daniel, and Amos Tversky. "Prospect Theory: An Analysis of Decisionmaking Under Risk." *Econometrica* 47 (1979): 263–291.

Kalyuzhnova, Yelena, Amy Myers Jaffe, and Dov Lynch, and Richard Sickles, eds. *Energy in the Caspian Region.* Hampshire, UK: Palgrave, 2002.

Kamenev, Sergei. "Ekonomika Turkmenistana na sovremennom etape." *Tsentral'naya aziya i kavkaz*, 3, 21 (2002): 194–205.

———. "Toplivno-energeticheskii kompleks Turkmenistana: sovremennoe sostoyanie i perspektivy razvitiya." *Tsenral'naya aziya i kavkaz* 6, 18 (2001): 178–192.

———. "Vneshnyaya politika Turkmenistana," *Tsentral'naya aziya i kavkaz* 4, 22 (2002): 90–101.

Kapstein, Ethan B., and Michael Mastanduno, eds. *Unipolar Politics: Realism and State Strategies after the Cold War.* New York: Columbia University Press, 1999.

Katzenstein, Peter J. "International Relations and Domestic Structures: Foreign Economic Policies of Advanced Industrial States." *International Organization* 30, 1 (Winter 1976): 1–45.

———, ed. *Between Power and Plenty.* Madison: University of Wisconsin Press, 1978.

Kaufman, Richard F., and John P. Hardt, eds. *The Former Soviet Union in Transition.* Armonk, NY: M.E. Sharpe, 1993.

Kazakhstanskaya pravda

Kent, Marian. *Oil and Empire: British Policy and Mesopotamian Oil, 1900–1920.* London: MacMillan, 1976.

Keohane, Robert O. ed. *Power and Governance in a Partially Globalized World*. London: Routledge, 2002.

———, and Helen V. Milner, eds. *Internationalization of Domestic Politics*. New York: Cambridge University Press, 1996.

———, and Joseph S. Nye. *Power and Interdependence*. 2nd ed. New York: HarperCollins, 1989.

Kertzer, David I. *Ritual, Politics, and Power*. New Haven: Yale University Press, 1988.

Khartukov, Eugene M. "Incomplete Privatization Mixes Ownership of Russia's Oil Industry." *Oil & Gas Journal* (18 August 1997).

———. "Low Oil Prices, Economic Woes Threaten Russian Oil Exports." *Oil & Gas Journal* (8 June 1998).

Khripunov, Igor. "MINATOM: Time for Crucial Decisions." *Problems of Post-Communism* 48, 4 (July/August 2001): 49–58.

———, and Maria Katsva. "Russia's Nuclear Industry: The Next Generation." *The Bulletin of Atomic Scientists* (March/April 2002): 51–57.

———, and Mary Matthews. "Russia's Oil and Gas Interest Group and Its Foreign Policy Agenda." *Problems of Post-Communism* (May/June 1996): 38–48.

Kiesling, Lynne, and Joseph Baker. "Russia's Role in the Shifting World Oil Market." *Caspian Studies Policy Brief,* Harvard University 8 (May 2002).

Kiewiet, D. Roderick, and Mathew D. McCubbins. *The Logic of Delegation: Congressional Parties and the Appropriations Process*. Chicago: University of Chicago Press, 1991.

Kingdon, John W. *Agendas, Alternatives, and Public Policies*. Boston: Little, Brown and Company, 1984.

Kirshner, Jonathan. *Currency and Coercion: The Political Economy of International Monetary Power*. Princeton: Princeton University Press, 1995.

Klare, Michael T. *Resource Wars*. New York: Metropolitan Books, 2001.

Klien, Benjamin, Robert G. Crawford, and Armen A. Alchian. "Vertical Integration, Appropriable Rents, and the Competitive Contracting Process." *Journal of Law and Economics* 21 (October 1979): 297–326.

Knapik, Michael, and Elaine Hiruo. "ITC Issues Decision in Kazakh Case." *Nuclear Fuel* (9 August 1999), www.mhenry.com.

———, Mark Hibbs, and Ann MacLachlan. "U.S. Open to Nuclear Pact with Conditions on International Spent Fuel Storage in Russia." *NuclearFuel,* (11 June 2001), www.mhenergy.com.

Kolossov, V., and N.S. Mironenko. *Geopolitika i politicheskaya geografiya*. Moskva: Aspekt Press, 2001.

Kommersant

Konoplyanik, Andrei. *Kaspiiskaya neft': na evraziiskom perekrestke*. Moscow: IGiRGI, 1998.

Krasner, Stephen D. *Defending the National Interest: Raw Materials Investments and U.S. Foreign Policy*. Princeton: Princeton University Press 1978.

———, ed. *Problematic Sovereignty: Contested Rules and Political Possibilities.* New York: Columbia University Press, 2001.

———, *Sovereignty: Organized Hypocrisy.* Princeton: Princeton University Press, 1999.

———. *Structural Conflict.* Berkeley: University of California Press, 1985.

Krasnov, Gregory v., and Josef C. Brada. "Implicit Subsidies in Russian-Ukrainian Energy Trade," *Europe-Asia,* 49, 5 (1997): 825–843.

Kubicek, Paul. "Regionalism, Nationalism, and Realpolitik." *Europe-Asia Studies* 49, 4 (1997): 637–656.

Kudrik, Igor. "Nuclear Fuel Producer Gets Nervous." *Bellona Report* (23 June 2000), www.bellona.no.

Lake, David A., and Robert Powell, eds. *Strategic Choice and International Relations.* Princeton: Princeton University Press, 1999.

Lane, David, ed. *The Political Economy of Russian Oil.* Lanham: Rowman & Littlefield Publishers, 1999.

Latsis, S.J., ed. *Method and Appraisal in Economics.* Cambridge: Cambridge University Press, 1976.

Laumulin, Murat. "Nonproliferation and Kazakhstani Security Policy." *The Nonproliferation Review* 5, 3 (Spring/Summer 1998): 126–133.

Lebow, Richard Ned. "What's So Different about a Counterfactual." *World Politics* 52 (July 2000): 550–585.

———, and Janice Gross Stein. "Rational Deterrence Theory: I Think, Therefore Deter." *World Politics* 41 (1989): 208–224.

Legvold, Robert, ed. *Thinking Strategically: The Major Powers, Kazakhstan, and the Central Asian Nexus.* Cambridge: The MIT Press, 2003.

Lelyveld, Michael. "Azerbaijan: LUKoil's Plan to Sell Azerbaijani Assets Raises Questions." *Radio Free Europe/Radio Liberty,* 29 October 2002.

Levy, Jack S. "Prospect Theory, Rational Choice, and International Relations." *International Studies Quarterly* 41, 1 (1997): 87–112.

Libecap, Gary. *Contracting for Property Rights.* New York: Cambridge University Press, 1989.

Liberman, Peter. "The Spoils of Conquest." *International Security* 18, 2 (1993): 125–153.

Lieven, Anatol. "The (Not So) Great Game." *The National Interest* 58 (Winter 1999–2000): 69–80.

Linden, Corina Herron. *Power and Uneven Globalization: Coalitions and Energy Trade Dependence in the Newly Independent States of Europe.* Ph.D. Dissertation, University of Washington, 2000.

Locatelli, C. "The Russian Oil Industry Restructuration: Towards the Emergence of Western Type Enterprises." *Energy Policy* 27 (1999).

Lomagin, Nikita. "Novye nezivisimaya gosudarstva kak sfera interesov Rossii i SshA." *Pro et Contra,* 5, 2 (Spring 2000): 65–87.

Long, William J. *Economic Incentives and Bilateral Cooperation.* Ann Arbor: University of Michigan Press, 1996.

Luong, Pauline Jones. *Institutional Change and Political Continuity in Post-Soviet Central Asia: Power, Perceptions and Pacts.* New York: Cambridge University Press, 2002.

―――, ed. *The Transformation of Central Asia: States and Societies from Soviet Rule to Independence.* Ithaca: Cornell University Press, 2004.

Lynch, Allen C. "The Realism of Russia's Foreign Policy." *Europe-Asia Studies* 53, 1 (2001): 7–31.

Lynch, Dov, ed. *The South Caucasus: A Challenge for the EU.* Institute for International Studies Chaillot Papers 65 (December 2003).

McFaul, Michael. "A Precious Peace: Domestic Politics in the Making of Russian Foreign Policy." *International Security* 22, 3 (Winter 1997/98): 5–35.

MacFarlane, Neil. *Western Engagement in the Caucasus and Central Asia.* London: The Royal Institute of International Affairs, 1999.

MacLachlan, Ann. "Rosenergoatom Change Raises Hopes for More Nuclear Reforms." *Nucleonics Week* 42, 37 (13 September 2001), http://www.mhenergy.com.

Makarov, I. "Itera, One of the Biggest Gas Suppliers in the CIS." *International Affairs* 46, 2 (2000): 69–76.

Maloney, Michael T., and Oana Diaconu. "Analysis of Privatization of Russia's Nuclear Ministry on the Nuclear Nonproliferation Objectives of the United States." The Strom Thurmond Institute, Clemson University Special Report, 2001, www.strom.clemson.edu.

Mandelbaum, Michael, ed. *The New Russian Foreign Policy.* New York: Council on Foreign Relations, 1998.

Manne, H.G., ed. *Economic Policy and the Regulation of Corporate Securities.* Washington, DC: American Enterprise Institute, 1969.

Mansfield, Edward D. "The Concentration of Capabilities and International Trade." *International Organization* 46, 3 (Summer 1992): 731–764.

―――, and Jon C. Pevehouse. "Trade Blocs, Trade Flows, and International Conflict." *International Organization* 54, 4 (Autumn 2000): 775–808.

Maoz, Zeev. "Framing the National Interest: The Manipulation of Foreign Policy Decisions in Group Settings." *World Politics* 43 (October 1990): 82–83.

―――, *Paradoxes of War: On the Art of National Self-Entrapment.* Boston: Unwin Hyman, 1989.

Marples, David R. "The Post-Soviet Nuclear Power Program." *Post-Soviet Geography* 34, 3 (1993): 172–184.

Martin, Lisa. *Coercive Cooperation: Explaining Multilateral Economic Sanctions.* Princeton: Princeton University Press, 1992.

―――. *Democratic Commitments.* Princeton: Princeton University Press, 2000.

Martin, Soizick, and Charles Digges. "Upcoming EU Enlargement Revives Long-Standing Nuclear Battle." *Bellona Working Paper* (10 February 2004), www.bellona.no.

————. "EC's 'Nuclear Package' to Harmonize Atomic Energy in Expanded EU—but Environmentalists Cry Foul." *Bellona Report* (17 October 2003), www.bellona.no.

Mastanduno, Michael. *Economic Containment: CoCom and the Politics of East-West Trade*. Ithaca: Cornell University Press, 1992.

McCubbins, Mathew D., Roger G. Noll, and Barry R. Weingast. "Administrative Procedures as Instruments of Political Control." *Journal of Law, Economics, and Organization* 3, 2 (Fall 1987): 243–277.

McDermott, Rose. *Risk-Taking in International Politics: Prospect Theory in American Foreign Policy*. Ann Arbor: University of Michigan Press, 1998.

McFarlane, Neil. *Western Engagement in the Caucasus and Central Asia*. London: The Royal Institute of International Affairs, 1999.

McPherson, Charles P. "Policy Reform in Russia's Oil Sector." *The World Bank* (June 1996).

Menon, Rajan. "In the Shadow of the Bear." *International Security* 20, 1 (Summer 1995): 149–181.

————, "The New Great Game in Central Asia." *Survival* 45, 2 (Summer 2003): 187–204.

————, Yuri E. Fedorov, and Ghia Nodia, eds. *Russia, the Caucasus and Central Asia*. Armonk, NY: M.E. Sharpe, 1999.

————, and Hendrik Spruyt. "The Limits of Neorealism: Understanding Security in Central Asia." *Review of International Studies*, 25 (1999): 87–105.

Milgrom, Paul, and John Roberts. "Predation, Reputation, and Entry Deterrence." *Journal of Economic Theory* 27: 2 (1982): 280–312.

Milner, Helen. *Interests, Institutions, and Interests: Domestic Politics and International Relations*. Princeton: Princeton University Press, 1997.

"Minatom Rising Updated." *NUKEM* (December 2002).

Mitchell, John V., Peter Beck, and Michael Grubb. *The New Geopolitics of Energy*. London: The Royal Institute of International Affairs, 1996.

————, Koji Morita, Norman Selly, and Jonathan Stern. *The New Economy of Oil: Impact on Business, Geopolitics, and Society*. London: The Royal Institute of International Affairs, 2001.

Miyamoto, Akira. *Natural Gas in Central Asia: Industries, Markets, and Export Options of Kazakhstan, Turkmenistan, and Uzbekistan*. London: The Royal Institute of International Affairs, 1997.

Moe, Arlid, and Valeriy A. Kryukov. "Joint Management of Oil and Gas Resources." *Post-Soviet Geography and Economics* 39, 7 (1998): 588–605.

Moe, Terry. "The New Economics of Organization." *American Journal of Political Science* 28, 4 (November 1984): 739–777.

Moltz, James Clay, Vladimir A. Orlov, and Adam N. Stulberg, eds. *Preventing Nuclear Meltdown: Managing the Decentralization of Russia's Nuclear Complex*. London: Ashgate Academic Publishers, 2004.

Morrow, James D. *Game Theory for Political Scientists.* Princeton: Princeton University Press, 1994.

Morse, Edward L., and James Richard. "The Battle for Energy Dominance." *Foreign Affairs* 81, 2 (March/April 2002): 16–31.

Moscow News

Mozaffari, Mehdi ed. *Security Politics in the Commonwealth of Independent States: The Southern Belt.* London: Macmillan Press/St. Martin's Press, 1997.

Nazar, Naz. "Turkmenistan: U.S. Official Discusses Trans-Caspian Pipeline," *Radio Free Europe/Radio Liberty*, 14 August 2000.

Newnham, Randall E. *Deutsche Mark Diplomacy: Economic Linkage in German-Russian Relations.* University Park: Penn State University Press, 2000.

———, "More Flies with Honey: Positive Economic Linkage in German Ostpolitik from Bismarck to Kohl." *International Studies Quarterly* 44, 1 (2000): 73–96.

New York Times

Nezavisimaya gazeta

Noreng, Oystein. *Crude Power: Politics and the Oil Market.* London: I.B. Taurus, 2002.

North, Douglas C. *Institutions, Institutional Change, and Economic Performance.* New York: Cambridge University Press, 1990.

Nuclear Energy Agency. *Trends in the Nuclear Fuel Cycle: Economic, Environmental, and Social Aspects.* Paris, OECD, 2001.

———, *Uranium 1999: Resources, Production, and Demand* Paris: OECD, 1999.

———, and the International Atomic Energy Agency, *Uranium 1999: Resources, Production, and Demand.* Paris, OECD, 1999.

"Nuclear Power in Russia." World Nuclear Association (March 2004), www.world-nuclear.org/info/printable_papers/inf45print.htm.

Nye, Joseph S. *The Paradox of American Power.* New York: Oxford University Press, 2002.

———, "Energy and Security in the 1980s." *World Politics* 35, 1 (October 1982).

———, "Soft Power." *Foreign Policy* 80 (Fall 1990): 153–171.

O'Sullivan, Meghan L. *Shrewd Sanctions: Statecraft and State Sponsors of Terrorism.* Washington, DC: The Brookings Institution Press, 2003.

Olcott, Martha Brill. *The Energy Dimension in Russian Global Strategy: Vladimir Putin and the Politics of Oil.* Houston TX: James A. Baker III Institute for Public Policy, Rice University, 2005.

———, *Kazakhstan: Unfulfilled Promise.* Washington, DC: Carnegie Endowment for International Peace, 2002.

———, *Central Asia's New States: Independence, Foreign Policy, and Regional Security.* Washington, DC: United States Institute of Peace, 1996.

————, Anders Aslund, and Sherman Garnett. *Getting It Wrong: Regional Cooperation and the Commonwealth of Independent States.* Washington, DC: Carnegie Endowment for International Peace, 1999.

Olson, Mancur. *The Logic of Collective Action.* Cambridge: Harvard University Press, 1965.

Orlov, Vladimir. "Chto bygodno Minatomu." *Pro et Contra*, 2 (Summer 1997).

Ottaway, Marina. *Democracy Challenged: The Rise of Semi-Authoritarianism.* Washington, DC: Carnegie Endowment for International Peace, 2002.

Ougrahm, Sonia Ben. *Study of Minatom's Organizational Structure.* Unpublished Paper, September 2000.

Papayoanou, Paul. "Economic Interdependence and the Balance of Power." *International Studies Quarterly* 41 (1997): 113–140.

Pape, Robert A. *Bombing to Win: Air Power and Coercion in War.* Ithaca: Cornell University Press, 1996.

Pappe, Yakov "Neftyanaya i gazovaya diplomatiya." *Pro et Contra*, 2, 3 (Summer 1997): 59–69.

Pauluis, Gerard. "The Global Nuclear Fuel Market to 2020." *The Uranium Institute Twenty Third Annual International Symposium,* London: 1998, www.uilondon.org/uilondon/uilondon/sym/1998/pauluis.htm.

Pavlov, Alexander, and Nigel Moe. *Uranium and Nuclear Energy 1997: Proceedings of the Twenty-Second Annual Symposium of the Uranium Institute.* London: Uranium Institute, September 1997.

PlanEcon. *Review and Outlook for the Former Soviet Republics.* Washington, DC: PlanEcon, September 1999.

Plater-Zyberk, Henry. *Kyrgyzstan-Focusing on Security.* Surrey, England: Conflict Studies Research Center (November 2003).

Pollack, Mark A. "Delegation, Agency, and Agenda Setting in the European Community." *International Organization* 51, 1 (Winter 1997): 99–134.

Pomfret, Richard. *Central Asia Turns South?* London: The Royal Institute of International Affairs, 1999.

————, *The Economies of Central Asia.* Princeton: Princeton University Press, 1995.

Puglisi, Rosario. "Clashing Agendas? Economic Interests, Elite Coalitions, and Prospects for Cooperation Between Russia and Ukraine." *Europe-Asia Studies* 55, 6 (2003): 827–845.

Putnam, Robert D. "Diplomacy and Domestic Politics: The Logic of Two Level Games." *International Organization* 42, 3 (Summer 1988): 427–460.

Quattrone, George A., and Amos Tversky. "Contrasting Rational and Psychological Analyses of Rational Choice." *American Political Science Review* 82, 3 (September 1988): 719–736.

Raczka, Will. "A Sea or a Lake? The Caspian's Long Odyssey," *Central Asian Survey* 19, 2 (2000): 189–221.

Rasizade, Alec. "Azerbaijan after a Decade of Independence: Less Oil, More Graft and Poverty." *Central Asian Survey*, 21, 4 (2002): 354–364.

————, "The Bush Administration and the Caspian Oil Pipeline." *Contemporary Review* 279: 1626 (July 2001), pp. 21–26.

————, "Mif ob uglevodorodnom izobilii kasniya i geopoliticheskaya strategiya 'truby.'" *Tsentral'naya aziya i kavkaz* 4,16 (2001): 19–32.

Reisinger, William M. *Energy and the Soviet Bloc: Alliance Politics after Stalin.* Ithaca: Cornell University Press, 1992.

Reuters

Roberts, John. *Caspian Pipelines.* London: the Royal Institute of International Affairs, 1996.

Rock, Stephen R. *Appeasement in International Politics.* Lexington, KY: University of Kentucky Press, 2000.

Rodman, Kenneth A. "Sanctions at Bay? Hegemonic Decline, Multinational Corporations, and U.S. Economic Sanctions Since the Pipeline Case." *International Organization* 49, 1 (Winter 1995): 105–37.

————, *Sanctions beyond Borders: Multinational Corporations and U.S. Economic Statecraft.* Lanham: Rowman & Littlefield Publishers, 2001.

Rogowski, Ronald. *Commerce and Coalitions: How Trade Affects Domestic Political Alignments.* Princeton: Princeton University Press, 1989.

Rose, Gideon. "Neoclassical Realism and Theories of Foreign Policy." *World Politics* 51, 1 (October 1998): 144–172.

Rotar, Igor. "Will Caspian Oil Flow over the Northern Variant?" *Prism* (Jamestown Foundation) 3, 12; part 3 (25 July 1997).

Rowe, David M. *Manipulating the Market: Understanding Economic Sanctions, Institutional Change, and the Political Unity of White Rhodesia.* Ann Arbor: University of Michigan Press, 2001.

Rumer, Boris Z., and Stanislav V. Zhukov, eds. *Central Asia: The Challenges of Independence.* New York: M.E. Sharpe, 1998.

Ruseckas, Laurent. "State of the Field Report: Energy and Politics in Central Asia and the Caucasus." *Access Asia Review*, 1, 2 (February 1998), www.nbr.org/publications/review/vol1no2/essay2.html.

Rutland, Peter, ed. *Economic Change in Post-Soviet Russia.* Westview Press, 1999.

————, "Lost Opportunities: Energy and Politics in Russia." *National Bureau of Asian Research Analysis* 8, 5 (1997).

Sabol, Steven. "Turkmenbashi: Going It Alone." *Problems of Post-Communism* 50, 5 (September/October 2003): 48–57.

Sagers, Matthew J. "Developments in Russian Crude Oil Production in 2000." *Post-Soviet Geography and Economics* 42, 3 (2001): 153–201.

————, "The Energy Industries of the former USSR: A Mid-Year Survey." *Post-Soviet Geography and Economics* 34, 6 (1993): 341–418.

————, "Russian Crude Oil Production in 1996: Conditions and Prospects." *Post-Soviet Geography and Economics* 37, 9 (1996): 523–587.

————, "The Russian National Gas Industry in the Mid-1990s." *Post-Soviet Geography and Economics* 36, 9 (1995): 521–586.

————, Igor A. Didenko, and Valeriy A. Kryukov. "Distribution of Refined Petroleum Products in Russia." *Post-Soviet Geography and Economic* 40, 5 (1999): 407–439.

————, Valeriy A. Kryukov, and Vladimir V. Shmat. "Resource Rent from the Oil and Gas Sector and the Russian Economy." *Post-Soviet Geography and Economics* 36, 7 (1995): 389–425.

Schmidt, Christian, ed. *Uncertainty in Economic Thought.* Cheltenham, UK: Edward Elgar, 1996.

Schultz, Kenneth A. *Democracy and Coercive Diplomacy.* Cambridge: Cambridge University Press, 2001.

————, "The Politics of Risking Peace: Do Hawks or Doves Deliver the Olive Branch." *International Organization* 59 (Winter 2005): 1–38.

Segodnya

Sevcik, Margarita. "Uranium Tailings in Kyrgyzstan: Catalyst for Cooperation and Confidence Building?" *The Nonproliferation Review* (Spring 2003): 147–154.

Shaffer, Brenda. "U.S. Policy toward the Caspian Region: Recommendations for the Bush Administration." Policy Brief, Caspian Studies Program, Harvard University (July 2001), www.harvard.edu/ksg/bcsia.

Shambaugh, George E. "Globalization, Sovereign Authority and Sovereign Control over Economic Activity." *International Politics* 37, 4 (Winter 2000): 403–431.

————, *States, Firms, and Power: Successful Sanctions in United States Foreign Policy.* Albany: State University of New York Press, 1999.

Shepsle, Kenneth A., and Barry R. Weingast. "The Institutional Foundations of Committee Power." *American Political Science Review* 81, 1 (1987): 85–104.

Shoumikhin, Andrei, "Developing Caspian Oil: Between Conflict and Cooperation," *Comparative Strategy* 16, 4 (Fall 1997): 337–351.

Shultse, P. "Kaspiiskaya neft': podkhody k probleme," *Kaspiiskaya neft' i mezhdunarodnaya bezopastnost'* 2 (1996): 1–10.

Skagen, Ottar. *Caspian Gas.* London: The Royal Institute of International Affairs, 1997.

Skalnes, Lars S. *Politics, Markets, and Grand Strategy.* Ann Arbor: University of Michigan Press, 2000.

Smolansky, Oles. "Fuel, Credit, and Trade: Ukraine's Economic Dependence on Russia." *Problems of Post-Communism* 46, 2 (March/April 1999): 49–58.

Socor, Vladimir. "Kazakhstan and Russia to Share Northern Caspian Oilfields." *The Monitor* (Jamestown Foundation) VII:94 (14 May 2002).

Soligo, Ronald, and Amy Jaffee. *Unlocking the Assets: Energy and the Future of Central Asia and the Caucasus.* Houston, TX: James A. Baker III Institute for Public Policy, Rice University (April 1998).

Solingen, Etel. "The Political Economy of Nuclear Restraint." *International Security* 19, 2 (Fall 1994): 126–169.

"Soveshchanie poslov Rossii v stranakh SNG," *Diplomaticheskii vestnik* 9 (September 1996).

Spector, Regine A. "The Caspian Basin and Asian Energy Markets." *The Brookings Institution Background Paper* (24 May 2001).

———, *The Transformation of Askar Akaev, President of Kyrgyzstan*. Berkeley Program in Soviet and Post-Soviet Studies Working Paper Series, Spring 2004.

Sperling, James, Sean Kay, and S. Victor Papacosma, eds. *Limiting Institutions?: The Challenge of Eurasian Security Governance*. Manchester: Manchester University Press, 2003.

Sprout, Harold, and Margaret Sprout. *Men-Milieu Relationship Hypotheses in the Context of International Politics*. Center for International Studies, Princeton University Research Monograph, 1956.

Srkar, Tina, Obut Avik, and Sankar Sunder. "Comparing Russian, and Western Major Oil Firms Underscores Problems Unique to Russia." *Oil and Gas Journal* 97, 5 (February 1, 1999): 20–25.

Starr, S. Frederick. "Power Failure: American Policy in the Caspian." *The National Interest* (Spring 1997): 20–31.

Stauffer, Thomas R. "Caspian Fantasy: The Economics of Political Pipelines." *Brown Journal of World Affairs* VII, 2 (Summer/Fall 2000), www.watsoninstitute.org.

Stein, Arthur A. *Why Nations Cooperate: Circumstance and Choice in International Relations*. Ithaca: Cornell University Press, 1990.

Steinbrunner, John. *The Cybernetic Theory of Decision*. Princeton: Princeton University Press, 1976.

Stern, Jonathan P. *Russian Natural Gas Bubble: Consequences for European Gas Markets*. London: The Royal Institute of International Affairs, 1993.

Strange, Susan. *The Retreat of the State: The Diffusion of Power in the World Economy*. New York: Cambridge University Press, 1996.

Strategy of Nuclear Power Development in Russia in the First Half of the 21st Century, Endorsed by the Russian Government on May 25, 2000 (Protocol No. 17).

Streifel, Shane S. *Review and Outlook for the World Oil Market*. Washington, DC: World Bank Discussion Paper, 1995.

Stulberg, Adam N. "The Federal Politics of Importing Spent Nuclear Fuel: Inter-Branch Bargaining and Oversight in the New Russia." *Europe-Asia Studies* 56, 4 (June 2004): 491–520.

———, "Moving Beyond the Great Game: The Geoeconomics of Russia's Influence in the Caspian Energy Bonanza." *Geopolitics* 10 (2005): 1–25.

———, "Nuclear Regionalism in Russia: Decentralization and Control in the Nuclear Complex." *The Nonproliferation Review* (Fall/Winter 2002): 31–46.

Suny, Ronald Grigor. "Provisional Stabilities: The Politics of Identities in Post-Soviet Eurasia." *International Security* 24, 3 (Winter 1999/2000); 139–178.

Taliaffero, Jeffrey W. *Balancing Risks: Great Power Intervention in the Periphery*. Ithaca: Cornell University Press, 2004.

Telegina, E.A. "Geopolicheskie i ekonomicheskie interesy Rossii v global'nom energeticheskom prostranstve," *Energiya: ekonomika, tekhnika, ekologiya* 1 (2001): 11–17.

Telhami, Shibley, and Michael Barnett, eds. *Identity and Foreign Policy in the Middle East*. Ithaca: Cornell University Press, 2002.

Tetlock, Philip E., and Aaron Belkin, eds. *Counterfactual Thought Experiments in World Politics: Logical, Methodological, and Psychological Perspectives*. Princeton: Princeton University Press, 1996.

Thomas, Timothy L. "Russian National Interests in the Caspian Sea." *Perceptions* 4, 4 (December 1999–February 2000): 75–96.

Toritsyn, Arkady, and Eric A. Miller. "From East to West and Back Again: Economic Reform and Ukrainian Foreign Policy." *European Security* 11, 1 (Spring 2002): 102–126.

Trenin, Dmitri. *The End of Eurasia: Russia on the Border Between Geopolitics and Globalization*. Washington, DC: Carnegie Endowment for International Peace, 2001.

Triesman, Daniel. "Rational Appeasement." *International Organization* 58 (Spring 2004): 345–373.

Tsnetr Strategicheskikh Razrabotok, *O vozmozhnyi napravleniyakh razvitiya infrastruktury po transportirovke rossiiskoi nefti* (12 November 2004), www.csr.ru.

Tsygankov, Andrei P. "Mastering Space in Eurasia: Russia's Geopolitical Thinking after the Soviet Break-up." *Communist and Post-Communist Studies* 36 (2003): 101–127.

———. *Pathways after Empire: National Identity and Foreign Economic Policy in the Post-Soviet World*. New York: Rowman & Littlefield Publishers, 2001.

"Uranium Imports to the USA from CIS Countries." *Uranium Institute Trade Briefing* 1 (1 August 1999).

"Uranium Production Figures, 1995–2003." *World Nuclear Association* (April 2004), www.world-nuclear.org/info/uprodhtm.

Vek

Vernon, Raymond. *Storm over the Multinationals: The Real Issues*. Cambridge: Harvard University Press, 1977.

Victor, David G., and Nadejda M. Victor. "Axis of Oil?" *Foreign Affairs* 82, 2 (March/April 2003): 47–61.

Viner, Jacob. "Power versus Plenty as Objectives of Foreign Policy in the Seventeenth and Eighteenth Centuries." *World Politics* 1, 1(October 1948): 1–29.

Wagner, R. Harrison. "Economic Interdependence, Bargaining Power, and Political Influence." *International Organization* 42, 3 (Summer 1988): 461–483.

Wakefield, Laura. "The Need for Comprehensive Legislation in the Russian Oil and Gas Industries." *Case Western Reserve Journal of International Law* 29, 1 (Winter 1997): 149–163.

Wall Street Journal

Waltz, Kenneth N. *Theory of International Politics.* Reading, MA: Addison-Wesley, 1979.

Watson, James "Foreign Investment in Russia: The Case of the Oil Industry." *Europe-Asia Studies* 48, 3 (May 1996): 429–456.

Weiss, Linda. *The Myth of the Powerless State: Governing the Economy in the Global Era.* Cambridge: Polity Press, 1998.

Wendt, Alexander, and Daniel Freidheim. "Hierarchy Under Anarchy: Informal Empire and the East German State." *International Organization* Vol. 49, No. 4 (Autumn 1995), pp. 689–721.

Williamson, Oliver E. *The Economic Institutions of Capitalism.* New York: Free Press, 1985.

Woodruff, David M. "It's Value That's Virtual: Bartles, Rubles, and the Place of Gazprom in the Russian Economy." *Post-Soviet Affairs* 15, 2 (1999): 130–148.

———, *Money Unmade: Barter and the Fate of Russian Capitalism.* Ithaca: Cornell UniversityPress, 1999.

The Word Bank. *Kazakhstan: Transition of the State.* Washington, DC, 1997.

———, *Russia Oil Transport and Export Study: Strategic Export Expansion Options and Legal, Contractual, and Regulatory Framework.* August 1997.

The World Fact Book, 2000 (http://.odci.gov/cia/publications/factbook/index-geo.html).

Yermukanov, Marat. "Nuclear Waste Disposal Scheme Sets Off Wave of Protests in Kazakhstan." *Central Asia-Caucasus Analyst Field Report* (15 January 2003), www.cacianalyst.org/2003–01–15/20030115_Kazakhstan_Nuclear_Waste.htm.

———, "CIS Leaders Meet at Summit in Kazakhstan." *Central Asia Caucasus Analyst,* 27 February 2002.

Yetiv, Steve A. *Crude Awakenings: Global Oil Security and American Foreign Policy.* Ithaca: Cornell University Press, 2004.

Zakaria, Fareed. *From Wealth to Power: The Unusual Origins of America's World Role.* Princeton: Princeton University Press, 1998.

Zeigler, Charles E., and Henry B. Lyon. "The Politics of Nuclear Waste in Russia." *Problems of Post-Communism* 49, 4 (July/August 2002): 33–42.

Index

319

322 *Index*

Export-Import Bank, 102, 144
Exxon, 120

Finlandization, 29

Gas, natural: alternatives for monetizing, 67; in Kazakhstan, 117–131;
liquefied (LNG), 67, 123, 124, 125;
regional structure of, 66; Trans-
Siberian natural gas pipeline, 14; in
Turkmenistan, 21, 99–117
Gas sector, Russian, 66–71, 93–132;
ability to sow uncertainty in Kazakhstan's gas sector regarding
market prospects, 129; attempts at
regional reintegration, 94; competitive advantages of, 66, 67, 106,
107, 124–129; corporatization of,
69; cost-competitive supply from,
67; costs of supplying as to industrial customers, 96; decision
authority in, 65; diplomacy in
Kazakhstan, 117–131; diplomacy
in Turkmenistan, 99–117;
economies of scale in, 66, 67;
efforts to dominate Eurasian
market, 130; *Energy Strategy
through 2020*, 97, 98; European
Union gas diplomacy, 106, 107;
exploitation of comparative advantages by, 122–124; exploitation of
economic difficulties in Turkmenistan, 105–110; exploration of
business with Iran, 108; fields in,
67; Gazprom in, 66, 68; lack of
institutional capacity to compel
domestic compliance in, 66; leading
supplier to Europe, 66; as mainstay
of economy, 66; mandates for royalty fees, excise taxes, and export
tariffs, 70; manipulation of risk in
Turkmenistan pipeline options,
110–115; market and institutional
components in, 94–99; market
strategy in, 106, 107; pipeline

infrastructure, 66; presentation of
risk in Kazakhstan gas options,
124–129; preservation of consolidated structure of industry, 68; privatization in, 69; reform of
domestic transmission system, 99;
regulatory authority in, 94–99;
reserves, 66; subsidization in, 94;
superpower status in regional markets, 65, 66–68
Gazexport, 69
Gazprom, 14, 94, 108, 112; attempts
to enlarge presence in emerging
markets, 95; authority to administer access, production, distribution
of natural gas, 68; commercial
incentives to cover insolvent customers, 94, 95, 128; constraints on
in foreign operations, 95; control of
Russian pipeline network, 69; dealings with Eurasian gas suppliers,
95; delivery to hard currency markets in Europe, 70; division of
ownership and regulatory responsibilities with state by, 71; dominant
market position of, 66, 130; dumping issues, 95, 96; expansion of
global market presence, 96, 97;
exploitation of authority by, 127;
foreign operations of, 14, 15; gas
subsidization by, 97; governmental
attempts to assert control over, 69;
in Kazakhstan energy development,
126–129; monopoly pipeline rents
and, 95; need for capital investment
in, 95; nonpayment crisis for, 70;
objections to relaxation of control
over European market, 98–99;
ownership of pipelines to European
market, 69; price discrimination by,
95, 96; price setting discretion, 68;
regulation of monopoly over internal transmission system of, 68;
rights to established gas fields, 68;
as state within a state, 68–71; tax

4, 17; effect of domestic regulatory authority on, 57; effect on policies of target, 17, 18; efficacy of, 31; expected utility principle in, 38–42; extending debate on, 219–226; findings concerning risk manipulation in securing strategic deference, 215–219; "hard" material instruments of, 26, 27, 28, 29; hard power outlooks in, 8; inducement as substitute for coercive diplomacy, 29; influence attempts assessed as static contests, 29, 30; influence of market power on, 57; information asymmetries and, 40, 41; in interdependent settings, 13; levels of compliance with, 57; making compliance advantageous for target, 17, 18; minimally acceptable stipulations in, 30; need for control of mobilization of resources in, 7; operationalization and, 54–57; positive inducement and, 24–31; relative power advantages in, 17, 18–19; resistance and, 27; restrictive characterization of positive inducement in, 28, 29, risk *versus* uncertainty and, 38–42; role of risk in, 8, 37; sectoral dimension of, 59; shaping situational contexts in, 27, 30, 31, 32, 37; situational rationality and, 43–44; "soft" power and, 13, 27; strategic manipulation in, 31–34, 37–61; structural power and, 17, 18, 20–22; success/failure decisions in, 30, 31, 31*fig*; theories of, 13–35; traditional assessment of, 26–31; use of diverse policy instruments in, 17

State(s): avenues for shaping interactions in, 32; capacity for manipulation of energy decisions, 7; capacity to exploit market power, 50; control of regulatory authority by, 7; mobilization of domestic energy

resources in, 7; power, 16; power in global energy market, 7; target manipulation by, 6

States, newly independent: advantages of compliance and resistance to Russia, 41; association with European Union, 60; asymmetrically dependent energy relations with Russia, 60; autonomous leaderships in, 60; compliance with Russian political demands in, 14; deference to Russia in gas supply policies, 14; differing compliance trends in, 16; disadvantages in meeting external energy demands, 67; diversified foreign energy ties for, 60; effect of Russian statecraft on energy policies, 17–24; energy security relations with extraregional actors, 14; ethnic and separatist conflicts in, 22; neopatrimonialism in, 23; policy choices circumscribed by Russian energy posture, 16; political insulation of regimes in, 60; regime survival and, 60; Russian leverage over, 18–19; strategic energy choices by, 16; trade with Russia by, 19, 20; uranium conversion market for Russia, 84, 85

Strategic manipulation, 37–61; ability to secure cooperation of domestic actors in, 43; advancing relative influence by foreign policy leaders in, 43; alignment choices in, 1; anticipation of adversarial countermoves in, 32; assumptions in, 42–46; central supervision of foreign policy and, 43; coercive diplomacy and, 1; common patterns of, 44, 45; containment of agency costs in, 48–54; decision situation in, 1; defining, 31–34; discretionary actions of domestic agents in, 49, 50; domestic dimensions of, 50, 51; and effectiveness of Russian energy

SUNY series in Global Politics
James N. Rosenau, Editor

List of Titles

www.ingramcontent.com/pod-product-compliance
Lightning Source LLC
Chambersburg PA
CBHW030637270326
41929CB00007B/112